Drug Discovery and Development

Contemporary Biomedicine

Drug Discovery and Development
Edited by ***Michael Williams*** and ***Jeffrey B. Malick***, 1987

Monoclonal Antibodies in Cancer
Edited by ***Stewart Sell*** and ***Ralph A. Reisfeld***, 1985

Calcium and Contractility: *Smooth Muscle*
Edited by ***A. K. Grover*** and ***E. E. Daniel***, 1984

Carcinogenesis and Mutagenesis Testing
Edited by ***J. F. Douglas***, 1984

The Human Teratomas: *Experimental and Clinical Biology*
Edited by ***Ivan Damjanov, Barbara B. Knowles***, and ***Davor Solter***, 1983

Human Cancer Markers
Edited by ***Stewart Sell*** and ***Britta Wahren***, 1982

Cancer Markers: *Diagnostic and Developmental Significance*
Edited by ***Stewart Sell***, 1980

DRUG DISCOVERY AND DEVELOPMENT

Edited by

MICHAEL WILLIAMS
Ciba-Geigy Corp., Summit, New Jersey

and

JEFFREY B. MALICK
Stuart Pharmaceuticals, Wilmington, Delaware

Humana Press · Clifton, New Jersey

Library of Congress Cataloging in Publication Data

Main entry under title:

Drug Discovery and development.

(Contemporary biomedicine)
Includes bibliographies and index.
1. Drugs—Research. 2. Drugs—Testing.
3. Pharmaceutical technology. I. Williams,
Michael, 1947- . II. Malick, Jeffrey B.
III. Series. [DNLM: 1. Drug Evaluation.
2. Drug Industry. 3. Pharmacology, Clinical.
4. Research Design. 5. Technology, Pharmaceutical. QV 771 D7942]
RM301.25.D78 1987 615.1'702 87-3343
ISBN 0-89603-108-X

Crescent Manor
PO Box 2148
Clifton, NJ 07015

Printed in the United States of America

Preface

The conceptual process of drug discovery is one that is often the result of an identified need in a defined disease area. This need represents a mandate from the marketing department of a pharmaceutical company or a breakthrough at the research level that has agreed applicability in response to a valid therapeutic demand. Although the intelligent design and development of new therapeutic entities, as evidenced by Sir James Black's H_2-receptor antagonist cimetidine (Tagamet), is intellectually satisfying, many novel drugs arise from serendipity, from the chance observation of the research scientist or the clinician, that a compound has unexpected actions of use for the treatment of human disease states. Drugs that have been identified by this route include the antipsychotic chlorpromazine and the putative anxiolytic buspirone.

The events surrounding the process of drug discovery and development are the theme of the present volume, which attempts to present, in a logical and lucid manner, the complexity of a process that is often naively assumed to represent nothing more than the identification of a new compound and its rapid introduction into humans, free of such complications as efficacy, selectivity, safety, bioavailability, toxicity, and need.

The volume is divided into three main sections: drug discovery from both the chemical and biological vantage points; drug development in terms of drug delivery, toxicological testing, and clinical evaluation; and examples of both the development of individual classes of therapeutic agents and new approaches to disease states of major importance.

The overview by Williams and Malick presents the overall concepts of the drug discovery process, outlining the logistics and pitfalls associated with the melding of science with business. Baldwin presents the chemical approaches to new compound synthesis and the structure–activity relationship, and Loftus et al. describe the increasing use of computers and molecular modeling to aid in this process. Goodman, Malick, Enna, and Fuller and Steranka describe some of the biological test procedures that form necessary steps in the drug discovery flowchart, whereas Young, Hong, and Khazan describe one of the more sophisticated techniques used in CNS drug research and brain electroencephalograms, for the networking of compounds and providing for the development of more predictive preclinical models of drug action. Gilman and Lewis review the area of immunology in relation to pharmacology, and the toxicological assessment of new chemical entities from the Ames test through to chronic whole animal studies is covered by Cavagnaro and Lewis. Bondi and Pope describe new developments in the area of drug delivery, a science that is radically changing many concepts in the process of drug design. The events related to the clinical assessment of a new chemical entity are cogently reviewed by Reines and Fong.

In the final chapters, Brimblecombe and Gannellin, Eison, Taylor, and Riblet, and Gould pull together the preceeding chapters and describe from both a historical and a research perspective the events underlying the discovery and development, respectively, of the H_2-receptor antagonist cimetidine, the psychotropic agents trazodone and buspirone, and the calcium entry blockers.

The individuals contributing to the present volume each bring with them a different perspective to the process of drug discovery and development, along with their personal experience in the context of the pharmaceutical industry. For this reason, it is anticipated that this volume will be a valuable reference source for persons in the industry involved in one or another of the separate phases described, for those in academia who depend on the success (and failures) of the industry for interesting new research tools, and for graduate students in the various branches of pharmacology whose interests and careers may impact with, or on, the industry.

Michael Williams
Jeffrey B. Malick

CONTENTS

OVERVIEW

COMPOUND DISCOVERY

TOXICOLOGICAL EVALUATION AND CLINICAL ASPECTS

THERAPEUTIC ENTITIES—FROM DISCOVERY TO HUMAN USE

CONTRIBUTORS

JOHN J. BALDWIN · *Merck Sharp & Dohme Research Laboratories, West Point, Pennsylvania*

JOSEPH V. BONDI · *Merck Sharp & Dohme Research Laboratories, West Point, Pennsylvania*

ROGER W. BRIMBLECOMBE · *Smith Kline and French Research Ltd., Hertfordshire, England*

JOY CAVAGNARO · *Hazleton Laboratories America Inc., Vienna, Virginia*

MICHAEL S. EISON · *Bristol-Myers Company, Wallingford, Connecticut*

S. J. ENNA · *Nova Pharmaceutical Corp., Baltimore, Maryland*

DICK FONG · *Merck Sharp & Dohme Research Laboratories, West Point, Pennsylvania*

RAY W. FULLER · *Lilly Research Laboratories, Eli Lilly and Company, Indianapolis, Indiana*

C. ROBIN GANELLIN · *Smith Kline and French Research Ltd., Hertfordshire, England*

STEVEN C. GILMAN · *Department of Experimental Therapeutics, Wyeth Laboratories, Inc., Philadelphia, Pennsylvania*

FRANK R. GOODMAN · *Pharmaceuticals Division, Ciba-Geigy Corp., Summit, New Jersey*

ROBERT J. GOULD · *Merck Sharp & Dohme Research Laboratories, West Point, Pennsylvania*

OKSOON HONG · *Department of Pharmacology and Toxicology, School of Pharmacy, University of Maryland at Baltimore, Baltimore, Maryland*

ROBERT F. HOUT, JR. · *Stuart Pharmaceuticals, ICI Americas, Inc., Wilmington, Delaware*

NAIM KHAZAN • *Department of Pharmacology and Toxicology, School of Pharmacy, University of Maryland at Baltimore, Baltimore, Maryland*

ALAN J. LEWIS • *Department of Experimental Therapeutics, Wyeth Laboratories, Inc., Philadelphia, Pennsylvania*

RICHARD M. LEWIS • *US Army Institute of Infectious Diseases, Fort Detrick, Maryland*

PHILIP LOFTUS • *Stuart Pharmaceuticals, ICI Americas Inc., Wilmington, Delaware*

JEFFREY B. MALICK • *Biomedical Research Department, Stuart Pharmaceuticals, ICI Americas Inc., Wilmington, Delaware*

D. G. POPE • *Merck Sharp and Dohme Research Laboratories, West Point, Pennsylvania*

SCOTT A. REINES • *Warner Lambert/Parke Davis, Ann Arbor, Michigan*

LESLIE A. RIBLET • *Preclinical CNS Research, Pharmaceutical Research and Development Division, Bristol-Myers Company, Evansville, Indiana*

LARRY R. STERANKA • *Nova Pharmaceutical Corp., Baltimore, Maryland*

DUNCAN P. TAYLOR • *Preclinical CNS Research, Pharmaceutical Research and Development Division, Bristol-Myers Company, Evansville, Indiana*

MARVIN WALDMAN • *Stuart Pharmaceuticals, ICI Americas Inc., Wilmington, Delaware*

MICHAEL WILLIAMS • *Drug Discovery Administration, Pharmaceuticals Division, Ciba-Geigy Corporation, Summit, New Jersey*

GERALD A. YOUNG • *Department of Pharmacology and Toxicology, School of Pharmacy, University of Maryland at Baltimore, Baltimore, Maryland*

Overview

Drug Discovery and Development

Reflections and Projections

Michael Williams and Jeffrey B. Malick

1. Introduction

The process of drug discovery involves a melding of many disciplines and interests, transcending the relatively simple process of identifying an active compound in the test tube. The discovery of a novel chemical entity that modulates some aspect of cell or tissue function is but the first step in the drug development process. Once shown to be selective and efficacious, a compound must also be relatively free of toxicity, bioavailable, and marketable before it can be considered to be a therapeutic entity.

In the last decade, dramatic changes have occurred in the way in which drugs are discovered and developed (Economist, 1985a). Federal legislation, primarily related to human safety issues, has made the process of introducing a drug a more time consuming and costly enterprise with a concomitant reduction in patent protection. Increased costs are resulting from marked changes in the world economy. Thus there is an increase in the actual costs of the research and development (R & D) process, and this is ultimately reflected in the increased cost of the end product to the consumer. This latter issue has ethical undertones and in countries such as Canada and the United Kingdom, the legislative concern in containing health care costs has compromised patent protection with increased generic competition. This in turn, has seriously, and in the case of Canada, fatally, affected the viability of the native pharmaceutical companies.

Changes in the basic sciences have also altered the process of drug discovery. Most notable are the major technological advances in the area of molecular biology, in which recombinant DNA and monoclonal antibody production have attracted much interest (Dibner, 1985a). Another factor that has influenced the nature of drug discovery is the existence of endogenous substances that mimic the actions of existing drugs. The enkephalins and endorphins are one such class of compounds (Snyder, 1975), and the search continues for the endogenous mediator(s) of anxiety (Williams, 1983).

The actual technical and conceptual aspects of the drug discovery process have also undergone revision, because of both initiatives within the pharmaceutical industry itself and the economic factors that govern the present climate of academic research (Maxwell, 1984). The former, to some extent, was prompted by the realization that the likelihood of the serendipitous discovery of major drugs such as chlorpromazine and the benzodiazepines could not be guaranteed or expected to create a basis for capital investment. The success of the beta-adrenoceptor blocker, propranolol (Black and Stephenson, 1962), the histamine-2 receptor blocker, cimetidine (Black et al., 1972; Brimblecombe and Gannelin, this volume), and the angiotensin-converting enzyme (ACE) inhibitor, captopril (Cushman and Ondetti, 1980; Sweet and Blaine, 1984), is evidence that a more rational, mechanistic approach to drug discovery can significantly contribute to the development of therapeutic entities. This approach has many positive aspects. As may be readily appreciated, however, the discovery of the opiate receptor and the enkephalins, and the elaboration of appropriate structure–activity relationships (SAR) still makes it highly improbable that the medicinal chemist would ever synthesize morphine or any of its congeners. As noted by Baldwin (this volume), drug discovery, despite much intellectual rationalization, is still very much dependent on serendipity.

The success associated with the mechanistic approach has, however, made pharmaceutical companies, even the smallest, acutely aware of the fact that to be too far distant from the "cutting edge" of science is to limit their chances of competing successfully in the marketplace. Thus many drug companies are expanding their R & D operations such that their expertise and potential for innovation is on a par with that of the best academic institutes. Furthermore, many companies have close associations with major academic centers, i. e., Harvard, Johns Hopkins, and Washington University. An additional facet of this process is that as the quality of science within the pharmaceutical industry has improved, so has the willingness of "pure" scientists to consider industry as a viable alternative to a career in academia. No longer is the industrial environment a refuge for the scientist who is unable or unwilling to compete for peer-evaluated grant funding. These changes must be viewed, however, as increasing the probability of success rather than guaranteeing it, a point echoed by Black (Economist, 1985b).

Although it is abundantly clear that the harnessing of the best expertise and the newest technologies are crucial to the interests of drug discovery, the road from concept to initiation, to discovery, to the clinic is a long one. To quote Testa (1984), "R and D . . . overlap considerably. Schematically, research is a long-range task with the potential for considerable scientific benefits and a high risk of failure. Development is a short-range task with the potential for considerable material profits and a lower risk of failure."

The pharmaceutical industry, both in the United States and Europe, has been considered to be in its "sunset years" (Williams, 1985). The successes associated with the Eisenhower era can no longer be taken for granted. Despite this, many of the larger companies are members of the "Fortune 500," and many nonpharmaceutical corporations see the area as worthy of investment (Dibner, 1985b). The harnessing of the newer biotechnologies, the acceptance of the Japanese pharmaceutical industry as a viable competitor, the eradication of an "ostrich" philosophy, typical of the US automobile industry in the seventies, and the setting of realistic goals will ensure the survival of an area of enterprise that is crucial to the well being, both physical and financial, of the West.

2. Drug Discovery

2.1. General Issues

Research management in the pharmaceutical industry is faced with many issues in regard to the drug discovery process. Calculated risk-taking, typical of the business world, can occur at two levels: radical new approaches to a known biological problem; or a "me-too" approach, taking a proven type of therapeutic entity to the marketplace in an area in which a company does not have an established foothold. A variety of factors—scientific, commercial, and social—need to be considered, however. Scientifically, the identification of therapeutic targets must be integrated within the context of advances in basic research knowledge. Experience and the existence of chemical leads are also crucial factors. The criteria established by the marketing departments, a need for innovation in a therapeutic area, the market size for a new agent, a sales force trained in a particular therapeutic area, current competitive activity, and the prevailing social environment are also important considerations. A new compound, regardless of its scientific credentials, presents a problem if it cannot be marketed in a professional manner. Viable therapeutic entities with marketing projections in the $5–25 million range are financially unappealing to the major pharmaceutical houses because of large administrative expenses. Smaller companies with little, if any, R & D costs and a small sales force may be able to market such agents at a profit, at the same time providing a needed entity in the marketplace. On a negative note, although a smaller company may be able to successfully promote an innovative new therapeutic entity, if the market is of sufficient interest to attract the attention of one of the larger, multinational companies with their huge R & D budgets and critical mass in terms of experienced personnel, the smaller company will probably be subjected to competition with a product of equal efficacy and attractiveness, but with more "clout."

2.2. Compound Sources

One source of new compounds for evaluation as potential therapeutic entities is the chemical laboratory. The generation of SARs can assist the chemist in the design of molecules (Baldwin, this volume; Topliss, 1983) that can, in conjunction with molecular modeling techniques (Loftus et al., this volume; Hopfinger, 1985), result in novel, active molecules that interact with a biologically relevant protein or have steric parameters that approximate those of known active entities.

Another source of compounds involving chemical synthesis is the random screen. The discovery of the central benzodiazepine receptor (Squires and Braestrup, 1977), coupled with the random assessment of in vitro activity at this site, has led to the discovery of compounds such as the pyrazoloquinoline, CGS 9896 (Yokoyama et al., 1982), which have anxiolytic activity and would not have been designed as such using a classical SAR approach.

A third source of compounds is that of natural products. This can extend from plant and marine products (Davies, 1985) to microbial fermentations (Albers-Schonberg et al., 1981). Although the concept of screening bacterial broths of unknown composition may be antithetical to a rational drug development program, many antibiotics—the antihelmintic, Avermectin (Campbell et al., 1983), and asperlicine, a peripheral cholecystokinin antagonist (Chang et al., 1985)—have been discovered by such strategies.

The importance of the contribution of the art of pharmacognosocy, or plant pharmacology, to modern drug development should not be ignored merely because of its "folk" associations. The efforts of the medicinal chemist and biologist should not blind them to the potential of viable products from herbs and other natural sources, especially in light of the known successes from Africa and the Far East. The ability to improve on nature's product should be no less challenging than *de novo* synthesis, especially when the SAR is far from obvious.

2.3. Goal Identification

Three factors are important to the task of drug discovery. These are serendipity, a factor that is welcome in all fields of endeavor, realistic goal identification, and the establishment of a multidisciplinary project team that will make the goal its immediate priority.

The issue of serendipity has been discussed briefly above and, as indicated, has contributed and will no doubt continue to contribute in a significant manner to the drug discovery process.

The identification of a viable research target involves many intangible factors that require consideration. For instance, the initial conceptualization of ACE inhibitors as potential antihypertensive agents by Cushman and Ondetti (1980) and the subsequent discovery of captopril® led to competitive programs at other companies. To date, enalapril® (MK 421; Patchett et al., 1980) is the major competitor for captopril® . Other companies active in this area that may have equally attractive compounds are faced with the possibility of being third or sixth on the marketplace, a difference that can represent many millions of dollars. In the final analysis, regardless of the quality of a product, its development may not be worthwhile because of its potential to cause a negative cash flow situation. A caveat, however, is that second-generation compounds may possess better side-effect profiles and may be very marketable in spite of having an identical mechanism of action as the initial agent. Furthermore, it is possible that as a compound is used by more and more of its targeted population, unanticipated side effects or toxicity may occur. This latter possibility underlines the need for a "backup" compound, ideally structurally unrelated to the marketed compound.

Another example of an area of intense competition is that of the leukotriene antagonist field. It is believed that over 25 companies are actively seeking LTC_4 and LTD_4 antagonists that are more potent than the benchmark compound, FPL 55712. Although the precise therapeutic role of such a compound remains to be de-fined, it would be absurd to expect all the companies involved in the area to reap the financial benefits from their R & D efforts because of the intense competition.

Not all areas of basic research endeavor are currently viable targets for drug development. There is intense interest in the industry for the potential offered by compounds that either mimic or antagonize the numerous peptide neurohumoral factors being discovered in ever-increasing numbers. This glamorous area of research, rightly considered to be on the ''cutting edge'' of biomedical research at the present time, is receiving significant support from the pharmaceutical industry. Although it may be too early to assess progress, it should be noted that the tremendous enthusiasm following the discovery of the endogenous opiates, the enkephalins and endorphins, has not resulted in a single useful analgesic, although nearly a decade has elapsed. More objective approaches to the assessment of the role(s) of these peptides in biological function and the development of therapeutic candidates are a shift to ''nonpeptide'' mimics or antagonists (Chang et al., 1985) and to inhibitors of degredative enzymes (such as enkephalinase) that will increase endogenous peptide levels by preventing their breakdown.

The finest example of the melding of innovative basic research with drug development is that of the histamine H_2-receptor antagonist, cimetidine (Black et al., 1972). Black, having identified a possible mechanism and a therapeutic area requiring a compound to replace surgery, together with a chemistry team, went through the intricate process of developing both an SAR and compounds for the gastric histamine H_2-receptor. The availability of cimetidine and similar products, such as ranitidine (Bradshaw et al., 1979; Humphrey et al., 1982) and MK 208 (Pendelton et al., 1983), has revolutionized ulcer treatment, both in terms of expense and time. The rational harnessing of basic research to a defined goal may be expected to provide compounds that will repeat the success story of cimetidine. Much is dependent, however, on the establishment of realistic, achievable research targets. This can be accomplished by establishing drug ''task forces'' to identify areas of interest to both the marketing sectors of the various pharmaceutical houses and to the scientists involved in the actual R & D process.

2.3.1. Research Strategy

Marketing and scientific issues related to a research project can be broken down into three phases (Jack, 1983); R & D policy, strategy, and tactics. The latter is one of the more crucial aspects and involves the formation of an interdisciplinary project team, a ''task force created to solve a set problem'' (Jack, 1983). The scientific aspects of a new project involve both traditional and basic research projects (Bartholini, 1983), the former related to the chemical modification of existing compounds, the ''me-too'' approach, and the latter revolving around a biological mechanism. For the basic approach it is essential that a pharmaceutical company have internal excellence in the form of its scientists and ''creative'' management. This encompasses the need for a critical mass, for the realistic setting of priorities for the various phases of the research process, and a review of progress. Once a project team capable of pursuing a given project has been put in place, it will take between 2 and 5 years to realistically assess progress. If the program is going well and encounters no serious setbacks, it may be 10–15 years or so before the goals of a project are achieved and realized in the marketplace. Only when a compound has passed through toxicological evaluation and clinical testing, does it have therapeutical potential. Thus in deciding on a research project, both management and the project team are undertaking what is hoped will be, at the very best, a minimal degree of calculated risk. This presents unavoidable political problems. Inevitably, the individuals initiating the project may not be the same as those finally taking credit for the final product. Conversely, the curtailing of a nonproductive project is difficult either because of vested interests or because of the inertia related to either an inability to make a decision or the consequent need to replace one program with another.

2.3.1.1. THE "ME-TOO" APPROACH. The ''me-too'' approach to drug development has two facets, one sterile and one innovative. In the former instance, a known chemical can be modified to establish a new patentable entity with the same biological characteristics of the original molecule. Although financially rewarding, this approach is intellectually arid. The conservative approach does however provide the financial security for innovative research. The second approach, akin to that of the ''me-too,'' is to take a novel mechanistic stance and evaluate other potential approaches. Returning once again to cimetidine, this highly successful H_2-receptor antagonist reduces ulcer formation by preventing the release of acid in the stomach. Another mechanistic approach given the proven utility of reduced acid secretion in ulcer prevention would be to inhibit the hydrogen/potassium ATPase that is the ''second messenger'' affected by receptor blockade (Sachs, 1984).

2.3.1.2. THE SCIENTIST IN INDUSTRY. The facilitation of any type of innovative approach to drug development is dependent ultimately on the degree of expertise present in the company and the willingness of management to support often long-term and frustratingly nonproductive research efforts. In order to achieve a balance between commercial factors and those required for good basic research, it is important for scientists to treat corporate management as colleagues rather than banking executives whose questioning of project goals is an infringement on their integrity. Conversely, there are inherent difficulties in accommodating often eccentric scientists in a business environment. In the pharmaceutical industry, one is melding the ''individual'' process of scientific expression with the necessary aspects of business (Badawy, 1985). Very often this degree of individuality is not compatible with the hierarchy of the corporate environment. Thus in many instances, the ''new blood'' that is so essential to basic research and, by extrapolation, to the drug discovery process, is often that which is the least appreciated. The need for scientists to manage is often a self-destructive process (Fitzgerald, 1983), and it is lamentable that many scientists in ''middle management'' who need to bridge the demands of the business world with those of their scientific careers are more familiar with the pages of *Fortune* and *Scrip* than they are with the *bona fide* scientific literature. For a scientist to compete in business requires a great deal of extra effort: not only does he or she need to learn new skills and compete with MBAs and their ilk, but the scientist must also maintain a good knowledge of the current literature and a viable *curriculum vitae*.

3. The Project Team

The project team is of paramount importance to the drug discovery process. It represents the core group of any drug discovery program. The ideal team should have members from the biology and chemistry departments, as well as representatives from marketing, clinical, and, where appropriate, safety evaluation personnel (toxicologists, pathologists). The responsibilities of the project team are to set the goals of the project at the tactical level; the team should define the flow chart and its attendant criteria, identify and request the resources necessary for its execution, identify "lead compounds," and ensure the follow-up procedures that will enable that compound to become a product candidate. In addition, the project team must keep abreast of new developments in their therapeutic area and also identify backup compounds for their leads. The project leader should have sufficient management support to ensure that his or her suggestions and needs are adequately dealth with, as well as peer respect in order to make the project team viable. In addition to the necessary management skills that veer between autocracy and democracy, the team leader must be a scientist capable of judging the quality and relevance of the data generated as part of the project. Such data should be of good quality (hopefully consistent with, if not conforming to, the FDAs Good Laboratory Practice standards) involving comparison with relevant controls, and all aspects relevant to determining the selectivity of a drug at the preclinical level should be addressed. The frustration associated with the scheduling of a compound for toxicological evaluation and IND (or equivalent report) compilation, only to find that data are incomplete or have not been generated is unacceptable. A team effort is absolutely essential for the project team concept to work. Too often, the chemist can assume an attitude that compounds would be active "if the test were more appropriate," whereas the biologist may be equally negligent by merely screening compounds, assuming a subservient role, failing to keep up with new developments in the literature, and not ascertaining whether in vitro activity has any in vivo corollary or selectivity.

The project team membership should be limited in number while maintaining viability (Jack, 1983), thus avoiding the inertia associated with large teams and allowing for the identification of key personnel in a project. The endless discussions and meetings that are the inevitable "curse" of large hierarchical organizations can only serve to dilute the progress of a bench-based R & D effort. Intellectual procrastination has no place in industry, and appreciation and respect for others' time should preclude lengthy meetings that are nothing more than "show and tell" sessions designed to impress the members and management of the importance of the individual presenting. A single experiment is worth ten meetings projecting its outcome (Brodie, 1986). Although project team meetings are useful for information flow, every effort should be made to encourage such flow at a more informal level. A team whose progress is reported and evaluated only at meetings is in poor shape. The team should also be able to make binding decisions, provided that they are not anarchical.

An excellent criterion for a project is that all data generated on a lead compound and its relevant standards should be publishable in a peer-reviewed scientific journal. Indeed, subject to patent constraints and the like, publication of data supporting the identification and development of a new chemical entity in a given therapeutic area should be a mandatory part of the drug development process. The project team also has political responsibilities, such as keeping lead compounds visible among the many others vying for budget support within the company. Accessing development and clinical support for a project is also a necessary part of compound development, as is involving those stages of the development process that are frequently thought of as support functions, e. g., toxicology, drug delivery, and formulation, in an active manner. Although not a current trend in the industry, the utilization of toxicological expertise at the research level can be very beneficial in alleviating a problem associated with a lead compound; by attempting to establish structure–activity relationships for the toxic effect, a backup compound can be more rapidly identified.

4. The Flow Chart

The costs involved in drug discovery make it very necessary to have a game plan to expedite the process. Although serendipity may still occur, the rational approach has much to recommend it from an intellectual vantage point. If the approach taken relies heavily on the use of biochemical methodologies, it is very necessary to have secondary and tertiary screens in place to rapidly substantiate the initial observations. The flow chart is a device borrowed from the business community that has two forms. The basic one is that in which a compound is identified in a primary screen and then put through a series of tests designed to test the efficacy and specificity of a new compound at the preclinical level. The second type of flow chart is a cost-effectiveness chart involving "go/no go" criteria. Schematic examples of these two types of chart are shown in Figs. 1 and 2. Quite clearly, one is independent of the other inasmuch as the compound flow chart has no finite time or cost boundaries. The superimposition of the chart in Fig. 2 on that in Fig. 1 is an additional method for ensuring that goal achievement is facilitated and that time constraints and costs are involved. The

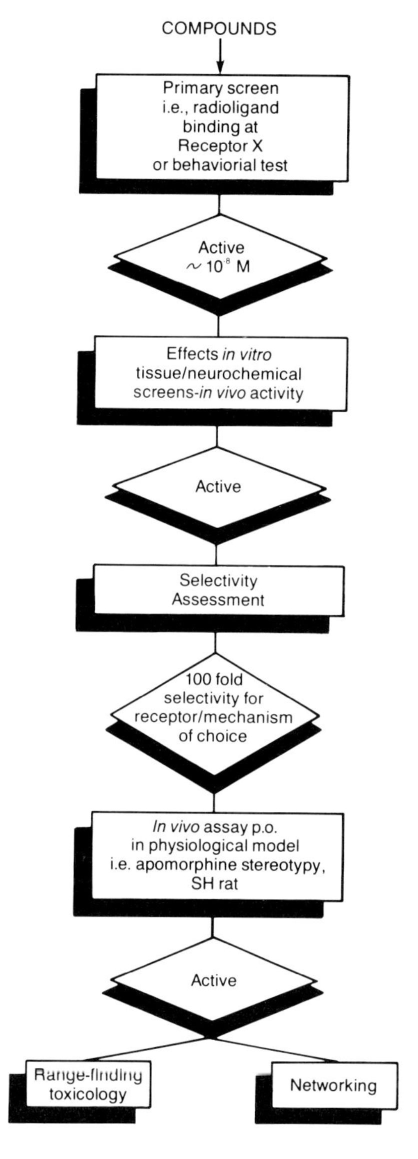

Fig. 1. Preclinical scientific project flow chart. A series of activity and selectivity criteria are established by the project team based on the current "state of the art" of the area under investigation. With such a flowchart in place, compound evaluation can proceed on a routine basis with periodic reevaluation dependent on the type of compound (*degree* of activity or selectivity) moving through the flowchart.

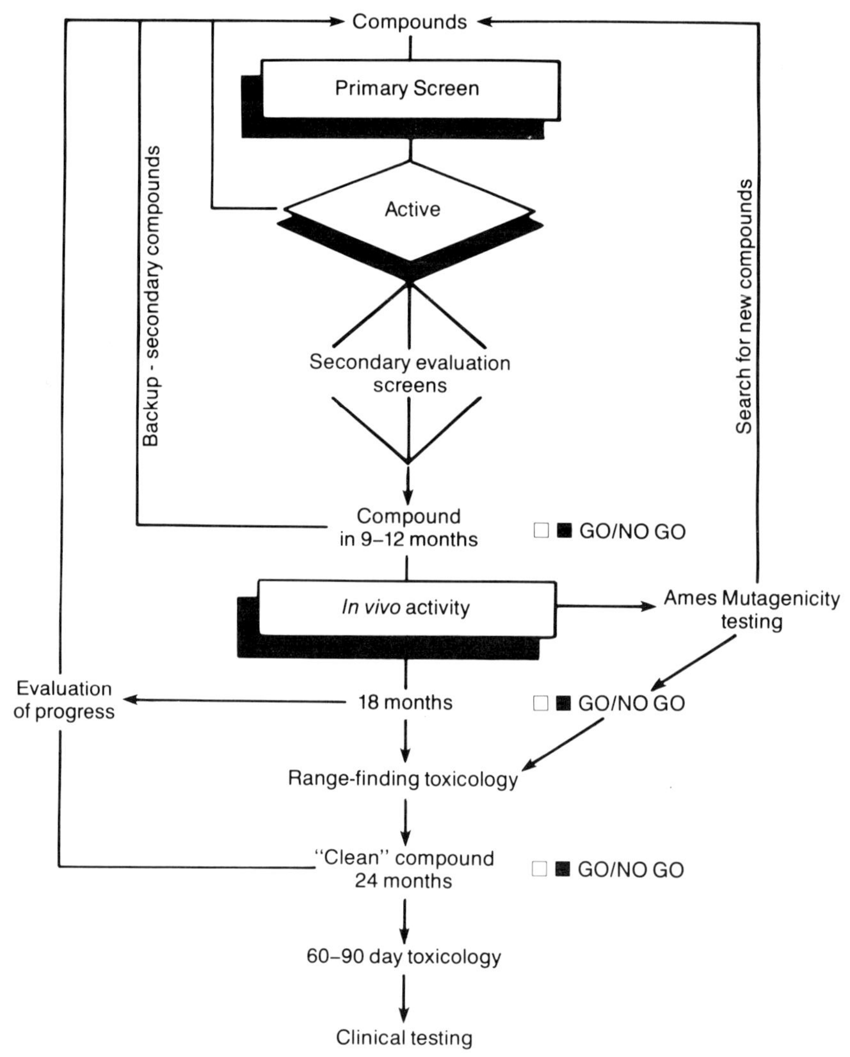

Fig. 2. Temporal flowchart (go/no go). A temporal flowchart such as that outlined here in schematic form can complement the preclinical flowchart (Fig. 1) and imposes time constraints on the drug discovery process. Such flowcharts can provide a realistic meeting point between the scientist and the budget/financial manager in terms of cost effectiveness and product expectations.

"go/no go" flow chart can facilitate decision-making by establishing criteria that are emotionally dissociated from the scientific aspects of the project. It would be of little competitive advantage to discover a potent analog of the LTD_4 antagonist FPL 55712 and take 30 years to do it. Although the goal may be achieved in that the compound is a viable therapeutic entity, the members of the project team will most likely be working for another company or be unemployed because of economic default.

4.1. The Primary Screen

The pharmacometric screen, involving the evaluation of new chemical entities in whole animal models, has been gradually replaced by more focused screens pertinent to a given discovery program. The classical concept that any new compound must do "something" if it is tested enough has given way to the more rational design of compounds (Baldwin, this volume). This in turn has resulted in the replacement of broad-based screening with more limited compound evaluation. There are pros and cons to this approach. In the first instance, the directed research program is more intellectually justified and possibly more economical. Random screening is inevitably dependent on models in which drugs already known have been discovered. This raises the possibility of only rediscovering "old compounds" or "me too" candidates. On the other hand, if a drug discovery program is only focused around a given enzyme inhibitor or neurotransmitter receptor, the likelihood of discovering a totally novel structural entity is remote. Other limitations of the pharmacometric screen involve the use of administration routes and time courses that relate to existing compounds and may very likely be of no relevance to new chemical entities. An additional negative in this random approach is the use of large numbers of animals. Such tests should not, however, be confused with those where animal usage is essential, as outlined below in sections 4.3 and 5.

The use of radioligand binding assays as primary screens is finding increasing favor (Snyder, 1983; Williams and U'Prichard, 1984; Williams and Wood, 1986; Creese, 1985; Williams and Enna, 1986), although these assays are not as widely accepted as may be thought or indeed as definitive as might be expected. Binding assays offer some very positive attributes in the drug discovery process (Enna, this volume). They require only small amounts (10–20 mg) of material, and by their very nature offer the most direct of all biological tests in terms of SAR generation since they involve the interaction of the compound directly with a recognition site. In addition they drastically reduce the number of animals used for compound evaluation, an issue of great importance at the present time. Binding assays do have their limitations, however, in that efficacy or physiological relevance cannot be measured. In addition, the delineation of compounds as agonists or antagonists is not always possible, although sodium ion (Pert and Synder, 1973) and GTP (Gavish et al., 1982) shifts can, in certain instances, identify compounds in this manner.

Once in vitro activity has been established, an in vivo correlation is necessary. The absence of activity in vivo may indicate either biotransformation or the lack of bioavailability. The converse, where a compound is inactive in vitro but shows activity in vivo, as in the case of the anticonvulsant/putative anxiolytic/central sympathomimeteic MK 801 (Clineschmidt et al., 1982), may be suggestive of a prodrug. Binding assays also involve their share of artifacts. The H_2-antagonists cimetidine and ranitidine, when radio-labeled, bind to imidazole and furane recognition sites rather than pharmacologically relevant H_2 receptors (Laduron, 1984), whereas the neuroleptic analog spiperone can bind to an anomalous spirodecanone site (Howlett et al., 1979). Another limitation of the binding technique is the assumption, based on the identification of the opiate receptor (Snyder, 1975) and recognition sites for the benzodiazepines (Squires and Braestrup, 1977) and tricyclic antidepressants (Raisman et al., 1980), that each drug has its own unique "receptor." This is very often a fallacious assumption, in that several compounds when labeled actually bind to known receptors. Binding assays have also been extended to the enzyme level. Compounds interacting with enkephalin convertase are potent displacers of the selective ligand ^{3}H-GEMSA (Strittmatter et al., 1984a), whereas radiolabeled captopril® can identify ACE-like binding sites in brain (Strittmatter et al., 1984b).

Binding assays may not always represent the optimal primary screen. Psychotherapeutic agents go through various vogues in terms of their postulated mechanisms of action and, since many such compounds tend to produce their effects only after chronic administration, it is necessary to predict their therapeutic potential following acute interactions with various receptor systems or in various animal models that have been shown to exhibit a relationship with active compounds. Inevitably, compounds with such in vitro activity require extensive in vivo evaluation, often extending to humans before the significance of such properties can be assessed. Ultimately, in many CNS-related disorders, it is the availability of novel compounds with defined mechanisms of action that allows for the testing of the hypothesis that resulted in such compounds being selected. At the present time, although classical antidepressants can cause selective decreases in the number of beta-adrenoceptors and serotonin-2 receptors (Charney et al., 1981), it is not well established as to whether this is a cause or effect phenomenon. In the area of excitatory amino acids, in which there is considerable interest in regard to the development of agonists or antagonists with potential use as anticonvulsants, for the treatment of ischemia or various neurological disorders including Huntington's disease (Shariff, 1984), newer binding assays have offered significant advantages (Foster and Fagg, 1984; Murphy et al., 1986).

Primary assays in these areas of research may, therefore, involve coupled systems such as adenylate cyclase, modulation of phospholipid metabolism, or where necessary, in vivo behavioral paradigms to assess compound activity. For antidepressants, the behavioral despair (Porsolt, 1981) or the muricidal rat (Vogel, 1971) models can be used as primary screens; in the area of antipsychotics, antagonism of dopamine-induced stereotype or hyperactivity is a routine screen (Martin et al., 1982; Williams et al., 1984), whereas in the anxiolytic area, shock-induced suppression of drinking behavior can be a high throughput screen (Patel and Malick, 1983).

The criteria for compound selection at the primary screening level are also important. In areas such as the benzodiazepine anxiolytics, where there are already compounds known that have affinity at the subnanomolar level for the central receptor and which have in vivo efficacy, activity in the $10^{-6}M$ range would not suffice to warrant compound selection.

4.2. The Secondary Screen

The secondary screen (or screens) serve to corroborate the activity identified in the primary screen and extend it by establishing some degree of functionality and, where appropriate, selectivity. A compound with activity at more than one receptor, a "polypharmic" compound, may also be of interest; in fact, many successful compounds have more than one activity and it is not known whether this profile is responsible for the efficacy of a compound or its side effects. In the antipsychotic area, the issue of extrapyramidal symptomatology and especially the induction of tardive dyskinesias is one of extreme concern. The availability of compounds that block dopaminergic hyperactivity and have ancillary properties that may prevent tardive dyskinesias has been a long and, to date, fruitless search. Dopamine-blocking agents with anticholinergic activity (Snyder et al., 1974) may be useful, effective neuroleptics. To date, however, all such relationships in terms of biological activity have been made with hindsight. The ultimate vindication of these hypotheses, as in the case of those related to the mode of action of antidepressants, relies solely on human data (Clineschmidt et al., 1979).

Efficacy and agonist/antagonist profiles at the secondary evaluation level can be assessed biochemically (adenylate cyclase, phospholipid turnover, modulation of receptor or enzyme-linked metabolite, or substrate formation; Enna, this volume; Fuller and Steranka, this volume). In the case of agonists, selectivity can be established using appropriate antagonists, whereas a compound without detectable activity could be evaluated as an antagonist. At this point IC_{50}, K_i, or p$A2$ values can be generated. Biological models for secondary evaluation would also include classical pharmacological models such as tissue bath preparations (rat vas deferens, guinea pig ileum or trachea) and whole animal models [spontaneously hypertensive rat (SH rat) and so on]. It is important at this stage that a potential lead compound be routinely compared with known standards to validate test procedures and extend the amount of information on the new compound.

4.3. The Tertiary Screen

Following evaluation in the secondary screening procedure, it is highly desirable to establish in vivo efficacy of the compound. Although this might be done as part of the secondary screen, before a compound assumes true ''lead'' status it must be shown to have in vivo activity before being considered a development candidate. Animal models that mimic the disease process are a valuable tertiary screen, since they should be capable of identifying active compounds regardless of chemical structure. In the area of hypertension, such models would be the SH rat (Antonaccio, 1984) and in the CNS, the behavioral despair (Porsolt, 1981), conflict models in rat (Cook and Davidson, 1973) or monkey (Patel and Malick, 1983), and muricidal rat (Vogel, 1971) models. Although these models have a relatively low throughput in terms of the number of compounds that can be studied, they are invaluable in assessing compound activity and efficacy. At the point that in vivo efficacy is tested, the aims of a given program and its overall importance will determine what precisely is sought. An aspirin-like analgesic that lacks oral activity is of little interest, whereas an opiate-like compound used in the treatment of the pain associated with terminal illnesses or an antiemetic used to prevent the emesis associated with chemotherapy may only need activity, regardless of the route of administration; this latter caveat may, however, limit the use of such a compound. Unless a specific reason has been identified for an oral route of administration, relative efficacy data on the lead compound can be generated using intraperitoneal (ip), intramuscular (im), subcutaneous (sc), and oral (po) administration routes. Duration-of-action studies and an acceptable side-effect profile can also be assessed at this point. With few exceptions, however, a development candidate must exhibit some degree of oral activity in several species before safety evaluation begins.

4.4. Networking

After preclinical evaluation, a compound then passes into toxicological and safety evaluation studies, and finally into clinical evaluation. These procedures are outlined in the present volume by Cavagnaro and Lewis and Reines and Fong.

In addition to these steps, a compound can at this stage be "networked," a term related to the substantiation of the primary properties identified in the molecule using more sophisticated biochemical and behavioral paradigms, as well as techniques such as electrophysiology, to assess the molecular mechanism of action of the compound. Quite clearly, as experience and serendipity show, there are many compounds in the clinic that have no clearly established mechanism of action, and the lack of such information should not preclude compound assessment through all stages of the process and flowchart. The availability of networking data and hopefully some idea of a compound's mechanism of action in comparison with a known entity (should there be one) can significantly enhance acceptance as a therapeutic agent. To this end, publication of preclinical data following patent filing can stimulate interest in a compound and enlist the support and interest of other biologists in establishing the usefulness of a compound. Unfortunately, many compounds that are developed have unacceptable side effects or toxicology that precluded their use as therapeutic entities. Nonetheless, such chemicals, which include 6-hydroxydopamine (Porter et al., 1962) and other chemical lesioning agents, are of inestimable value in basic research. The tragic but nonetheless serendipitous discovery that MPTP (*N*-methyl-4-phenyl-1,2,3,4-tetrahydropyridine) can in humans cause lesions similar to those associated with parkinsonism (Langston et al., 1983) has made available a new model for the study of anti-parkinsonian drugs.

4.5. Backup Compounds

A single chemical entity is unlikely to be the sole lead compound since, in developing a new structure through the primary and secondary assays, a structure–activity profile is inevitably established. Thus certain substitutions will increase activity, and others will reduce it. Following lead compound identification from a given chemical series, backups are required—either analogs in the same series or preferably other, novel chemical entities with similar profiles. Most preclinical testing paradigms do not objectively address the issues of bioavailability or pharmacokinetics, and it is possible that a given compound will be less favorable than anticipated because of unforeseen in vivo problems. In addition, toxicity is not always a predictable event. A compound found to be negative in the Ames mutagenicity test procedure may still have the potential to cause carcinogenic toxicity in mammals. Ideally, the Ames test should be performed as soon as possible after the identification of a lead compound to prevent the workup of a chemical entity that has no future.

5. Toxicological Evaluation

Once a compound has been taken through a relevant, objective flow chart designed to detect therapeutic potential and has passed the criteria established, it is then ready for toxicity evaluation, a series of tests dealt with in the present volume by Cavagnaro and Lewis. Initially, a compound is put into limited toxicity studies (1–6 wk) followed by a full 90-d study, involving extensive biochemical and histological examination of animal tissue (Traina, 1983). Another factor of importance in the future development of a compound as a therapeutic entity is the incidence of chronic effects unique to humans that do not show up readily in preclinical toxicity tests. This may reflect discrete species differences ascribable to tissue differences or to different metabolites being formed. At this point, there is still the possibility of the development of unacceptable side effects. To enter this stage of the drug development process without identified backups can be construed, at best, as shortsighted. The elimination of a compound because of toxicity problems can involve the loss of many millions of dollars, and although it is very necessary to establish such properties before a compound enters phase I—human trials—there has long been a need to develop cheaper, more rapid ways to assess compound toxicity. Liver toxicity tests comparable to the Ames test would be most welcome, and it may be anticipated given the increased emphasis on reducing animal usage and the need for rapid, reliable screens that biotechnology has an important contribution to make in regard to these aspects of drug evaluation. Hepatocyte cultures that can reflect the liver damage associated with heptatotoxic agents are just one potential in vitro approach to this issue.

6. Clinical Evaluation

The process of clinical evaluation has been dealt with in the excellent monographs by Pocock (1983) and Hamner (1982) and is dealt with in the current volume by Reines and Fong.

Clinical trials are broken down into four distinct categories:

Phase I: Clinical pharmacology and toxicity. Evaluation of drug safety in human volunteers using drug-escalation paradigms. Drug metabolism and bioavailability are also studied at this stage.

Phase II: Initial clinical investigation for treatment effect. Effectiveness and safety of the drug are evaluated.

Phase III: Full scale evaluation of treatment.

Phase IV: Postmarketing surveillance.

Such studies can take anywhere between 3 and 5 yr, depending on the therapeutic area and the social environment. In the area of Acquired Immune Deficiency Syndrome (AIDS), promising treatments would no doubt be "fast-tracked" to ensure that patients suffering from this insidious disease do not suffer unnecessarily because of administrative delays. In the area of baldness, it is unlikely that drugs to "treat" this essentially cosmetic disease, i. e., minoxidol, would be given priority.

7. Newer Aspects Of Drug Development

7.1. The Mechanistic Approach

As newer developments in science impact on the drug development process, commonly held conceptions become open to challenge. Tischler and Denkewalter (1966) reported some 20 yr ago that 95% of all pharmaceutical research ever undertaken had occurred since 1935. Many of the concepts that are currently in vogue in regard to drug discovery have arisen in the subsequent 20 yr. Although the receptor concept goes back to the turn of the century, it is only recently, with the development of cimetidine (Black et al., 1972) and the radioligand binding assays (Snyder, 1983; Williams and U'Prichard, 1984; Williams and Wood, 1986), that this cornerstone of pharmacology has been realistically addressed from a purely mechanistic viewpoint. The generation of SARs for the various classes of receptors and enzymes, although adding valuable knowledge, has not in reality resulted in a drug design program that is purely computer based (Goodford, 1984). If drug design were that easy, the content of this chapter and of the monograph itself would be far shorter.

Drug development in the future will be increasingly dependent on the manipulation of events at the molecular level, involving both membrane components and nucleic acids.

7.2. Drug Delivery

Newer methods of drug delivery, the subject of the chapter by Bondi and Pope in this volume, have also changed some of the criteria related to drug administration. The emesis associated with the administration of dopamine agonists (Van Rooyen and Offermeier, 1981) may be circumvented by the use of cutaneous patch administration. Slow release forms of various drugs and the consequent avoidance of bolus-type effects may circumvent many of the drawbacks associated with the present forms of drug administration, the unfortunate effects seen by Merck with their slow-release form of Indomethacin notwithstanding. The use of ''soft'' drug formulations (Bodor and Farag, 1982) and liposome-type delivery systems may radically change perceptions related to the properties required of a compound for successful use in humans if and when such techniques become applicable to the marketplace. The anticipated input from newer biotechnology concepts (Casement, 1984) will also change and hopefully improve the tactical approach to drug discovery. Hopes for the vectored delivery of CNS drugs across the blood–brain barrier may not be realized for another 10 yr or so. The tremendous impact that physicochemical approaches to drug delivery have had, however, reflected in a huge investment in basic research in this area by several major pharmaceutical companies, requires that the drug delivery specialist should be a necessary member of the drug discovery project team, having input at the preclinical discovery phase, rather than helping to put out forest fires that develop in the human trials.

8. The Future

8.1. The Industry . . .

As outlined in the section 1, the world in which the pharmaceutical industry now operates is different from that in which its fortunes were made. Increasingly the major corporations will have to adjust to the newer biotechnology conceptualizations of how research should be done. Although manpower allocations can be effective, it is increasingly apparent that critical masses in terms of intellect are of equal importance. Large companies are going to compete with smaller, "leaner," less-stratified organizations and although the corporate coffers can be used for licensing from, and marketing for, such entities, these major corporations will have to consider themselves as either research and development or development companies. If the latter, then names like Mitsubishi, Chase Manhattan, and Barclays will validly join the roster of the pharmaceutical industry and it in turn will become a service industry whose impact on society will solely be legislated by the bottom line. Good research requires large sums of money and a long-range commitment to success. Unless the industry continues its support of excellence in science, its position in society will be seriously undermined.

8.2. . . . and Society

The year 2000 has supplanted 1984 as an epochal time point in the evolution of Western civilization. Accordingly, the subject of pharmaceuticals at the turn of the century has been addressed (Bezold, 1983). Projections have been made that cancer and cardiovascular disease will be controllable by the end of the century

(Schwartz, 1983). Balanced against such optimism is the knowledge that health care costs and the geriatric population will increase such that the incidence of arthritis and rheumatism will increase, as will the occurrence of senile dementia and Alzheimer's disease. History has indicated that as one disease is eliminated or controlled, another takes it place, becoming prominent by virtue of the changes in life expectancy that accompany the gradual eradication of diseases that precluded the expression of the "new" disease. AIDS is a recent example of a disease that has emerged, whereas obesity has made the progression from a cosmetic preoccupation to a topic of major concern (Kolata, 1985).

The fact that antibiotics, contraceptives, and psychotherapeutic agents have all dramatically and irreversibly changed society is indicative of the impact that the pharmaceutical industry has had, and will continue to have, on society. The development of cognitive enhancers, or nootropics, to aid the elderly will similarly impact the fabric of society. On the whole, within a materialistic framework, these advances may be considered to have benefited humankind (Stetler, 1985). There is a negative side, however, and this may be more aptly viewed from an age-old ethical standpoint; an increasing concern with the quantity of life at the expense of the quality (Lasch, 1978). The addiction and dependence liability associated with the use of benzodiazepines, the extrapyramidal side effects associated with neuroleptic therapy, the life-negating properties of most current chemotherapeutic agents, the ability to control stress-induced diseases without addressing their genesis, and the ability to prolong life in a society that has difficulty in allocating its resources equitably at the present time and has yet to make old age synonymous with dignity, are all issues that the pharmaceutical industry will have to join society in addressing in a meaningful manner. Also of importance are the increasing ethical concerns related to the use of laboratory animals in drug evaluation and the often fanatical behavior of animal rights groups; all sectors of the industry are committed to finding ways of assuring drug safety and efficacy with the use of fewer animals.

Clearly, as the 20th century draws to its close and society becomes increasingly complex, the pharmaceutical industry must seek to assist in the search for knowledge and improvement in society that epitomizes the scientific ethic and in doing so avoid the scenario visualized by Sanders (1975) in his novel *The Tomorrow File* of a state-condoned, drug-regulated society.

Acknowledgments

The authors would like to thank their many friends and colleagues in the pharmaceutical industry who have aided, knowingly and unknowingly, the genesis of this overview.

References

Albers-Schonberg, G., Arison, B. H., Hirshfield, J. M., and Hoogsteen, K. (1981) The absolute stereochemistry and conformation of Avermectin B_2a aglycone and avermectin B_1a. *J. Amer. Chem. Soc.* **103**, 4221–4225.

Antonaccio, M. (1984) *Cardiovascular Pharmacology* 2nd Ed., Raven, New York.

Badawy, M. K. (1985) Career advancement. *Chem. Eng. News* **63**, 28–38.

Bartholini, G. (1983) Organization of Industrial Drug Research, in *Decision Making In Drug Research* (Gross, F., ed.) Raven, New York.

Bezold, C. (1983) *Pharmaceuticals In The Year 2000. The Changing Context for Drug R & D.* Institute for Alternative Futures, Alexandria, Virginia.

Black, J. W. and Stephenson, J. S. (1962) Pharmacology of a new adrenergic beta-receptor-blocking compound (nethalide). *Lancet* **ii**, 311–314.

Black, J. W., Duncan, W. A. M., Durant, C. J., Ganellin, C. R., and Parsons, E. M. (1972) Definition and antagonism of histamine H_2 receptors. *Nature* **236**, 385–390.

Bodor, N. and Farag, A. H. (1983) Improved delivery through biological membranes 13. Brain specific delivery of dopamine with a dihydropyridine $\rightleftharpoons$ pyridinium salt type redox delivery system. *J. Med. Chem.* **26**, 528–534.

Bradshaw, J., Brittain, R. T., Clitherow, J. W., Daly, M. J., Jack, D., Price, B. J., and Stables, R. (1979) AH 19065. A new potent, selective histamine H_2 receptor antagonist. *Br. J. Pharmacol.* **66**, 464p.

Brodie, B. B. (1986) *Apprentice to Genius. The Making of a Scientific Dynasty* (Kanigel, R., ed.) Macmillan, New York.

Campbell, W. C., Fisher, M. H., and Stapley, E. O. (1983) Ivermectin: A potent new antiparasitic agent. *Science* **221**, 823–828.

Casement, R. (1984) The Bug Business Becomes a Gold Rush, in *Man Suddenly Sees To The Edge Of The Universe* Open Court, La Salle, Illinois.

Chang, R. S. L., Lotti, V. J., Monaghan, R. L., Birnbaum, J., Stapley, E. O., Geotz, M. A., Albers-Schonberg, G., Patchett, A. A., Liesch, J. M., Hensens, O. D., and Springer, J. P. (1985) A potent non-peptide cholecystokinin antagonist selective for peripheral tissues isolated from *Aspergillus alliaceus*. *Science* **230**, 177–179.

Charney, D. S., Menkes, D. B., and Heninger, G. R. (1981) Receptor sensitivity and the mechanism of action of antidepressant treatment. *Arch. Gen. Psychiat.* **38**, 1160–1180.

Clineschmidt, B. V., Mckendry, M. A., Papp, N. L., Pflueger, P. B., Stone, C. A., Totaro, J. A., and Williams, M. (1979) Stereospecific antidopaminergic and anticholinergic actions of the enantiomers of (±)-1-cyclopropylmethyl-4-(3-trifluoromethylthio-5H-dibenzo[a,d]-cyclohe;ten-5-ylidene)piperidine (CTC), a derivative of cyproheptadine. *J. Pharmacol. Exp. Ther.* **208**, 460–467.

Clineschmidt, B. V., Williams, M., Witoslawski, J. J., Bunting, P. R., Risley, E. A., and Totaro, J. A. (1982) Restoration of shock-suppressed behavior by treatment with (+)-5-methyl-10,11 dihydro-5H-dibenzo[a,d]cyclohepten 5,10-imine (MK 801), a substance with potent anticonvulsant, central sympathomimimetic and apparent anxiolytic properties. *Drug Develop. Res.* **2**, 123–134.

Cook, L. P. and Davidson, A. B. (1973) Effect of Behaviorally Active Drugs in a Conflict-Punishment in Rats, in *The Benzodiazapines* (Garrattini, S., Mussini, R., and Randall, L. O., eds.) Raven, New York.

Creese, I. (1985) Receptor Binding as a Primary Drug Screen, in *Neurotransmitter Receptor Binding* (2nd Ed.) (Yamamura, H. I., Enna, S. J., and Kuhar, M. J., eds.) Raven, New York.

Cushman, D. W. and Ondetti, M. A. (1980) Inhibitors of angiotensin-converting enzyme. *Prog. Med. Chem.* **17**, 43–104.

Davies, L. P. (1985) Pharmacological studies on adenosine analogs isolated from marine organisms. *Trends Pharmacol. Sci.* **6**, 143–146.

Dibner, M. D. (1985a) The changing pharmaceutical industry: Impacts of biotechnology. *Trends Pharmacol. Sci.* **6**, 343–346.

Dibner, M. D. (1985b) Biotechnology in pharmaceuticals: The Japanese challenge. *Science* **229**, 1230–1235.

Economist (1985a) The medicine-men spend more to discover less. *Economist* **295** (7394), 93–94.

Economist (1985b) The primrose path. *Economist* **295** (7391), 104.

Fitzgerald, J. D. (1983) Reflections on Some Problems in the Management of Drug Discovery, in *Decision Making in Drug Research* (Gross, F., ed.) Raven, New York.

Fagg, G. E. and Foster, A. C. (1984) Acidic amino acid binding sites in mammalian neuronal membranes. Their characteristics and relationship to synaptic receptors. *Brain Res. Rev.* **7**, 103–164.

Gavish, M., Goodman, R. R., and Snyder, S. H. (1982) Solubilized adenosine receptors in the brain; regulation by guanine nucleotides. *Science* **215**, 1633–1635.

Goodford, P. J. (1984) Drug design by the method of receptor fit. *J. Med. Chem.* **27**, 557–564.

Hamner, C. E. (1982) *Drug Development*, CRC Press, Boca Raton, FL.

Hopfinger, A. J. (1985) Computer-assisted drug design. *J. Med. Chem.* **28**, 1133–1139.

Howlett, D. R., Morris, H. M., and Nahorski, S. R. (1979) Anomalous properties of [^{3}H] Spiperone binding sites in various areas of the rat limbic system. *Mol. Pharmacol.* **15**, 506–514.

Humphray, J. M., Daly, M. J., and Stables, R. (1982) Inhibition of gastric acid secretion by AH 22216, a new long-lasting histamine H-2 receptor antagonist. *Gut* **23**, A899.

Jack, D. (1983) Project Teams in Pharmaceutical Research, in *Decision Making in Drug Research* (Gross, F., ed.) Raven, New York.

Kolata, G. (1985) Why do people get fat? *Science* **227**, 1327–1328.

Laduron, P. (1984) Criteria for receptor sites in binding studies. *Biochem. Pharmacol.* **33**, 833–839.

Langston, J. W., Ballard, P., Tetrud, J. W., and Irwin, I. (1983) Chronic Parkinsonism in humans due to product of meperidine-analog synthesis. *Science* **219**, 979–982.

Lasch, C. (1978) *The Culture Of Narcissism, American Life in an Age of Diminishing Expectations.* Norton, New York.

Martin, G. E., Williams, M., and Haubrich, D. R. (1982) A pharmacological comparison of TL-99 and 3-PPP with selected dopamine agonists. *J. Pharmacol. Exp. Ther.* **223**, 298–304.

Maxwell, R. A. (1984) The state of the art of the science of drug discovery—an opinion. *Drug Dev. Res.* **4**, 375–389.

Murphy, D. E., Schneider, J., Lehmann, J., and Williams, M. (1986) Binding of [^{3}H]CPP (3(2-carboxypiperazin-4-yl)propyl-1-phosphonic acid): A selective high affinity ligand for *N*-methyl-D-aspartate receptors. *Abst. Soc. Neurosci.* **12**, 21.4.

Patchett, A., Harris, E., Tristan, E., Wyvratt, M., Wu, M. T., Taub, D., Peterson, E., Ikeler, T., ten Broeke, J., Payne, L., Oneyka, D., Thorsett, E., Greenlee, W., Lohr, N., Maycock, A., Hoffsommer, R., Joshua, H., Ruyle, W., Rothrock, J., Aster, S., Robinson, F. M., Sweet, C. S., Ulm, E. H., Gross, D. M., Vassil, T. C., and Stone, C. A. (1980) A new class of angiotensin converting enzyme inhibitors. *Nature* **288**, 280–283.

Patel, J. B. and Malick, J. B. (1983) Neuropharmacological Profile of an Anxiolytic, in *Anxiolytics—Neurochemical Behavioral and Clinical Per-*

spectives (Malick, J. B., Enna, S. J., and Yamamura, H. I., eds.) Raven, New York.

Pendelton, R. G., Torchiana, M. L., Chung, C., Cook, P., Weise, S., and Clineschmidt, B. V. (1983) Studies on MK 208 (YM-1170) a new slowly dissociable H_2 receptor antagonist. *Arch. Int. Pharm. Ther.* **266**, 4–11.

Pert, C. B. and Snyder, S. H. (1973) Properties of opiate receptor binding in rat brain. *Proc. Natl. Acad. Sci. USA* **70**, 2243–2247.

Pocock, S. J. (1983) *Clinical Trials. A Practical Approach.* Wiley, New York.

Porsolt, R. D. (1981) Behavioral Despair, in *Antidepressants—Neurochemical, Behavioral and Clinical Perspectives* (Enna, S. J., Malick, J. B., and Richelson, E., eds.) Raven, New York.

Porter, C. C., Watson, L. S., Titus, D. C., Totaro, J. A., and Byer, S. S. (1962) Inhibition of dopa decarboxylase by the hydrazine analog of α-methyldopa. *Biochem. Pharmacol.* **11**, 1067–1071.

Raisman, R., Briley, M., and Langer, S. Z. (1980) Specific tricyclic antidepressant binding sites in rat brain characterized by high affinity ^{3}H-imipramine binding. *Eur. J. Pharmacol.* **61**, 363–370.

Sachs, G. (1984) Pump blockers and ulcer disease. *N. Eng. J. Med.* **310**, 785–786.

Sanders, L. (1975) *The Tomorrow File.* Putnam, New York.

Schwartz, H. (1983) Images of the Year 2000: Alternative Pharmaceutical Futures, in *Pharmaceuticals In The Year 2000. The Changing Context for Drug R & D* (Bezold, C., ed.) Institute for Alternative Futures, Alexandria, Virginia.

Sharif, N. (1984) Excitatory amino acid receptors. *Handb. Neurochem.* **6**, 239–259.

Snyder, S. H. (1975) Neurotransmitter and drug receptors in the brain. *Biochem. Pharmacol.* **24**, 1371–1374.

Snyder, S. H. (1983) Neurotransmitter binding and drug discovery. *J. Med. Chem.* **26**, 1667–1672.

Snyder, S. H., Greenberg, D., and Yamamura, H. I. (1974) Antischizophrenic drugs and brain cholinergic receptors; affinity for muscarinic sites predicts extrapyramidal effects. *Arch. Gen. Psych.* **31**, 58–61.

Squires, R. F. and Braestrup, C. (1977) Benzodiazepine receptor in rat brain. *Nature* **266**, 732–734.

Stetler, C. J. (1985) An antidote for misconceptions. *Pharm. Exec.* **5**, 54–55.

Strittmatter, S. J., Lo, M. M. S., Javitch, J. A., and Snyder, S. H. (1984a) Autoradiographic visualization of antiotensin-converting enzyme in rat brain with [^{3}H] captopril located to a striatonigral pathway. *Proc. Natl. Acad. Sci. USA* **81**, 1599–1605.

Strittmatter, S. J., Lynch, D. R., and Snyder, S. H. (1984b) [^{3}H] Guandino ethylmercaptosuccinic acid binding to tissue homogenates. *J. Biol. Chem.* **259**, 11812–11817.

Sweet, C. S. and Blaine, E. H. (1984) Angiotensin-Converting Enzyme and Renin Inhibitors, in *Cardiovascular Pharmacology* (2nd Ed.) (Antonaccio, M., ed.) Raven, New York.

Testa, B. (1984) Drugs? Drug research? Advances in drug research? Musings of a medicinal chemist. *Adv. Drug. Res.* **13**, 1–58.

Tischler, M. and Denkewalter, R. G. (1966) Drug Research: Whence and Whither? in *Progress in Drug Research* (Jucker, E., ed.) Basel, Switzerland.

Topliss, J. (1983) *Quantitative Structure–Activity Relationships Of Drugs*, Academic, New York.

Traina, V. M. (1983) The role of toxiology in drug research and development. *Med. Res. Rev.* **3**, 43–72.

Van Rooyen, J. M. and Offermeier, J. (1981) Peripheral dopaminergic receptors. Physiological and pharmaceutical aspects of therapeutic importance. *S. Afr. J. med.* **59**, 329–332.

Vogel, J. R. (1971) Antidepressants and Mouse Killing (Muricide) Behavior, in *Antidepressants, Industrial Pharmacology* vol. 2 (Fielding, S., and Lal, H., eds.) Futura, Mt. Kisco, New York.

Williams, M. (1983) Anxioselective anxiolytics. *J. Med. Chem.* **26**, 619–628.

Williams, M. and Enna, S. J. (1986) The receptor: from concept to function. *Ann. Rep. Med. Chem.* **21**, 211–235.

Williams, M. and U'Prichard, D. C. (1984) Drug discovery at the molecular level: A decade of radioligand binding in retrospect. *Ann. Rep. Med. Chem.* **19**, 283–292.

Williams, M. and Wood, P. L. (1986) Receptor Binding as a Tool in Drug Screening and Evaluation. *Neuromethods* vol. 4 (Boulton, A. A., Baker, G. B., and Hrdina, P., eds.) Humana, New Jersey.

Williams, M., Jones, J. H., and Watling, K. J. (1983) Biochemical evaluation of the enatiomers of the novel ergoline 6-ethyl-9-oxaergoline (EOE). *Drug Dev. Res.* **3**, 573–579.

Williams, M., Martin, G. E., Taylor, D. A., Yarbrough, G. G., Bendesky, R. J., King, S. W., Robinson, J. L., Totaro, J. A., and Clineschmidt, B. V. (1984) L-646, 462, a cyproheptadine related antagonist of dopamine and serotonin with selectivity for peripheral systems. *J. Pharmacol. Exp. Ther.* **229**, 775–781.

Williams, W. (1985) Glory days end for pharmaceuticals. *New York Times* 24 February 1985, sec. 3, p. F1.

Yokoyama, N., Ritter, B., and Neubert, A. D. (1982) 2-Arylpyrazolo[4,3c] quinolin-3-ones: A novel agonist, a partial agonist and an antagonist of benzodiazepines. *J. Med. Chem.* **25**, 337–339.

Compound Discovery

Drug Design

JOHN J. BALDWIN

1. Drug Design

The discovery of drugs depends on a number of interdependent factors, including insight, serendipity, and persistence. A common characteristic of the successful program is a clearly delineated strategy based on quantitative pharmacological assays. The particular problem chosen often will influence this strategy and dictate the type of approach; that is, either a rational or an empirical one. Depending on this selection, the synthetic targets then will either be derived from drug design concepts for the former or developed around systematic variation of a lead compound for the latter.

The choice of drug design, as the problem-solving technique, implies the selection of a rational approach that is based on detailed information concerning the structure of the target receptor or enzyme. The term has been broadened, however, to include a hypothesis-based lead development effort that evolves from a consideration of a suspected or known biological mechanism (Williams and Malick, this volume). Success under this broader definition will depend on the depth of biochemical knowledge, the validity of the hypothesis, and the probability of finding compounds that act by the required mechanism. An empirical strategy for drug discovery has been termed practical drug design or systematic drug design and is an iterative process of chemical modification emphasizing a stepwise improvement.

The number of books and reviews describing the drug design process must approximate the number of compounds that have made the transition from test tube to pharmacy shelf. In spite of this, there are few examples of successes that have been based solely on the rational approach, i. e., from a basic understanding of the macromolecular structure. With the current techniques, concepts, and databases, however, design is becoming possible, and even the more empirical methods are being influenced by the philosophy of rational strategy.

The question is not, can new drugs be discovered, but, can a rational approach do so more expeditiously. Toward this end, it is clear that medicinal chemistry, along with molecular pharmacology and computer-assisted design, forms the new basis for drug discovery (Hopfinger, 1984; Cohen, 1983; Loftus et al., this volume). The former of these, molecular pharmacology, may be defined as the application of molecular biology and biochemistry to drug discovery and involves the incorporation of receptor technology and biochemical pathways into the research strategy. By starting the drug design process at the level of receptors and enzymes, the variables of absorption, distribution, metabolism, and excretion are set aside temporarily, thus allowing the optimization of affinity or potency to become the key issue. The ability to understand, even at a primitive level, the interaction of a ligand with a macromolecule is the central issue in drug research.

In spite of the recent progress in molecular pharmacology to open new windows for discovery, and although advances in design are being made, drug discovery today remains largely empirical, a mixture of intuition, experience, and serendipity. The guide to design under the empirical approach is the structure–activity relationship, which is often subjected to intuitive surveys rather than statistical analysis. Nevertheless, such relationships remain central to the process of lead optimization (Albert, 1971, 1982; Hathway, 1982).

Two aspects of the drug discovery process will be emphasized in the following pages; they are the critical issues of lead selection and development. Both rational and empirical approaches will be discussed, highlighting those options and strategies that have the potential to improve the efficiency of the discovery effort.

2. Lead Discovery by Design

Drug design in its purest form requires detailed structural knowledge of the receptor or the active site of an enzyme. In certain cases, a thorough understanding of the reaction mechanism may serve as the basis on which to design specific inhibitors. The type of information needed for *de novo* design can best be obtained by X-ray crystallographic techniques. Although this method has been successful with model enzymes, it has not been applied to receptors. Members of this latter class of macromolecules are structurally more complex, often being composed of subunits, and are associated with membranes (Meunier et al., 1972; Defau et al., 1975). In addition, none of the solubilized receptor complexes have been amenable to crystallographic determinations.

With a few enzymes the technical problems of quantity, purity, and crystallinity have been overcome, and X-ray structures have been generated. The most widely studied is dihydrofolate reductase, whose X-ray crystallographic structure has been used in the drug design process and for which novel high-affinity inhibitors have been produced (Burchall and Hitchings, 1965; Baker et al., 1981; Kuyper et al., 1982). A second example is renin; a three-dimensional model of this enzyme has been developed using a combination of techniques. In this approach, X-ray crystallographic data for the related protein, pepsin, were combined with interactive computer graphics to build a working model from which surfaces defining hydrophobicity, charge distribution, and hydrogen bonding can be generated. In a drug design mode, molecular models of proposed substrate analogs can be compared with the surfaces of the active site searching for physiochemical and structural compatibility (Tickle et al., 1984).

Thus far there are no examples of drugs whose design was based on the X-ray structure of the macromolecule. All have failed one or more of the other requirements, i. e., transport, metabolic stability, and biological half-life. This observation had led Goodford to emphasize that in the receptor fit approach to drug design, not only must the macromolecule be a therapeutically relevant target, but the appropriate physiochemical properties must be incorporated into the design using the principles of Hammett and Hansch (Goodford, 1984). Regarding this combined approach, Hansch has concluded that although crystallography will play an important role in drug discovery, it will not replace information from well-designed structural probes of the receptor (Hansch, 1982). Although such a dual-design process can, in principle, lead directly to compounds of high affinity, bioavailability, and specificity, it may be more realistic to incorporate those critical ancillary properties of a drug into a separate development phase after affinity has been optimized. An integrated approach of target selection, X-ray determination, and lead design followed by optimization may be the most expeditious way to proceed using this totally rational approach.

Although the ideal starting point in designing an inhibitor may be the X-ray crystal structure of the enzyme, a sophisticated understanding of how the target enzyme carries out its transformation also can serve as the primary database. Using such a mechanism-based approach, the target selection criteria must be expanded; not only must the enzyme be therapeutically relevant, but the biochemical pathway and underlying enzymatic mechanism must be understood. It is usually preferable for the selected enzyme to be the one that is rate limiting in the sequence. In addition there should be a convenient source of material and a relatively straightforward assay.

Once the enzymatic target has been chosen, the design of inhibitors can be initiated. The overall strategy can evolve from a consideration of either the proposed transition state or the ground-state structure. The aim can be to achieve enzyme inhibition through reversible or irreversible mechanisms.

Among the reversible classes of inhibitors, the transition state analog approach has received the most attention and is the subject of several reviews (Wolfenden, 1969a,b, 1972, 1976, 1978; Lienhard, 1972, 1973; Douglas, 1983; Stark and Bartlett, 1983; Andrews and Winkler, 1984). This approach, pioneered by Wolfenden and Lienhard, seeks to take advantage of binding interactions available to the transition state structure. Usually such an inhibitor will structurally resemble the transition state for the enzymatic process and is expected to bind more tightly than substrate, although it is reversible and competitive.

Examples include the inhibition of adenosine deaminase by 1, a compound that mimics the proposed transition state 2 (Wolfenden, 1969a; Cha et al., 1975). The antibiotic norjirimycin 3 has been viewed as a transition state inhibitor of glycosidase. Although the structural relationship of the dehydro form 4 to the putative intermediate 5 is clear, the transition state mechanism has not been established. Methionine sulfoximine 6 inhibits glutamine synthetase after phosphorylation generates the transition state analog 7 of the proposed intermediate 8. Replacement of a peptide carbonyl by a tetrahedral phosphorus as in the phosphonamidate analogs leads to effective inhibitors of peptidases. A recent application of this strategy is seen in the design of the angiotensin converting enzyme (ACE) inhibitor 9 (Thorsett et al., 1982).

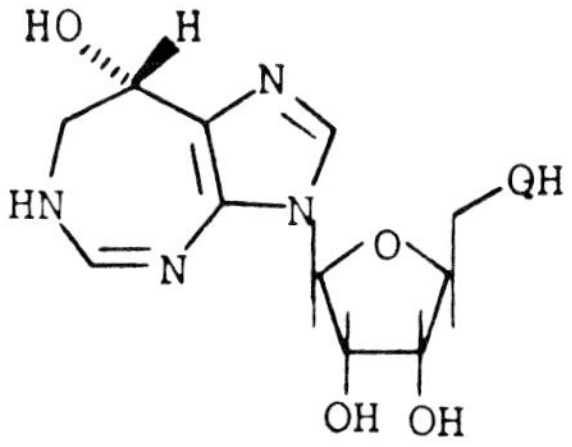

1 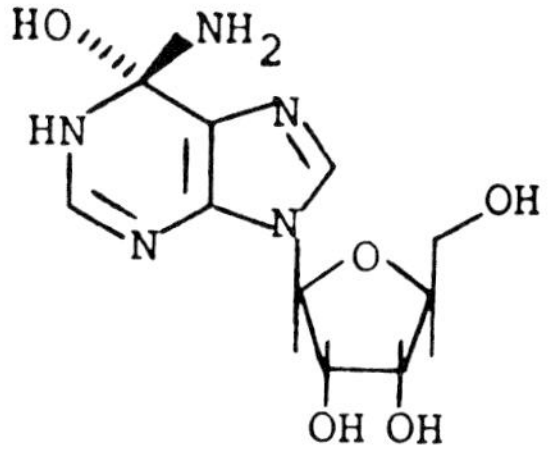 2

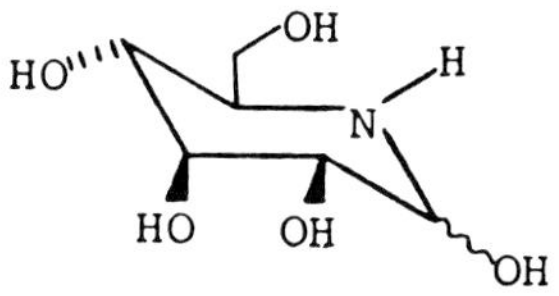

3 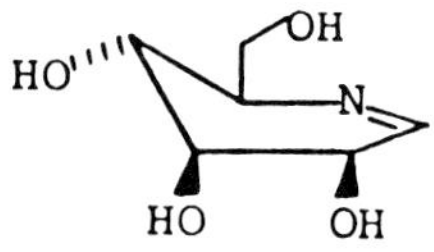 4

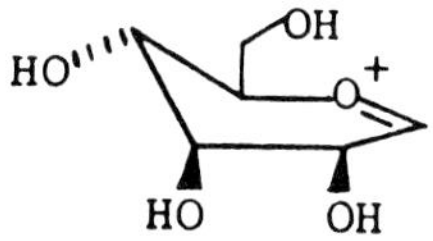

5

An analogous strategy based on the hypothesis that the *S*-3-hydroxyl group of the statine residue in pepstatin serves a key role in the transition state mechanism for inhibition of hydrolysis at the peptide scissile bond (Marciniszyn et al., 1976; Schmidt et al., 1982; Tewksbury et al., 1981) had led to the design of the highly potent renin inhibitor 10 (Boger et al., 1983; Boger, 1983). Similarly, the aminoalcohol modification in 11 was designed to mimic the putative transition state at the scissile bond in the ACE substrate *N*-benzoyl-Phe-Ala-Pro (Gordon et al., 1985).

Although there are numerous examples of the successful design of transition state analogs, few have proceeded past the level of biochemical interest. The strategy itself focuses on the design of novel, potent inhibitors and does not consider metabolic and transport requirements. This is the same concern expressed by Goodford (1984) and Hansch (1982) in their considerations of designs based on macromolecular structure. Thus the design segment must be considered as only the first step in the process of drug discovery.

The second favored design concept based on enzyme inhibition is the mechanism-based approach. This strategy has been the focus for several reviews (Kalman, 1981; Abeles and Maycock, 1976; Walsh, 1978; Stark and Bartlett, 1983; Silverman and Hoffman, 1984). In this scheme the enzyme must first act on the inhibitor to generate a species that is capable of blocking the enzyme either reversibly or irreversibly. The latter type is the more common and examples operating through this mechanism have been termed suicide inhibitors. It can be argued that irreversible mechanism-based inhibitors, because of the required enzymatic activation, will be free of the nonspecific alkylation seen with the haloacetyl type of active site-directed irreversible agent (Burger, 1983).

This approach has found broadest application for those enzymes requiring pyridoxal phosphate or flavin as necessary cofactors. Examples include the inhibition of GABA aminotransferase by 4-amino-5-fluoropentanoic acid (Silverman and Levy, 1981) and by gabaculine (Rando, 1977), the inhibition of 2-hydroxy acid oxidase by 2-hydroxy-3-butynoate (Cromartie and Walsh, 1975), the inhibition of D-amino acid transaminase by D-vinylglycine (Soper et al., 1977), and the inhibition of aromatic amino acid decarboxylase by α-vinyl DOPA (Metcalf and Jund, 1977). In addition, a number of drugs have been determined retrospectively to be mechanism-based inhibitors; these have been summarized by Silverman and Hoffman and include chloramphenicol, tranylcypromine, 5-fluorouracil, norethindrone, 5-trifluoromethyl-2-deoxyuridine, and allopurinol (Silverman and Hoffman, 1984).

3. Lead Discovery by Random and Directed Screening

When drug design through the use of X-ray crystal structure or enzymatic mechanisms is not an option, the traditional method of lead discovery can be used. In this approach the structural basis for the optimization effort is supplied either from a screening program or developed around a known agent (Maxwell, 1984; Testa, 1984; Wooldridge, 1984). Figure 1 schematizes an idealized pathway illustrating the central role of lead discovery in drug development.

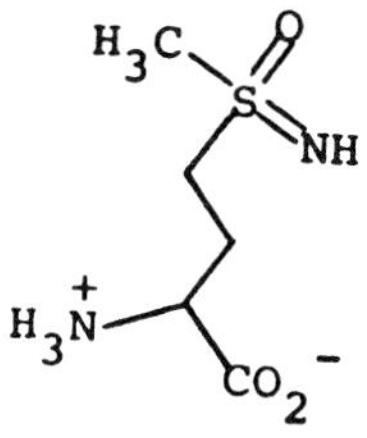

H_3C O S NH $H_3\overset{+}{N}$ CO_2^-

6

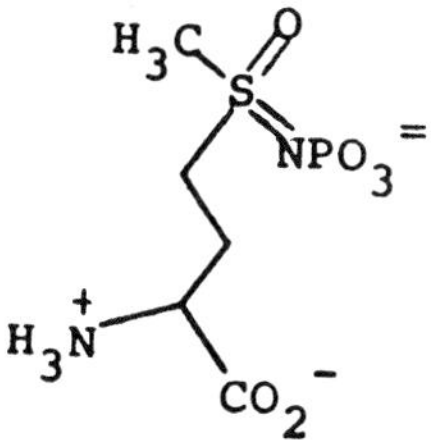

H_3C O S $NPO_3^=$ $H_3\overset{+}{N}$ CO_2^-

7

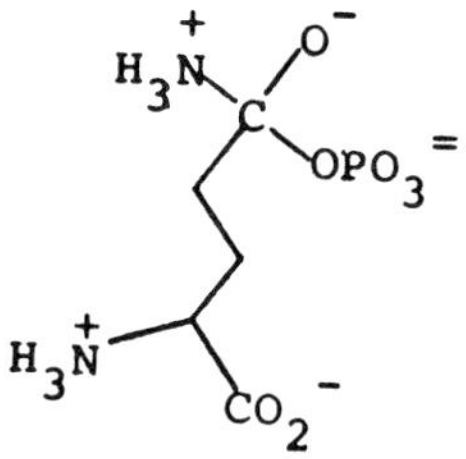

$H_3\overset{+}{N}$ O^- C $OPO_3^=$ $H_3\overset{+}{N}$ CO_2^-

8

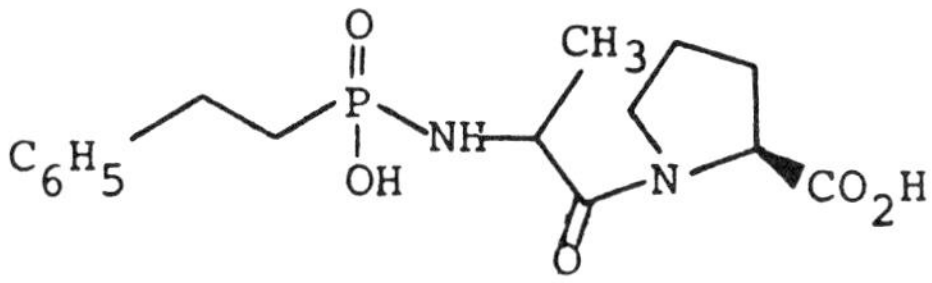

C_6H_5 O P OH NH CH_3 O N CO_2H

9

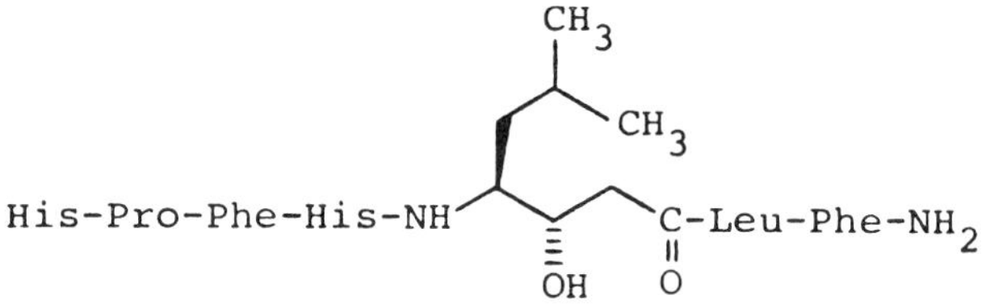

CH_3 CH_3 His-Pro-Phe-His-NH OH C-Leu-Phe-NH_2 O

10

C_6H_5 C_6H_5 NH O H OH NH O N CO_2H

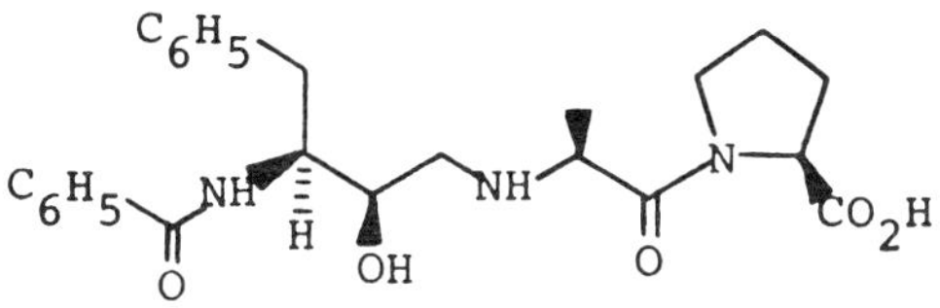

11

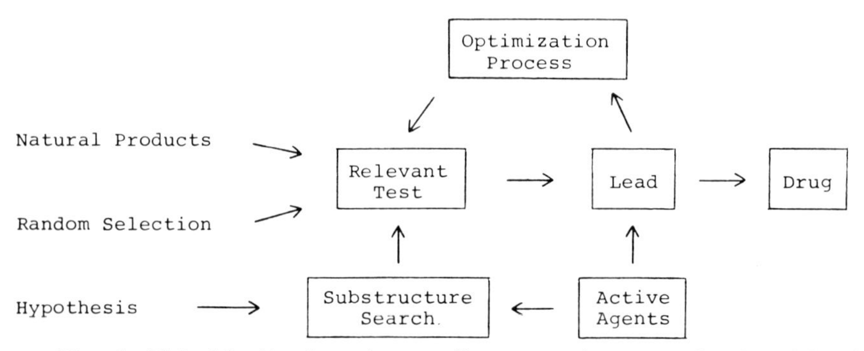

Fig. 1. This idealized pathway illustrates the central role of lead discovery and optimization in the drug discovery process.

Screening refers to a random testing of compounds or mixtures in a relevant, high-throughput assay. Pharmacological evaluation of either substances that look like the substrate, i. e., the antimetabolite concept, or natural products has been historically the most successful path to lead discovery. Enzymatic- or radioligand-based protocols are examples of rapid test systems that are suitable for such initial evaluations. The test compounds can be secured from numerous sources, including synthesis (Westwood, 1981), proprietary sample collections, arrangements with other laboratories, microbiological sources (Conover, 1971), toxins, and plant and marine extracts. The structurally novel, innovative leads are most often found in the latter sources; that is, among naturally occurring compounds. Recent examples of successes based on the random screening of fermentation broths include the antiparasitic avermectins (Albers-Schonberg et al., 1981) and the HMG-CoA reductase inhibitors, mevinolin (Alberts et al., 1980) and compactin (Endo et al., 1976). The problems and successes concerned with finding leads from plant sources have been described in detail by Farnsworth and Bingel (1977). Imidazobenzoxazines and triazoisoquinazolinones are examples of novel lead structures discovered by evaluating unanticipated products formed during chemical reactions (Westwood, 1981).

To improve the chances of successful lead discovery from screening programs, especially those following the natural product approach, a major commitment is required in terms of equipment and personnel. The cost of high-volume production, testing, isolation, and characterization of natural products is high and the chance for success low (Robinson, 1974). This random approach to lead discovery has been criticized as inefficient (Editorial, *Lancet*, 1981).

A rational, science-based research effort centered on molecular biology has been proposed as the replacement. For the foreseeable future, however, the chemist will need a basis for design and a structure to explore; the advances in biology likely will serve to provide more sophisticated, novel test systems for the evaluation of potential leads. Such new protocols are critical for major therapeutic advances, since the use of established models will likely provide new compounds acting by old mechanisms.

A middle ground between the extremes of design and chance is the hypothesis-dependent screen. This can serve as an alternative for or supplement to the totally random approach. This strategy uses a mechanistic hypothesis or structural information about the natural ligand, substrate, or inhibitor as the starting point for compound selection. As illustrated in Fig. 1, substructure searching is the approach of choice to find compounds for testing that contain the particular arrangement of atoms thought to be important for the biological response. When large data sets must be searched, computer-assisted substructure matching capabilities are a near mandatory requirement (Gund, 1984). Active compounds found from substructure screening serve as the basis for optimization.

A second method of lead selection is simply to use a known active compound. Two of the common rationales put forward to justify a program built around a known agent are (1) to improve the parent in terms of potency or side effects, and (2) to exploit an unexpected biological activity. When the objective is to improve the known agent, for example, in terms of subreceptor selectivity or pharmacokinetic properties, then assays are needed that will quantitatively distinguish between compounds in the targeted area. The approach can be hypothesis-based or depend on serendipity. Maxwell (1984) has divided such molecular modification of known agents into two classes. The "enlightened" approach is hypothesis-based and directed toward overcoming deficiencies of the parent; the "unenlightened" approach is one whose main goal is to discover a biologically active analog that may provide by chance an improvement over the parent. It is recognized that historically this latter approach has led to improved therapy (Halberstam, 1979). Examples superior to the parent have frequently been produced by intuitive modifications. The ultimate economic success of the "unenlightened" approach depends on producing a drug candidate within a narrow time frame; that is, during the period between the discovery of the parent and the over-exploitation of the class. To achieve this objective, the molecular modification program must be initiated as soon as the significance of the parent is recognized. The strategy should allow for a rapid development program; a critical mass of chemists, biochemists, and pharmacologists must be assembled quickly in a well-coordinated program. Without decisiveness and rapid implementation, such a program will produce, as is usually the case, too little too late.

4. Quantitative Structure–Activity Relationships

When direct information about the receptor is lacking, or when the enzymatic mechanism is not fully understood, more indirect approaches must be used to optimize the selected lead structure. Under these conditions, lead development depends on building a meaningful structure–activity relationship. Two tools are available to search for such a relationship and, when found, to predict the biological response; they are quantitative structure–activity relationships (QSAR) and pattern recognition.

The first of these, QSAR (Martin, 1979, 1981, 1983; Tute, 1971; Hansch and Blaney, 1984; Fujita, 1984), has become known as the Hansch approach in recognition of the pioneering role of Corwin Hansch. This method is based on the premise that within a series of related compounds the potency of a biological response can be expressed in terms of a function of quantifiable physiochemical parameters, such as hydrophobic, electrostatic, steric, and dispersion interactions. It is assumed that within a cogeneric series the compounds are exerting their effects by identical mechanisms. It follows that if the mode of interaction with the macromolecule for each analog is the same, then differences in biological activity are related to changes in physical parameters induced by structural variation. If the three-dimensional structure is considered, then the variable substituent is likely to affect activity from a noncritical position on the molecular surface.

The parameters (Hansch and Leo, 1979) that have found widest application are measures of hydrophobicity, such as the octanol–water partition coefficient, the electronic characteristics monitored as the pK_a or Hammett values, the molar refractivity (M_R), which is a measure of the tendency to participate in Van der Waal's interactions, and steric parameters (Charton, 1983; Motoc, 1983; Charton and Motoc, 1983; Kier, 1980), such as the Taft value E_s and Verloop parameters.

For the Hansch analysis to be applicable, the biological response must be relevant and quantitative. In addition, data must be generated on all compounds, not just the few most active. This requirement for dose–response curves and not simply single-point determinations has been reviewed by Martin (Martin, 1978) and has proven to be a major stumbling block in the application of QSAR. The biological result (e. g., ED_{50}, LD_{50}) is usually expressed as log $1/C$, where C is the molar concentration of the compound eliciting the standardized response.

Since it is assumed that the data are continuous, a least-squares multiple regression analysis can be performed for determination of the coefficient. Such an analysis yields the basic Hansch equation

$$\log 1/C = K_i\pi - K_2\pi^2 + K_3\sigma + K_4$$

This equation reflects the underlying assumption that the effects of the substituents on the molecular interaction are related to the biological response and are equal to an additive combination of the effects of these substituents on model systems. In this example, π and σ are illustrated as parameters. The π value is a substituent constant formulated by Hansch and is a reflection of hydrophobicity. It is defined by the Hammett-like relationship

$$\pi = \log P_X/P_H = \log P_X - \log P_H$$

when P_X and P_H are the octanol-water partition coefficient for the substituted and unsubstituted examples, respectively. A lower π value denotes greater hydrophilicity and a higher value indicates greater lipophilicity. The π value allows the partition coefficient of compounds to be calculated. Methods for complex molecules have been described and computer programs written (Chou and Jurs, 1979).

The second term in the Hansch equation, the π^2 term, reflects the parabolic relationship that is often seen between lipophilicity and biological activity. Since the equation represents a linear free-energy relationship, parameters (Martin, 1979) other than σ may be used, such as dipole moments, polarizability, inductive-field effects, and resonance effects.

In order for the regression equation to be predictive, certain statistical criteria must be met. These conditions have been described by Wold and Dunn (1983) for each of the common models used in QSAR. Simply stated, the number of indicator variables used in the analysis must be less than the number of compounds used in the set. When the number of parameters (P) equals or exceeds the degrees of freedom (DOF), the predictive ability of the model becomes no better than chance. The relationship between P and DOF can also be affected by bias causing the $P << DOF$ rule to be violated. A common example of this type of bias is the selection of compounds or variables to be used in the analysis; for example, by using only those compounds that fit a preconceived model and classifying others as exceptions, or by including more variables in the calculation of confidence limits than are actually used in the analysis. These problems have been reviewed by Topliss (Topliss and Edwards, 1979).

Another difficulty associated with QSAR is the retrospective nature of the method. The large database needed for the analysis makes the technique more appropriate in the final stages of the design process (Austel, 1984). Associated with this criticism is the observation that in practice QSAR has not been able to replace the more intuitive approaches (Burger, 1983). Because of this need for a large database, the method has been viewed and applied as a method for optimization of well-characterized series. Indeed the two examples of marketed products that resulted from a QSAR approach and were cited by Hansch (1984) and Fujita (1984) were based on well-exploited areas. As pointed out by Hansch (1984), however, because of the complexity of the problems these optimizations probably could not have been achieved without the application of QSAR.

In spite of such criticism, it is clear that the Hansch method has changed the way drug design problems are viewed and attacked. Most strategies include the Hansch approach as a critical component in the overall plan. More important than this nearly general acceptance is the fact that the Hansch approach represents a vibrant and developing field. The method is being incorporated into other strategies and new applications are being developed. A simple example of this integration of the linear free-energy approach with other methods is illustrated by a combination of the Hansch and the Free and Wilson methods (Martin, 1979, 1981). In the method of Free and Wilson (1964), only the presence or absence of a substituent is noted; a measurement of physical properties is not required. The biological activity is expressed as log $1/C$ and is related to the number and position of the substituents by the equation:

$$\log 1/C = X + \sum^{mn} a_{ij} G_{ij}$$

where m is the number of substituted positions; n is the number of substituents; a_{ij} is the presence or absence of a substituent j at position i, and G_{ij} is the group contribution obtained by multiple regression analysis. This method applies best to multisubstituent problems; Martin (1979, 1981) has pointed out that for validity each substituent that occurs at any position must be present at that position in at least two examples. By combining the Hansch and Free and Wilson methods, it is possible to include in the analysis an indicator variable, i. e., one that denotes the presence or absence of a particular substituent, atom, ring, and so on, that is believed to effect biological activity by a process other than through its physiochemical properties (Martin, 1979, 1981).

This growth in the applicability of QSAR has been summarized by Martin with examples from receptor mapping using local properties, the treatment of stereoisomers, and the design of test series (Martin, 1979, 1981; Martin and Panas, 1979).

5. Pattern Recognition

A second method to search for those features that are critical to biological activity is pattern recognition. This technique is receiving increased attention and has been the subject of in-depth reviews (Kirschner and Kowalski, 1979; Jurs et al., 1981; Jurs, 1983; Ramiller, 1984).

Thus far, the principal use of this technique has been the classification of compounds according to the presence or absence of a particular biological activity (Ramiller, 1984). Large historic databases have been analyzed in such fields as carcinogens (Jurs et al., 1983; Rose and Jurs, 1982; Yuta and Jurs, 1981), antitumor agents (Henry et al., 1982), anticonvulsants (Klopman and Contreras, 1985), olfactory stimulants (Brugger and Jurs, 1977), and monoamine oxidase inhibitors (Martin et al., 1974), among others.

The method of pattern recognition as applied to structure–activity relationships has evolved from the field of artificial intelligence; such a technique is, in principle, ideally suited to unravel the complex interdependences found at the chemical–biological interface. When this method is applied to a series of structurally related compounds, the local linearization of complicated functional relationships is assumed (Wold and Dunn, 1983). The goal is to identify those properties, present in each member of a series, that are directly related to the biological activity. These classifiers, once defined, should provide a means to discriminate between those proposed synthetic targets that will be biologically active from those that will not.

An advantage of pattern recognition over QSAR is that it lends itself to the use of qualitative rather than quantitative biological data. In practice, members of a series of known compounds are divided into two sets, the training set and the test set. The training set is used to define those features of the molecule that may be related to the biological activity. Obviously the accuracy and relevance of both the molecular parameters and the pharmacological data are critical to a successful and meaningful analysis. After the parameters separating the active from inactive compounds have been defined, the test set can be used to validate the pattern recognition through a comparison of the predicted biological response with the observed.

The molecular descriptors receiving the widest application include substructure coding, position coding that defines the location of a substructure, electronic coding, such as partial charges, dipole moments, and bond strengths, mass spectral coding, physiochemical coding, and three-dimensional coding. The typical geometric descriptors include molecular volume, surface area, and moments of inertia.

In the ADAPT computer program of Jurs (1983), the structures are entered into the program, which converts them to three-dimensional strain-minimized conformers. A search is then initiated for those descriptors that separate the set into two subsets within multidimensional space. Once discriminants have been found they can be tested for predictive ability and used for design of new analogs.

The CASE computer program for pattern recognition has been described by Klopman and Contreras (1985). This program is capable of selecting the factors that are relevant to the biological activity and automatically generating suitable descriptors. The method is based on molecular fragments of the chemical structure that are formed by breaking the molecule into substructures containing 3–10 nonhydrogen atoms. Fragments belonging to active and inactive examples are separated. The advantage of CASE over ADAPT is in the automatic selection of descriptors. The disadvantage is the lack of three-dimensional parameters and a term indicative of intramolecular interactions. A related method based on the Soltzberg-Wilkens topological molecular transform has been used to divide compounds into active and inactive clusters and several model series analyzed (Gabayani et al., 1982).

Esaki (1982) has described a computer program based on searching for a pattern match between a candidate compound and 261 selected pharmacophors. The information base on pharmacophoric patterns covers 25 drug classes and is derived from crystallographic data. The match is based on a comparison of geometry, electron density, and frontier orbital energy levels. The approach has the same limitations as others that use X-ray crystal structure; that is, the solid-state structure may not adequately represent the receptor-bound conformation. This pattern-recognition method is suitable for finding new structures related to established and well-defined agents, and would not be useful in the development of ''breakthrough''-type drugs.

A similar program has been described by Rozenblit (1982). The software includes a database of approximately 9000 compounds covering 57 classes of biological activity. Each compound is defined by a set of structural descriptors. Comparison of this set with a test structure can be used to predict the activity profile, or new structural types can be designed using an interactive mode.

A note of caution concerning pattern recognition analyses in general has been raised. In many of the reported studies, the number of parameters equals or exceeds the number of examples, thus yielding spurious results. The causes for such serious errors in analysis have been described (Wold and Dunn, 1983).

6. Lead Development

6.1. Systematic Empirical Approach

Lead development can occur via one of three possible pathways: a systematic empirical approach, a hypothesis-based approach, or some combination of the two.

In the systematic empirical approach, test compounds or series are selected based on parameter variation of the lead, taking into account synthetic accessibility. Using this optimization method, test compounds or a series of compounds are prepared and evaluated; the interpretation of the results leads to a more refined series. This iterative process continues until no further improvement is obtained.

The most straightforward procedure of this type is the Topliss tree. This operational scheme for the optimization of the substituents on aromatic rings and side chains has been clearly described and usefully illustrated by Topliss and Martin (1975) and Martin (1979). The approach, based on the Hansch method, uses π and σ values to construct a selection grid. In the case of optimization of aromatic substitution, the 4-chloro derivative of the lead compound is first prepared. The biological testing results with this compound set into motion a selection process that continues until optimization has been achieved. A similar scheme has been devised for side-chain optimizations.

The simplicity of this approach is the foundation for its appeal to medicinal chemists and pharmacologists, and its value has been amply demonstrated. Examples of series development using the Topliss method come from a variety of drug classes, including anti-inflammatory, cardiovascular, CNS, and antibacterial. The method is not universally applicable to design problems, however. Only substituent variations are involved; other properties that may be critical to biological activity, such as topological features, are not included. In addition, the hierarchical approach does not provide the flexibility that may be needed when synthetic accessibility becomes the limiting factor.

A more individualized optimization procedure is the sequential simplex method. In the simple two-dimensional case, the parameters, e. g., π and σ, are plotted. Three examples are prepared and evaluated. The point describing the least active of the three is reflected along the midpoint of the line joining the two more active examples. The line is extended to the next nearest substituent. Synthesis and testing of this compound provides the information necessary for movement to the next point, i. e., by projection of a line from the least active of the three through the midpoint of the line connecting the two more active examples. A potential disadvantage of the sequential simplex method, as pointed out by both Austel (1984) and Martin (1979), is that the procedure may fail by entrapment in a local optimum.

The Fibonacci series forms the basis for another method of target selection involving a stepwise optimization procedure (Martin, 1979; Austel, 1984). In this approach a search space is defined and the parameter under study organized according to the Fibonacci series. The second-from-last and next-to-last Fibonacci numbers are converted to the parameter series and the two indicated compounds prepared. In the series analyzed by Bustard (1974), the carbon chain from 7 to 20 atoms in a series of alkylammonium compounds served as the parameter space as indicated:

Fibonacci series number	0	1	2	3	[5]	[8]	13
Parameter series number	7	8	9	10	[12]	[15]	20

The Fibonacci numbers 5 and 8 translate to the two examples with carbon chains of 12 and 15 atoms. The compounds are prepared and the least active serves as the start of a new series. This second series is made up of the elements from the least active through the more active example to the end of the series. The selection process is repeated. Each step leads to a new series until optimization is complete.

The systematic empirical approach is applicable to series selection, as well as to the stepwise optimization process discussed above. A series based on broad parameter variation and synthetic accessibility is selected, and each example made and tested. Regression analysis of the test series leads to a more refined second series. Repetition of the process would be expected to result in optimization. Several methods have been developed that allow for the selection of a balanced series, i. e., one that is dissimilar and orthogonal. A generally applicable systematic approach to series design has been suggested by Austel (1982). An ideal series will be based on all important physical properties with uniform coverage of the relevant space. Since it is not known which of the parameters are correlated with the biological activity, analyses in multidimensional space increase the probability of developing a meaningful series. This manual method is based on a 2^n factorial design. Indicators of (+), (0), and (−) are assigned to each descriptor based on high, mean, or low values or (+) and (−) to indicate the presence or absence of a feature. In a two-parameter study, for example, one in which the values of π and σ are being varied, the matrix illustrated on the left side of Fig. 2 would be used. Four compounds having substituents that satisfy the matrix would be selected after considering synthetic accessibility. With a three-parameter system, eight compounds would be required, as illustrated by the matrix on the upper right-hand side.

In these multiparameter problems, however, the number of required test compounds can be reduced by the use of cross terms. As indicated on the lower left of Fig. 2, the three-parameter study requires only four compounds if the cross term is used as the third indicator. Similarly for a four-parameter variation, only eight com-

2^n FACTORIAL DESIGN - INDICATOR VARIABLES

Indicators
Two Parameter Problem

-	-
+	-
-	+
+	+

Indicators
Three Parameter Problem

-	-	-
+	-	-
-	+	+
-	+	-
+	+	-
-	-	+
+	-	+
+	+	+

Indicators
Three Parameter Problem
With a Cross Term

-	-	(+)
+	-	(-)
-	+	(-)
+	+	(+)

Indicators
Four Parameter Problem
With a Cross Term

-	-	-	(-)
+	-	-	(+)
-	+	-	(+)
-	+	+	(-)
+	+	-	(-)
-	-	+	(+)
+	-	+	(-)
+	+	+	(+)

Fig. 2. Indicators are assigned + for high and − for low values of each parameter used in the analysis. The calculated values for the cross terms are shown in parentheses (modified and reproduced with permission from Austel, 1982).

pounds are required if the fourth descriptor is determined by the cross term of the first three. This is illustrated on the lower right side of Fig. 2. Using several examples, Austel has analyzed the test sets selected by this method and found that they meet the requirements of dissimilarity and orthogonality.

Hansch et al. (1973) has used cluster analyses to divide substituents into groups depending on their position in space defined by lipophilicity (π and π^2), the Swain and Lupton inductive-field, and resonance constants ($\mathcal{F}$ and $\mathcal{R}$), molar refractivity, and molecular weight. In a design process, a substituent from each cluster would be selected for synthesis and evaluation. The successful use of such clusters has been described (Dunn et al., 1976). A similar, fully computerized and interactive approach has been developed by Wootton et al. (1975), and recently models by Borth et al. (1985) have introduced a synthetic difficulty factor into the analysis with only a moderate decline in predictive ability.

6.2. Hypothesis-Based Lead Development

6.2.1. Active Analog Approach

Some design problems are better approached by developing a working hypothesis; for example, one that attempts to define the proposed interaction of the drug with the macromolecule. This approach operates in drug discovery at various levels of refinement. One of the most sophisticated is that advocated by Garland Marshall. This hypothesis-centered method depends heavily on computational chemistry and computer graphics.

The computer-assisted representation of molecular structures is a powerful design tool (Gund, 1984; Loftus et al., this volume). Perhaps its most useful asset is to provide a view of the three-dimensional structure; such graphic representations allow structural overlays in a manner that can only be accomplished by computer graphics. This simple ability to model structures and test structural hypotheses has provided insights into bioactive conformations.

Computer graphics may be at its most powerful when combined with computational chemistry. Energy minimized conformations can be computer generated and viewed in three dimensions. Although the problem of achieving the global minimum has not been solved, its value has been questioned since extrapolation to the receptor-bound conformation is tenuous at best. To overcome this problem of the global minimum and its significance, Marshall uses the concept of the pharmacophore, i. e., those features of a molecule that in their three-dimensional arrangement are recognized by the macromolecule (Marshall et al., 1979; Marshall, 1984). In defining a hypothesis based on a proposed pharmacophore, it is recognized that the minimum energy structure as determined by X-ray analysis or computations may not represent the three-dimensional array recognized by the target receptor or enzyme. Thus, all allowed conformations are considered and each resulting arrangement of the pharmacophore is represented in ''orientation space.'' Through the analyses of a series of active compounds the common three-dimensional patterns are determined. Once a pattern has been selected, other active compounds can be tested against the proposed pharmacophore; failure to fit the pattern would compromise the hypothesis.

Once defined, the pharmacophore allows for calculation of the volume available for binding to the macromolecule, i. e., the excluded volume. Subtraction of the excluded volume from the volume common to inactive analogs provides the receptor essential volume. With this information new compounds can be defined and tested, and the results used to further refine the hypothesis.

Development of the hypothetical pharmacophore is dependent on a background of information on active and inactive analogs that operate by the same mechanism. The addition to the analysis set of compounds that function by another pharmacological mechanism will compromise any conclusions. Problems without such a background of data are not readily adaptable to this approach.

Flexible molecules constitute another problem, since the analysis of their allowed conformations becomes rate limiting. Problems with four rotatable bonds are currently common and those with eight rotatable bonds are becoming approachable. Attempts to circumvent the problem of flexibility involve modifications to reduce the number of variables. Reduction of flexibility is a common strategy in drug design, which is used both as an approach to increase receptor affinity and as a method to establish likely conformations of the pharmacophore. The application of molecular constraints to deduce the bioactive conformation is perhaps best illustrated with the highly flexible peptides. With this class such information should allow a more rational approach to modifications. Veber (1982; Freidinger and Veber, 1984) has incorporated backbone constraints in analog design as a method to determine the active conformation of somatostatin. Aided by computer modeling and solution-state nuclear magnetic resonance (NMR) studies and using covalent bridges, proline residue, D-amino acids, and *N*-methyl amino acids, the naturally occurring hormone was simplified to a metabolically stable cyclic hexapeptide. The key to the design process was the recognition that the bioactive moiety in somatostatin was the tetrapeptide Phe-Trp-Lys-Thr held in a β-turn. Other examples illustrating the concept of conformational constraint will be described in detail later in the chapter.

When the active analog approach cannot be applied because of these limitations, the design strategy can proceed at a lower level of approximation. An analysis based on three-dimensional structural information derived from X-ray analysis or computational methods can serve as the starting point. Because of the recognized limitations of these methods for predicting the conformation required for ligand–macromolecule interaction, a combination of the crystal structure analysis and molecular mechanics has been recommended. A successful application of this approach is illustrated in attempts to better define the pharmacophore for the opiate receptor (Duchamp, 1979).

6.2.2. Bioisosterism

A classic hypothesis-based strategy in drug design involves the use of bioisosteric replacements. Such replacements define groups that, when substituted for a moiety present in the parent, produce a compound having similar chemical, physical, and biological properties. This approach has been broadly applied and is the subject of several reviews (Burger, 1970, 1983; Foye, 1974; Korolkovas and Burckhalter, 1976; Ariens, 1971; Hansch, 1974; Thornber, 1979).

Since no two substituents are exactly alike, any substitution will result in changes in size, shape, electronic distribution, partition coefficient, solubility, pK_a, chemical reactivity, susceptibility to metabolism, and hydrogen bonding capabilities. The rationale for the bioisosteric approach is that the total change induced by the substituent replacement will serendipitiously result in improved potency, selectivity, duration of action, bioavailability, and/or a reduction in toxicity. For programs using known agents as the lead structure, an additional potential benefit would be that the chance improvement along with the resulting structural novelty would form the basis for patentability.

A wide range of bioisosteric replacements has been used in the optimization process and is summarized in the reviews cited above and references therein.

Isosteric groups can also be defined in terms of similarity in physiochemical parameters such as σ and π. Homogeneous grouping of substituents by individual descriptors constitutes a common method used in analog selection. This definition of isosterism brings the approach into the realm of series design. As discussed previously, when more complex interdependencies between several variables define biological activity, more sophisticated approaches to selection are needed.

The value of the bioisosteric approach and its integration into the drug design process may be best illustrated with the discovery of cimetidine (Ganellin, 1982). The isosteric replacement of the thiourea moiety of metiamide 12 by a cyanoguanidine to give the H_2-receptor antagonist cimetidine (13) was based on the similarity in the physiochemical characteristics between the two groups. This simple application of bioisosterism overcame the low incidence of granulocytopenia seen with metiamide, and resulted in one of the most successful drugs to be produced by medicinal chemistry.

6.2.3. Conformational Restriction

A favored strategy in drug design is to restrict conformational mobility of the pharmacophore, often with the aim of introducing subreceptor selectivity, enhancing affinity, or improving the working model of the "receptor map." With the numerous ways of building rigidity into a molecule, a clever application of the approach is also likely to lead to structural novelty and patentability.

The concept of conformational restraint through incorporation of the pharmacophoric elements into a cyclic structure has met with frequent success. One of the clearest examples is based on the neurotransmitter dopamine. Molecular restriction has led directly to the proposal that D_1 and D_2 dopamine receptors (Seiler and Markstein, 1984) accept the fully extended β-transoid geometry 14 rather than the α conformer 15 or the two cisoid possibilities. This assignment is based on studies of conformationally restricted analogs by Cannon and others using the dopamine agonist ADTN, 16, and related compounds (Cannon et al., 1980; Kohli et al., 1979; Volkman et al., 1977; Olson et al., 1981).

More recent semirigid analogs include the neuroleptic candidate centbutindole 17, which contains an *N*-arylpiperazine in a conformationally restrained structure. Modeling studies indicate that centbutindole, with its extended *trans*-arylethylamine geometry, can bind to the dopamine receptor, whereas the butyrophenone chain occupies an accessory binding site (Anand, 1983).

Modeling studies have been used to design conformationally restricted bicyclic vasopressin analogs. Such an example, 5,8-cyclo (1-β-mercaptopropionic acid-2-phenylalanine-5-aspartic acid-8-lysine)vasopressin, not only exhibits selectivity for the vasopressin antidiuretic receptor, but functions as a nearly pure antagonist (Skala et al., 1984).

The angiotensin-converting enzyme inhibitor 18 illustrates the use of conformational restriction as applied to enzyme inhibitors. This design was based on a working model of ACE that postulated two binding regions, S_1 and S_2. The incorporation of conformational constraints was used to efficiently orient the zinc-binding carbonyl ligand and resulted in a 1000-fold increase in potency over the acyclic acyldipeptide analog (Weller et al., 1984).

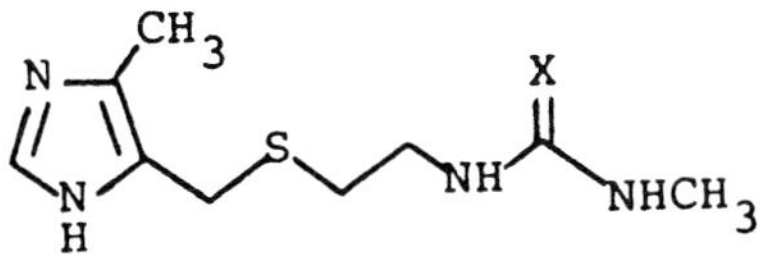

CH_3 N N H S NH X $NHCH_3$

12 X = S

13 X = NCN

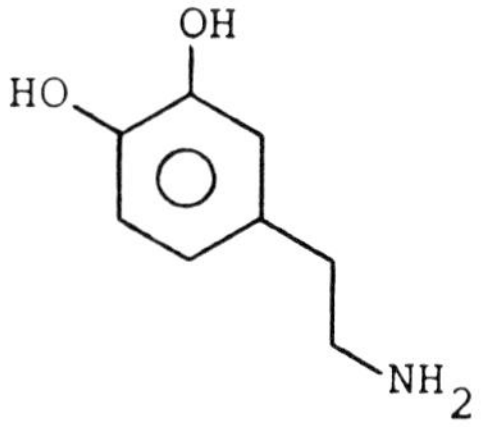

OH HO NH_2 β-trans 14

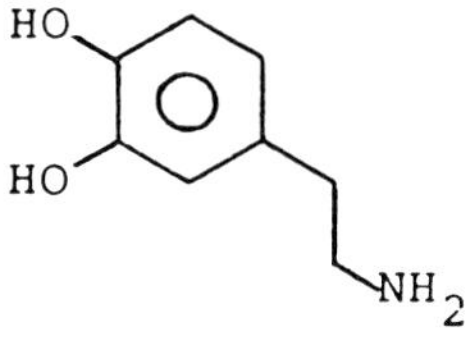

HO HO NH_2 α-trans 15

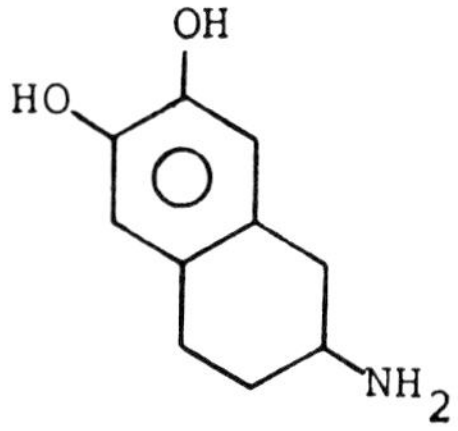

OH HO NH_2 16

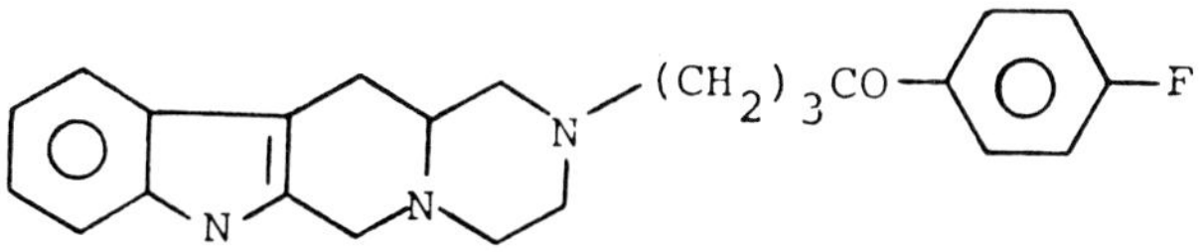

N N N $(CH_2)_3CO$ F

17

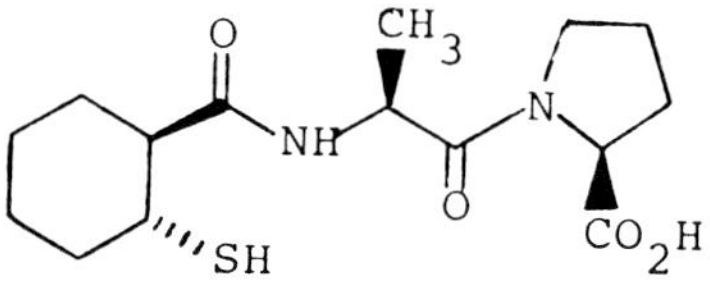

O CH_3 NH N O SH CO_2H

18

6.2.4. Symbiotic Approach

The symbiotic approach (Baldwin et al., 1979) to drug design or medicinal chemical hybridization involves the molecular combination of two mutually complimentary biological activities into a single entity. As most often used, the approach involves either introduction of both pharmacophores into structurally allowed areas of the molecule or the optimization of a lead found to exhibit both activities and selected by computerized matching of biological profiles.

The strategy requires a carefully chosen pharmacological target and, to be useful, the resulting compound must express both activities in an appropriate balance. The main advantage of such a hybrid over the concomitant administration of two drugs is in pharmacokinetics. With two drugs, each pharmacological activity depends on individual absorption, metabolism, and excretion profiles, whereas the hybrid does not.

This approach, which Gross (1984) has cited as a useful design strategy for novel antihypertensive agents, has recently been reviewed (Nicolaus, 1983). The most successful application has been in the cardiovascular area, in which a variety of hybrids have been produced and studied. These include the α,β-adrenoceptor antagonist labetalol (19) and the vasodilator-β-adrenergic blocking agents MK-717 (20), prizidilol (21), USF 3382 (22), and BM 14190 (23).

6.2.5. Metabolic Conversions

Metabolic conversion can be made a part of the drug design process for either activation or destruction of the administered dosage form. A widely used example of metabolic activation is found in the prodrug approach. This strategy calls for the bioconversion of inactive or weakly active species to the therapeutic agent. Such bioconversions can involve removal of a carrier or protecting group by cleavage of a labile bond, or by a more complex enzymatic alteration of a bioprecursor to the active species (Oelschlager, 1982). Many variations and combinations on these themes have been played with the aim of improving bioavailability and tissue selectivity.

Examples of prodrugs include the converting enzyme inhibitor enalapril 24, which has greater bioavailability than its therapeutically active acid form. Pivampicillin 25 and bacampicillin 26 further illustrate the strategy where the polar carboxylic acid group has been masked to enhance gastrointestinal absorption. Esterification of the

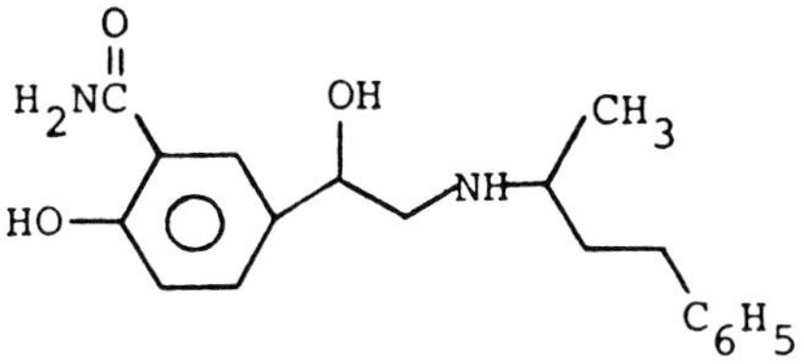

19

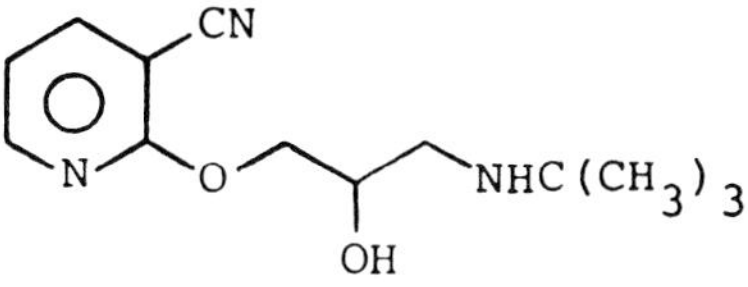

20

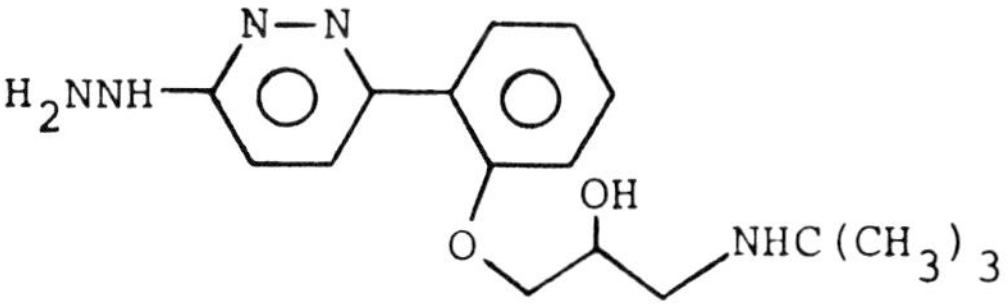

21

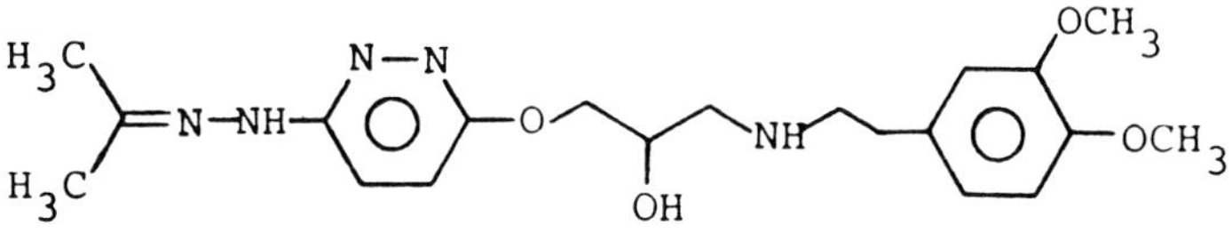

22

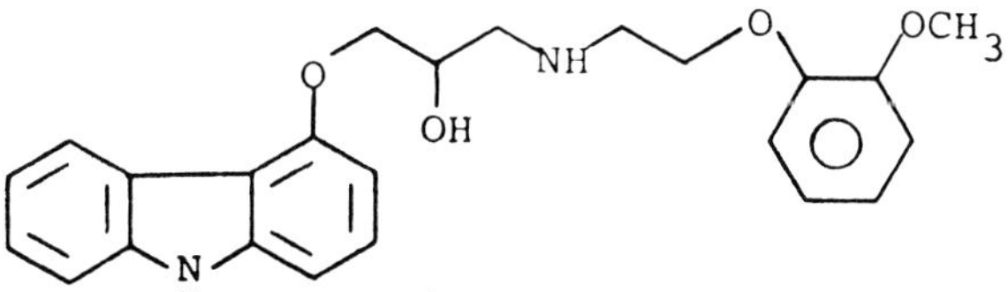

23

phenolic groups has been used to improve the penetration of epinephrine across the cornea, as in the dipivaloyl derivative 27, and of ADTN across the blood–brain barrier, as in the dibenzoyl derivative 28.

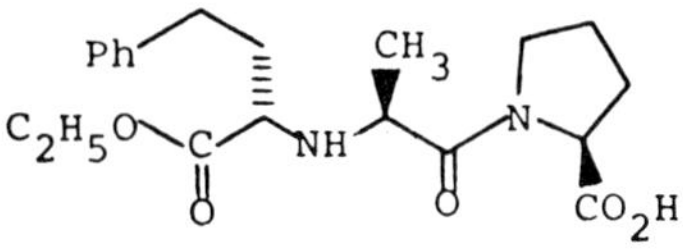

24

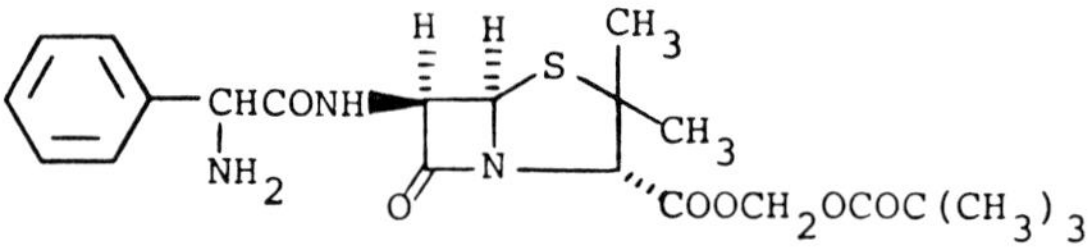

25

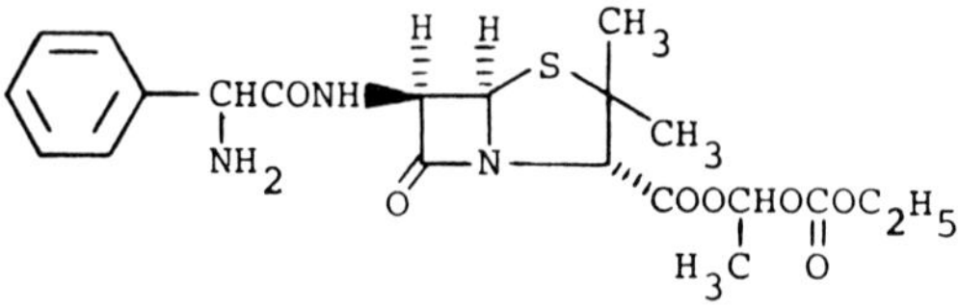

26

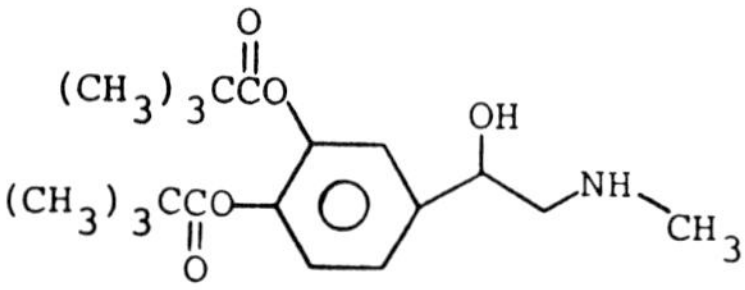

27

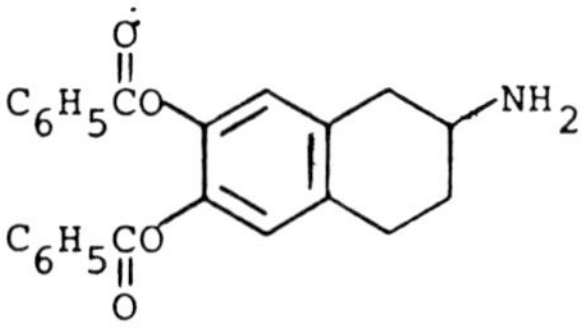

28

DOPA, after inhibition of peripheral decarboxylase with carbidopa, is converted to dopamine within the CNS, an example of enforced tissue selectivity through a controlled conversion of the prodrug. Many other intriguing examples that illustrate the breadth in the application of the concept can be found in several recent reviews (Wermuth, 1984; Pitman, 1981; Notari, 1981).

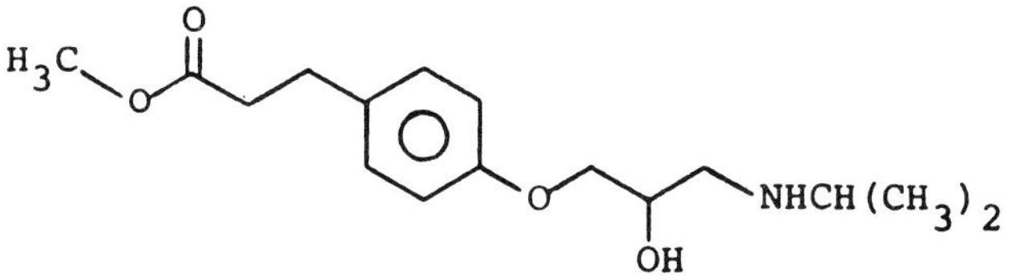

29

When the design includes the introduction of labile groups for rapid metabolic conversion to less active species, the strategy is termed the soft drug approach and has been reviewed by Bodor (1982, 1984a,b). Soft drugs are defined as "biologically active, therapeutically useful chemical compounds characterized by a predictable and controlled in vivo destruction to nontoxic moieties" (Bodor, 1984a). To achieve the objective of predictable and controlled metabolism, oxidative processes are usually avoided because of slow rates and sensitivity to saturation. Hydrolytic conversions are the first choice for the deactivation process. A recent example of a potentially useful therapeutic agent that fits the soft drug definition is the β-adrenergic blocking agent esmolol 29. This compound has inhibitory potency as the ester, but hydrolysis to the acid by esterase results in a dramatic decrease in β-adrenoceptor affinity. The compound can be administered by infusion; when the infusion is stopped, blockade of the β-receptor rapidly disappears. Similarly, esters of prednisolonic acid and its dehydro analog behave as soft drugs; that is, they possess local antiinflammatory activity without systemic toxicity due to hydrolysis to the inactive acid metabolite (Lee et al., 1984). As exemplified by esmolol and the steroid esters, the soft drug concept can be used to produce useful agents, especially when local action or short duration are desirable attributes.

Just as design strategies based on metabolic conversion can be used to guide biodegradation toward predictable and nontoxic pathways, approaches can also be directed toward avoiding metabolism entirely. Implementation of this concept leads to metabolically stable compounds termed hard drugs, which in principle should be safer and have optimal pharmacokinetics. Minimizing metabolism should result in long duration of action, decreased variability, and low toxicity. The improved toxicity profile is directly related to stability. Metabolic processes are usually associated with oxidative pathways and lead to alkylation of nucleophilic groups in proteins and DNA by chemically reactive electrophilic species such as epoxides and *N*-hydroxyarylamines.

The methods to achieve metabolic stability have been reviewed (Ariens, 1980; Ariens and Simonis, 1982) and include the substitution of fluorine for hydrogen on olefins sensitive to epoxide formation and the introduction of steric constraints to prevent enzymatic attack.

This tactic for designing less toxic and longer-acting drugs has not been universally accepted (Bodor, 1982), mainly because of the likelihood that metabolism cannot be totally avoided. It has been suggested that the fraction of the hard drug subjected to bioconversion may be more prone to generate highly reactive intermediates.

6.2.6. Set Theory and Drug Design

An integrated approach to drug design based on set theory has been illustrated by Kutter and Austel in their approach to novel cardiotonic agents (Austel and Kutter, 1981) and has been incorporated into a general philosophy for drug discovery (Austel, 1984). By definition, compounds possessing all of the defined attributes of the ideal drug can be viewed as a subset Z of all compounds that meet at least one of the requirements. In a second step the various options aimed at achieving the objective are evaluated. The current agents and biochemical approaches are listed and viewed as sets. Each is rated on the probability that it will form a nonempty intersection with the subset Z, i. e, the subset containing compounds having all the desired features. Based on this evaluation, certain lead structures or biochemical approaches are eliminated or given low priority.

Those sets chosen for evaluation may be viewed independently or examples representing the union of sets may be considered, i. e., compounds presenting structural features that are present in more than one of the sets may be considered as candidates for synthesis. The set or sets chosen for synthetic probes form a new set. From this new set structural subsets are defined based on synthetic accessibility, structural dissimilarity, and intuition. A small number of test compounds representing each subset are then designed, prepared, and evaluated. Those examples expressing the desired pharmacological activity are defined as leads and an optimization program initiated.

Those features of the ideal drug that are not met by the lead, e. g., duration of action or bioavailability, become the focus in the development phase. Both topological- and hypothesis-based approaches are used to eliminate subsets. As the structural features necessary for activity narrow, physiochemical properties combined with cluster analysis are used to exclude further subsets. This method of successive exclusion results in lead optimization defining the most active structural type with optimum physiochemical properties.

6.2.7. Antibodies

An application of molecular biology to rational drug discovery, as an alternative to classic approaches, has been reviewed by Haber (1983). This strategy is based upon antibodies as a source of potentially useful therapeutic agents. The advantage lies in the specificity of the interaction between antibody and target protein.

Two approaches are envisioned. The first would involve antibodies raised to a target protein. In the second, antibodies are raised to a ligand followed by the production of antibodies for the interacting site of the first antibody that is the antiidiotypic antibody. This latter method has been applied to the β-adrenergic receptor. The antiidiotype behaves as a competitive antagonist (Haber, 1983).

Formidable problems remain, however, before this technology can produce agents competitive with more traditional drugs. The problems are the same as those that plague the application of peptides, but are magnified by the molecular size of the antibody or antigen-binding fragment Fab, e. g., the molecule must be metabolically stable, nonimmunogenic, and absorbable, preferably from the gastrointestinal tract. Overcoming such major hurdles will not be trivial. Use of this technology to produce specialty agents such as the digoxin specific antibody (Curd et al., 1971; Smith et al., 1982) is already underway, however.

6.2.8. Immobilized Drugs

Another nonclassical approach to drug discovery is through the use of immobilized drugs; this strategy has been reviewed by Venter (1982). One application of the concept involves covalently attaching a bioactive molecule to a target-specific carrier. For example, various anticancer drugs have been linked to estrogens, antibodies, dextran, synthetic polymers, and DNA in an attempt to achieve selective biodistribution and pharmacological specificity. Covalently joining daunomycin and DNA produced a compound with reduced toxicity, while retaining antitumor properties. A second application involves the use of immobilized enzymes on solid support in an extracorporeal shunt to correct metabolic disorders. The technique has proven successful in animals with experimentally induced hyperuricemia and phenylketonuria. As indicated, this approach is primarily directed toward improving known bioactive molecules, but could, equally as well, be applied to new agents as part of a development program. The imaginative application of immobilization has the potential for interesting successes.

7. Conclusion

Despite advances in design strategies, chance has not been eliminated as an element in drug discovery. From the search for a lead structure through optimization most approaches rely, to some degree, on serendipity. Finding suitable lead compounds remains the most difficult phase of the entire process and still depends in many instances on random screening as the source for biologically active, novel structures.

Many of the hypothesis-based optimization methods, i. e., receptor mapping and the active analog approach, are aimed at generating new structures with the hope for chance improvement. Others, such as the hard, soft, and prodrug concepts, are somewhat closer to true design, being directed toward overcoming particular problems through a directed approach.

Important advances have been made, especially in systematic optimization and series design. The techniques are associated primarily, however, with the exploration of cogeneric series and require large information bases for application. These potency-oriented strategies have resulted in focusing much of medicinal chemistry on the improvement of lead structures. The principal advantage of systematic design may not lie in reducing the element of chance, but in increasing the odds for success over intuition and experience. Such enhanced efficiency may result from improved organization, which allows for better conceptualization and facilitates the formulation of testable questions.

Design strategies such as transition-state analogs and mechanism-based enzyme inhibitors are closer to a more purely rational approach. Unlike the systematic methods, these depend on a fundamental understanding of the biological process, rather than on a structure–activity database. In the near term, it is likely that therapeutic advances will result from applying this level of rational design, along with systematic strategies, to new areas in which advances in biology have exposed novel and unexploited control mechanisms. In the longer view, the gradual growth in the library of macromolecular structures coupled with further advances in computer technology offer the promise for design from first principles and the real possibility for stepping beyond the limitations of current approaches.

References

Abeles, R. H. and Maycock, A. L. (1976) Suicide enzyme inactivators. *Acc. Chem. Res.* **9**, 313–319.

Albers-Schonberg, G., Arison, B. H., Chabala, J. C., Douglas, A. W., Eskola, P., Fisher, M. H., Lusi, A., Mrozik, H., Smith, J. L., and Tolman, R. L. (1981) Avermectins. Structure determination. *J. Am. Chem. Soc.* **103**, 4216–4221.

Albert, A. (1971) Relations between molecular structure 6501 and biological activity: Stages in the evolution of current concepts. *Ann. Rev. Pharmacol.* **11**, 13–36.

Albert, A. (1982) The long search for valid structure–action relationships in drugs. *J. Med. Chem.* **25**, 1–5.

Alberts, A. W., Chen, J., Kuron, G., Hunt, V., Huff, H., Hoffman, C., Rothrock, J., Lopez, M., Joshua, H., Harris, E., Patchett, A., Monaghan, R., Currie, S., Stapley, E., Albers-Schonberg, G., Hensens, O., Hirshfield, J., Hoogsteen, K., Liesch, J., and Springer, J. (1980) Mevinolin: A highly potent competitive inhibitor of hydroxymethylglutaryl-coenzyme A reductase and a cholesterol-lowering agent. *Proc. Natl. Acad. Sci. USA* **77**, 3957–3961.

Anand, N. (1983) Molecules with restricted conformational mobility—an approach to drug design. *Proc. Indian Natn. Sci. Acad.* **49**, A, 233–255.

Andrews, P. R. and Winkler, D. A. (1984) The Design and Medicinal Applications of Transition State Analogues, in *Drug Design: Fact or Fantasy?* (Jolles, G. and Wooldridge, K. R. H., eds.) Academic, New York.

Ariens, E. J. (1971) A General Introduction to the Field of Drug Design, in *Drug Design* vol. I (Ariens, E. J., ed.) Academic, New York.

Ariens, E. J. (1980) Design of Safer Chemicals, in *Drug Design* vol. IX (Ariens, E. J., ed.) Academic, New York.

Ariens, E. J. and Simonis, A. M. (1982) Optimalization of Pharmacokinetics—An Essential Aspect of Drug Development—by "Metabolic Stabilization," in *Strategy in Drug Research* (Keverling Buisman, J. A., ed.) Elsevier, Amsterdam.

Austel, V. (1982) A manual method for systematic drug design. *Eur. J. Med. Chem.* **17**, 9–16.

Austel, V. (1984) Design of test series in medicinal chemistry. *Drugs of the Future* **9**, 349–365.

Austel, V. and Kutter, E. (1981) The theory of sets as a tool in systematic drug design. *Arzneimittelforsch./Drug Res.* **31**, 130–135.

Baker, D. J., Beddell, C. R., Champness, J. N., Goodford, P. J., Norrington, F. E. A., Smith, D. R., and Stammers, D. K. (1981) The binding of trimethoprim to bacterial dihydrofolate reductase. *FEBS Lett.* **126**, 49–52.

Baldwin, J. J., Lumma, W. C., Lundell, G. F., Ponticello, G. S., Raab, A. W., Engelhardt, E. L., Hirschmann, R., Sweet, C. S., and Scriabine, A. (1979) Symbiotic approach to drug design: Antihypertensive β-adrenergic blocking agents. *J. Med. Chem.* **22**, 1284–1290.

Bodor, N. (1982) Soft drugs: Strategies for Design of Safer Drugs, in *Strategy in Drug Research* (Keverling Buisman, J. A., ed.) Elsevier, Amsterdam.

Bodor, N. (1984a) Novel Approaches to the Design of Safer Drugs: Soft Drugs and Site-Specific Chemical Delivery Systems, in *Advances in Drug Research* vol. 13 (Testa, B., ed.) Academic, New York.

Bodor, N. (1984b) Soft drugs: Principles and methods for the design of safe drugs. *Med. Res. Rev.* **4**, 449–469.

Boger, J. (1983) Renin Inhibitors. Design of Angiotensinogen Transition-State Analogs Containing Statine, in *Peptides: Structure and Function* (Hruby, V. J. and Rich, D. H., eds.) Pierce Chemical Company, Rockford, Illinois.

Boger, J., Lohr, N. S., Ulm, E. H., Poe, M., Blaine, E. H., Fanelli, G. M., Lin, T. Y., Payne, L. S., Schorn, T. W., LaMont, B. I., Vassil, T. C., Stabilito, I. I., Veber, D. F., Rich, D. H., and Bopari, A. S. (1983) Novel renin inhibitors containing the amino-acid statine. *Nature* **303**, 81–84.

Brugger, W. E. and Jurs, P. C. (1977) Extraction of important molecular features of musk compounds using pattern recognition techniques. *J. Agric. Food Chem.* **25**, 1158–1164.

Burchall, J. J. and Hitchings, G. H. (1965) Inhibitor binding analysis of dihydrofolate reductases from various species. *Mol. Pharmacol.* **1**, 126–136.

Burger, A. (1970) Hallucinogenic Agents, in *Medicinal Chemistry* 3rd ed. (Burger, A., ed.) Wiley-Interscience, New York.

Burger, A. (1983) *A Guide to the Chemical Basis of Drug Design.* Wiley, New York.

Bustard, T. M. (1974) Optimization of alkyl modifications by Fibonacci search. *J. Med. Chem.* **17**, 777–778.

Cannon, J. G., Lee, T., Goldman, D., Long, J. P., Flynn, J. R., Verimer, T., Costall, B., and Naylor, R. J. (1980) Congeners of the β conformer of dopamine derived from *cis-* and *trans*-octahydrobenzo[f]quinoline and *trans*-octahydrobenzo[g]quinoline. *J. Med. Chem.* **23**, 1–5.

Cha, S., Agarwal, R. P., and Parks, R. E., Jr. (1975) Tight-binding inhibitors. II. Non-steady state nature of inhibition of milk xanthine oxidase by allopurinol and alloxanthine and of human erythrocytic adenosine deaminase by coformycin. *Biochem. Pharmacol.* **24**, 2187–2197.

Charton, M. (1983) The Upsilon Steric Parameter-Definition and Determination, in *Steric Effects in Drug Design* (Charton, M. and Motoc, I., eds.) Springer-Verlag, New York.

Charton, M. and Motoc, I. (1983) Introduction, in *Steric Effects in Drug Design* (Charton, M. and Motoc, I., eds.) Springer-Verlag, New York.

Chou, J. T. and Jurs, P. C. (1979) Computer-assisted computation of partition coefficients from molecular structures using fragment constants. *J. Chem. Inf. Comput. Sci.* **19**, 172–178.

Cohen, N. C. (1983) Towards the rational design of new leads in drug research. *TIPS* 503–506.

Conover, L. H. (1971) Discovery of drugs from microbiological sources. *Adv. Chem. Ser.* **108**, 33–80.

Cromartie, T. H. and Walsh, C. (1975) Mechanistic studies on the rat kidney flavoenzyme L-alpha-hydroxyacid oxidase. *Biochemistry* **14**, 3482–3489.

Curd, J., Smith, T. W., Jaton, J. C., and Haber, E. (1971) The isolation of digoxin-specific antibody and its use in reversing the effects of digoxin. *Proc. Natl. Acad. Sci. USA* **68**, 2401–2406.

Douglas, K. T. (1983) Transition-state analogues in drug design. *Chem. Ind.* 311–315.

Duchamp, D. J. (1979) Molecular mechanics and crystal structure analysis in drug design. *ACS Symp. Ser.* **112**, 79–102.

Dufau, M. L., Ryan, D. W., Baukal, A. J., and Catt, K. J. (1975) Gonadotropin receptors. Solubilization and purification by affinity chromatography. *J. Biol. Chem.* **250**, 4822–4824.

Dunn, W. J., Greenberg, M. J., and Callejas, S. S. (1976) Use of cluster-analysis in development of structure–activity relations for antitumor triazenes. *J.Med. Chem.* **19**, 1299–1301.

Editorial (1981) Drug licensing or innovation. *Lancet* **II**, 788.

Endo, A., Kuroda, M., and Tsujita, Y. (1976) ML-236A, ML-236B, and ML-236C, new inhibitors of cholesterogenesis produced by *Penicillium citrinum. J. Antibiotics* **29**, 1346–1348.

Esaki, T. (1982) Quantitative drug design studies. V. Approach to lead generation by pharmacophoric pattern searching. *Chem. Pharm. Bull.* (Tokyo) **30**, 3657–3661.

Farnsworth, N. R. and Bingel, A. S. (1977) Problems and prospects of discovering new drugs from higher plants by pharmacological screening. *New Nat. Prod. Plant Drugs Pharmacol. Biol. Ther. Act.*, Proc. Int. Cong., 1st, 1–22.

Foye, W. O. (1974) *Principles of Medicinal Chemistry* pp. 93–102. Lea & Febiger, Philadelphia.

Free, S. M. and Wilson, J. W. (1964) A mathematical contribution to structure–activity studies. *J. Med. Chem.* **7**, 395–399.

Freidinger, R. M. and Veber, D. F. (1984) Design of novel cyclic hexapeptide somatostatin analogs from a model of the bioactive conformation. *ACS Symp. Ser.* **251**, 169–187.

Fujita, T. (1984) The Role of QSAR in Drug Design, in *Drug Design: Fact or Fantasy?* (Jolles, G. and Wooldridge, K. R. H., eds.) Academic, New York.

Gabayani, Z., Surjan, P., and Naray-Szabo, G. (1982) Application of topological molecular transforms to rational drug design. *Eur. J. Med. Chem.-Chim. Ther.* **17**, 307–311.

Ganellin, C. R. (1982) Cimetidine, in *Chronicles of Drug Discovery* vol. 1 (Bindra, J. S. and Lednicer, D., eds.) Wiley, New York.

Goodford, P. J. (1984) Drug design by the method of receptor fit. *J. Med. Chem.* **27**, 557–564.

Gordon, E. M., Godfrey, J. D., Pluscec, J., VonLangen, D., and Natarajan, S. (1985) Design of peptide derived amino alcohols as transition-state analog inhibitors of angiotensin converting enzyme. *Biochem. Biophys. Res. Commun.* **126**, 419–426.

Gross, F. (1984) Antihypertensive therapy: Modern concepts, future aspects in research. *Triangle* **23**, 25–32.

Gund, P. (1984) Present and future computer aids to drug design. *X-Ray Crystallogr.* Drug Action, Course Int. Sch. Crystallogr. 9th, 495–506.

Haber, E. (1983) Antibodies as models for rational drug design. *Biochem. Pharmacol.* **32**, 1967–1977.

Halberstam, M. J. (1979) Too many drugs? *Forum on Medicine* **2**, 170–291.

Hansch, C. (1974) Bioisosterism. *Intrascience Chem. Rep.* **8**, 17–25.

Hansch, C. (1982) Dihydrofolate reductase inhibition. A study in the use of X-ray crystallography, molecular graphics, and quantitative structure–activity relations in drug design. *Drug. Intel. Clin. Pharm.* **16**, 391–396.

Hansch, C. (1984) On the state of QSAR. *Drug. Infor. J.* **18**, 115–122.

Hansch, C. and Blaney, J. M. (1984) *The New Look to QSAR*, in *Drug Design: Fact or Fantasy?* (Jolles, G. and Wooldridge, K. R. H., eds.) Academic, New York.

Hansch, C. and Leo, A. J. (1979) *Substituent Constants for Correlation Analysis in Chemistry and Biology*. Wiley, New York.

Hansch, C., Unger, S. H., and Forsythe, A. B. (1973) Strategy in drug design. Cluster analysis as an aid in the selection of substituents. *J. Med. Chem.* **16**, 1217–1222.

Hathway, D. E. (1982) Structure–activity considerations; a synthesis of ideas. *Chem. Biol. Interact.* **42**, 1–26.

Henry, D. R., Jurs, P. C., and Denny, W. A. (1982) Structure–antitumor activity relationships of 9-anilinoacridines using pattern recognition. *J. Med. Chem.* **25**, 899–908.

Hopfinger, A. J. (1984) Computational chemistry, molecular graphics and drug design. *Pharm. Int.* **5**, 224–228.

Jurs, P. C. (1983) Studies of relationships between molecular structure and biological activity by pattern recognition methods. *Struct.*-Act. Correl. Predict. Tool Toxicol. (Golberg, L., ed.) Hemisphere, Washington, DC.

Jurs, P. C., Ham, C. L., and Brugger, W. E. (1981) Computer-assisted studies of chemical structure and olfactory quality using pattern recognition techniques. *ACS Symp. Ser.* **148**, 143–160.

Jurs, P. C., Hasan, M. N., Henry, D. R., Stouch, T. R., and Whalen-Pedersen, E. K. (1983) Computer-assisted studies of molecular structure and carcinogenic activity. *Fund. Appl. Toxicol.* **3**, 343–349.

Kalman, T. I. (1981) Enzyme inhibition as a source of new drugs. *Drug Dev. Res.* **1**, 311–328.

Kier, L. B. (1980) Molecular Connectivity as a Description of Structure for SAR Analyses, in *Physical Chemical Properties of Drugs* (Yalkowsky, S. H., Sinkula, A. A., and Valvani, S. C., eds.) Dekker, New York.

Kirschner, G. L. and Kowalski, B. R. (1979) The Application of Pattern Recognition to Drug Design, in *Drug Design* vol. VIII (Ariens, E. J., ed.) Academic, New York.

Klopman, G. and Contreras, R. (1985) Use of artificial intelligence in structure–activity correlations of anticonvulsant drugs. *Mol. Pharmacol.* **27**, 86–93.

Kohli, J. D., Goldberg, L. I., and Nand, N. (1979) *1*-Aminomethyl isochromans: New vascular dopamine agents. *Pharmacologist* **21**, 202.

Korolkovas, A. and Burckhalter, J. H. (1976) *Essentials of Medicinal Chemistry*, pp. 23–26. Wiley, New York.

Kutter, E. and Austel, V. (1981) Application of the theory of sets to drug design. *Arzneimittelforsch./Drug Res.* **31**, 135–141.

Kuyper, L. F., Roth, B., Baccanari, D. P., Ferone, R., and Beddell, C. R. (1982) Receptor-based design of dihydrofolate-reductase inhibitors—comparison of crystallographically determined enzyme binding with enzyme affinity in a series of carboxy-substituted trimethoprim analogs. *J. Med. Chem.* **25**, 1120–1122.

Lee, H. J., Khalil, M. A., and Lee, J. W. (1984) Antedrug—a conceptual basis for safer anti-inflammatory steroids. *Drugs Under Experimental and Chemical Research* **10**, 835–844.

Lienhard, G. E. (1972) Transition state analogs as enzyme inhibitors. *Ann. Rep. Med. Chem.* **7**, 249–258.

Lienhard, G. E. (1973) Enzyme catalysis and transition-state theory. *Science* **180**, 140–154.

Lindquist, R. N. (1975) The Design of Enzyme Inhibitors: Transition State Analogs, in *Drug Design* vol. V (Ariens, E. J., ed.) Academic, New York.

Marciniszyn, J., Hartsuck, J. A., and Tang, J. (1976) Mode of inhibition of acid proteases by pepstatin. *J. Biol. Chem.* **251**, 7088–7094.

Marshall, G. R. (1984) Computational Chemistry and Receptor Characterization, in *Drug Design: Fact or Fantasy?* (Jolles, G. and Wooldridge, K. R. H., eds.) Academic, New York.

Marshall, G. R., Barry, C. D., Bosshard, H. E., Dammkoehler, R. A., and Dunn, D. A. (1979) The conformational parameter in drug design: The active analog approach. *ACS Symp. Ser.* **112**, 205–226.

Martin, Y. C. (1978) *Quantitative Drug Design. A Critical Introduction*. Dekker, New York.

Martin, Y. C. (1979) Advances in the Methodology of Quantitative Drug Design, in *Drug Design* vol. VIII (Ariens, E. J., ed.) Academic, New York.

Martin, Y. C. (1981) A practitioner's perspective of the role of quantitative structure–activity analysis in medicinal chemistry. *J. Med. Chem.* **24**, 229–237.

Martin, Y. C. (1983) Studies of relationships between structural properties and biological activity by Hansch analysis. *Struct.-Act. Correl. Predict. Tool Toxicol.* (Golberg, L., ed.), Hemisphere, Washington, DC.

Martin, Y. C. and Panas, H. N. (1979) Mathematical considerations in series design. *J. Med. Chem.* **22**, 784–791.

Martin, Y. C., Holland, J. B., Jarboe, C. H., and Plotnikoff, N. (1974) Discriminant analysis of the relationship between physical properties and the inhibition of monoamine oxidase by aminotetralins and aminoindans. *J. Med. Chem.* **17**, 409–416.

Maxwell, R. A. (1984) The state of the art of the science of drug discovery—an opinion. *Drug Dev. Res.* **4**, 375–389.

Metcalf, B. W. and Jund, K. (1977) Synthesis of beta$_1$ gamma-unsaturated amino acids as potential catalytic irreversible enzyme inhibitors. *Tetrahedron Lett.* 3689–3692.

Meunier, J. C., Olson, R. W., Menez, A., Fromageot, P., Boquet, P., and Changeux, J. P. (1972) Some physical properties of the cholinergic receptor protein from Electrophorus electricus revealed by a tritiated α-toxin from *Naja nigricollis* venom. *Biochemistry* **11**, 1200–1210.

Motoc, I., ed. (1983) Molecular Shape Descriptors, in *Steric Effects in Drug Design* (Charton, M. and Motoc, I., eds.) Springer-Verlag, New York.

Nicolaus, B. J. R. (1983) Symbiotic Approach to Drug Design, in *Decision Making in Drug Research* (Gross, F., ed.) Raven, New York.

Notari, R. E. (1981) Prodrug design. *Pharmacol. Ther.* **14**, 25–53.

Oelschlager, H. (1982) Drug Biotransformation as a Source of Drug Development, in *Strategy in Drug Research* (Keverling Buisman, J. A., ed.) Elsevier, Amsterdam.

Olson, G. L., Cheung, H., Morgan, K. D., Blount, J. F., Todaro, L., Berger, L., Davidson, A. B., and Boff, E. (1981) A dopamine receptor model and its application in the design of a new class of rigid pyrrolo[2,3-g]isoquinoline antipsychotics. *J. Med. Chem.* **24**, 1026–1034.

Pitman, I. H. (1981) Pro-drugs of amides, imides and amines. *Med. Res. Rev.* **1**, 189–214.

Ramiller, N. (1984) Computer-assisted studies in structure–activity relationships. *Am. Lab.* 78–88.

Rando, R. R. (1977) Mechanism of irreversible inhibition of gamma-aminobutyric acid alpha-ketoglutaric acid transaminase by neurotoxin gabaculine. *Biochemistry* **16**, 4604–4610.

Robinson, F. A. (1974) Therapeutic innovation—the end or a new beginning? *Chem. Brit.* **10**, 129–136.

Rose, S. L. and Jurs, P. C. (1982) Computer-assisted studies of structure–activity relationships of *N*-nitroso compounds using pattern recognition. *J. Med. Chem.* **25**, 769–776.

Rozenblit, A. B. (1982) Computer-Assisted Drug Design. Strategy and Algorithms, in *Strategy in Drug Research* (Keverling Buisman, J. A., ed.) Elsevier, Amsterdam.

Schmidt, P. G., Bernatowicz, M. S., and Rich, D. H. (1982) Pepstatin binding to pepsin-enzyme conformation changes monitored by nuclear magnetic resonance. *Biochemistry* **21**, 6710–6716.

Seiler, M. P. and Markstein, R. (1984) Further characterization of structural requirements for agonists at the striatal dopamine D_2 receptor and a comparison with those at the striatal dopamine D_1 receptor. *Mol. Pharmacol.* **26**, 452–457.

Silverman, R. B. and Hoffman, S. J. (1984) The organic chemistry of mechanism-based enzyme inhibition: A chemical approach to drug design. *Med. Res. Rev.* **4**, 415–447.

Silverman, R. B. and Levy, M. A. (1981) Mechanism of inactivation of gamma-aminobutyric acid alpha-ketoglutaric acid aminotransferase by 4-amino-5-halopentanoic acids. *Biochemistry* **20**, 1197–1203.

Skala, G., Smith, C. W., Taylor, C. J. and Ludens, J. H. (1984) A conformationally constrained vasopressin analog with antidiuretic antagonistic activity. *Science* **226**, 443–445.

Smith. T. W., Butler, V. P., Haber, E., Fozzard, H., Marcus, F. I., Bremner, W. F., Schulman, I. C., and Phillips, A. (1982) Treatment of life-threatening digitalis intoxication with digoxin-specific Fab antibody fragments. *N. Eng. J. Med.* **307**, 1357–1362.

Soper, T. S., Manning, J. M., Marcotte, P. A., and Walsh, C. T. (1977) Inactivation of bacterial D-amino acid transaminases by olefinic amino acid-D-vinylglycine. *J. Biol. Chem.* **252**, 1571–1575.

Stark, G. R. and Bartlett, P. A. (1983) Design and use of potent, specific enzyme inhibitors. *Pharmacol. Ther.* **23**, 45–78.

Testa, B. (1984) Drugs? Drug research? Advances in drug research? Musings of a Medicinal Chemist, in *Advances in Drug Research* vol. 13 Academic, New York.

Tewksbury, D. A., Dart, R. A., and Travis, J. (1981) The amino terminal amino acid sequence of human angiotensinogen. *Biochem. Biophys. Res. Commun.* **99**, 1311–1315.

Thornber, C. W. (1979) Isosterism and molecular modification in drug design. *Chem. Soc. Rev.* **18**, 563–580.

Thorsett, E. D., Harris, E. E., Peterson, E. R., Greenlee, W. J., Patchett, A. A., Ulm, E. H., and Vassil, T. C. (1982) Phosphorus-containing inhibitors of angiotensin converting enzyme. *Proc. Natl. Acad. Sci. USA* **79**, 2176–2180.

Tickle, I. J., Sibanda, B. L., Pearl, L. H., Hemmings, A. M., and Blundell, T. L. (1984) Protein crystallography, interactive computer graphics, and drug design. *X-Ray Crystallogr.* Drug Action, Course Int. Sch. Crystallogr. 9th, 427–440.

Topliss, J. G. and Edwards, R. P. (1979) Chance factors in studies of quantitative structure–activity relationships. *J. Med. Chem.* **22**, 1238–1244.

Topliss, J. G. and Martin, Y. C. (1975) Utilization of Operational Schemes for Analog Synthesis in Drug Design, in *Drug Design* vol. V (Ariens, E. J., ed.) Academic, New York.

Tute, M. S. (1971) Principles and Practice of Hansch Analysis: A Guide to Structure–Activity Correlation for the Medicinal Chemist, in *Advances in Drug Research* (Harper, N. J. and Simmonds, A. B., eds.) Academic, New York.

Veber, D. F. (1982) Peptide analogue design based on conformation. *Special Publication of the Royal Society of Chemistry* **42**, 309–319.

Venter, J. C. (1982) Immobilized and insolubilized drugs, hormones, and neurotransmitters: Properties, mechanisms of action and applications. *Pharmacol. Rev.* **34**, 153–180.

Volkman, P. H., Kohli, J. D., Goldberg, L. I., Cannon, J. G., and Lee, T. (1977) Conformational requirements for dopamine-induced vasodilation. *Proc. Natl. Acad. Sci. USA* **74**, 3602–3606.

Walsh, C. (1978) Chemical approaches to study of enzymes catalyzing redox transformations. *Ann. Rev. Biochem.* **47**, 881–931.

Weller, H. N., Gordon, E. M., Rom, M. B., and Pluscec, J. (1984) Design of conformationally constrained angiotensin-converting enzyme inhibitors. *Biochem. Biophys. Res. Commun.* **125**, 82–89.

Wermuth, C. G. (1984) Designing Prodrugs and Bioprecursors, in *Drug Design; Fact or Fantasy?* (Jolles, G. and Wooldridge, K. R. H., eds.) Academic, New York.

Westwood, R. (1981) The synthesis of novel heterocyclics as one approach to drug discovery. *Bull. Soc. Chim. Belg.* **90**, 777–780.

Wold, S. and Dunn, W. J., III (1983) Multivariate quantitative structure–activity relationships (QSAR): Conditions for their applicability. *J. Chem. Inf. Comput. Sci.* **23**, 6–13.

Wolfenden, R. (1969a) On the rate-determining step in the action of adenosine deaminase. *Biochemistry* **8**, 2409–2415.

Wolfenden, R. (1969b) Transition state analogs for enzyme catalysis. *Nature* **223**, 704–705.

Wolfenden, R. (1972) Analog approaches to the structure of the transition state in enzyme reactions. *Acc. Chem. Res.* **5**, 10–18.

Wolfenden, R. (1976) Transition-state analog inhibitors and enzyme catalysts. *Ann. Rev. Biophys. Bioeng.* **5**, 271–306.

Wolfenden, R. (1978) Transition-State Affinity as a Basis for the Design of Enzyme Inhibitors, in *Transition States of Biochemical Processes* Plenum, New York.

Wooldridge, K. R. H. (1984) The Virtues of Present Strategies for Drug Discovery, in *Drug Design: Fact or Fantasy?* (Jolles, G. and Wooldridge, K. R. H., eds.) Academic, New York.

Wootton, R., Cranfield, R., Sheppey, G. C., and Goodford, P. J. (1975) Physicochemical-activity relationships in practice. 2. Rational selection of benzenoid substituents. *J. Med. Chem.* **18**, 607–613.

Yuta, K. and Jurs, P. C. (1981) Computer-assisted structure–activity studies of chemical carcinogens. Aromatic amines. *J. Med. Chem.* **24**, 241–251.

Computer-Based Approaches to Drug Design

Philip Loftus, Marvin Waldman, and Robert F. Hout, Jr.

1. Introduction

Two major factors have influenced the dramatic growth in the use of computer-modeling techniques as an integral aspect of the drug discovery process. The first of these has been the increased availability of powerful, but relatively inexpensive, computer systems in both academic and industrial research laboratories. This provided researchers with direct access to a level of computational capacity that had previously been available only on large and prohibitively expensive mainframe computers. The second important factor was the availability of high-speed, high-resolution, graphical display systems.

The numeric, and frequently complex, input requirements of many theoretical chemistry programs had long been a formidable obstacle to their widespread acceptance by most chemists. By enabling researchers to "build" and manipulate chemical structures in a direct and simple manner and to transform the output of theoretical calculations into visual images, graphical display devices provided a powerful visual interface between the computer and the researcher working at the laboratory bench. The ability to display theoretical results in a visual form also provided a common channel of communication between theoretical, physical, and synthetic chemists, facilitating the development of effective multidisciplinary approaches to the area of drug design.

2. Theoretical Calculations

A wide variety of computational approaches to the determination of chemical properties is currently available. In terms of computer-aided drug design, the chemical aspects of most interest include determinations of molecular geometries, atomic interactions, heats of formation, relative conformational energies, and electron density distributions.

2.1. Molecular Mechanics-Based Approaches

Conceptually among the most straightforward of the approaches employed in this area, molecular mechanics programs are also among the most widely used. In this type of approach the energy of a molecule is partitioned into a number of discrete components that can be represented in a form similar to that of Eq. (1).

$$E_{\text{molecule}} = E_{\text{steric}} + E_{\text{bond length}} + E_{\text{bond angle}} + E_{\text{torsional}} \quad (1)$$

Individual energy terms are calculated between all significant pairs of atoms, then summed to give the overall energy of the molecule.

The molecular mechanics approach was initially applied to the study of conformational stabilities in hydrocarbon molecules (Hendrickson, 1961; Wiberg, 1965; Lifson and Warshel, 1968), in which good overall agreement between the calculated and experimentally determined values was obtained. Since carbon and hydrogen do not differ greatly in terms of electronegativity, it was not necessary to take account of polar interactions in these calculations. In the case of most drug molecules, however, significant differences in elecronegativity occur because of the presence of hetero-atoms, and it becomes important to take explicit account of the polar component of the energy. This has been accomplished in a number of ways, including the assignment of discrete bond dipoles and the evaluation of the resultant dipolar interactions (Abraham and Parry, 1970), but the most common approach has been the assignment of a partial charge to each atom in the molecule. The polar interaction energy may then be defined as the sum of the electrostatic interactions, as shown in Eq. (2)

$$E_{\text{polar}} = \sum \frac{Cq_iq_j}{r_{ij}} \quad (2)$$

where q_i and q_j represent the partial charges on two atoms i and j, r_{ij} is the internuclear separation in angstroms, C is a constant, and the summation is carried out over all appropriate pairs of atoms.

A considerable effort has been invested in empirically determining partial charge values for use in calculations of this type (Del Re, 1958; Momany, 1978; Abraham and Hudson, 1984), but the sensitivity of polar interactions to subtle changes in molecular geometry and to neighboring group effects, particularly in conjugated systems, imposes significant limitations on the applicability of such a highly simplified approach. The issue is also complicated by the fact that the approach of one polar molecular close to another causes a mutual polarization of charge—an effect that may also occur intramolecularly in conformationally flexible molecules in which the spacial orientation of one polar group relative to another exhibits a significant conformational dependence. Consequently an approach based on the use of formalized partial charges will always be subject to significant constraints as to its applicability, particularly in the case of highly polarizable molecules such as those that are frequently found among extensively conjugated drug molecules.

The principle attraction of the molecular mechanics technique is its computational speed. The highly simplified nature of the calculation means that it is possible to perform localized energy minimizations on many typical drug molecules within a few minutes, generating optimized molecular geometries and conformations. Consequently, molecular mechanics is frequently used to perform initial "refinement" of structures entered directly by means of a graphics terminal. The MMx series of programs, developed by Allinger (1976), is among the most widely used for this purpose.

2.2. General Limitations of Theoretical Calculations

Used appropriately, molecular mechanics routines provide a powerful computational tool, but, as with all theoretical techniques, it is important to appreciate the fundamental limitations of the approach adopted. Almost all such calculations are performed on a single isolated structure producing what is commonly known as a "gas phase" energy. This means that no account is taken of intermolecular interactions, particularly those between a compound and neighboring solvent molecules. Similarly, polar interaction energies are normally calculated *in vacuo,* with no allowance being made for the influence of solvent on the dielectric constant. Consequently, even with the most rigorous of calculations, care must be used in "transferring" the results to an experimental environment, since, although a conformational preference for a polar drug molecule in a highly hydrophobic environment may approximate fairly closely to that of the "gas phase" calculation, the conformational preference in a hydrophilic environment, such as that encountered in the plasma or cytoplasm, may be markedly different from the calculated value.

In most computational applications, an initial starting geometry for the structure under study is derived either from X-ray data or tables of standard bond lengths and bond angles, and the program attempts to optimize all or part of the structure in order to minimize its energy. Although a variety of minimization procedures are available, varying in terms of both speed and overall accuracy, most minimization methods converge on the first significant minimum encountered. Consequently, great care must be taken, particularly when attempting to study the conformational properties of highly flexible compounds, to ensure that the calculation does not become trapped in a "local" or "false" minimum.

Molecular mechanics approaches inherently assume the additivity of atomic interactions and normally assume all interactions to a given atom to be radially symmetric (e.g., no specific allowance for the directionality of lone pair interactions is made). Another assumption that is frequently inherent in this type of approach is that partial charges, and hence polar interaction energies, are independent of both molecular conformation and geometrical distortions. Clearly, the validity of such an assumption must be questioned very closely for molecules with extensive conjugation pathways.

Finally, being a purely empirical approach, the validity of a molecular mechanics calculation is directly determined by the appropriateness of the parameters employed. Generally, this means that routines parameterized using a specific series of compounds, such as the prostaglandins, for example, would typically give good results when applied to other closely related series of compounds, but may give very poor results when applied to unrelated systems such as heterocyclic aromatic compounds. Careful parameterization is critical to the performance of any molecular mechanics routine, and several pharmaceutical companies have already developed substantial "in-house" collections of parameters for particular applications.

2.3. Quantum Mechanics-Based Approaches

One of the earliest and simplest methods of calculating the resonance energy for a delocalized system involved the use of Extended Huckel Theory (EHT). Although now rarely used directly, several major quantum mechanical methods have evolved from this approach.

A number of quantum mechanics programs, at the semi-empirical level, have been applied to problems of drug design. The most common of these is probably the CNDO/2 (complete neglect of differential overlap) method of Pople and Segal (1966). Although still ultimately dependent on empirical parameterization, these programs calculate atomic interactions in terms of the overlap of localized atomic orbitals and take explicit account of electron delocalization effects such as conjugation and aromaticity. In addition, they are already fully parameterized (usually covering at least first and second row elements) so that the extensive parameterization problems encountered with molecular mechanics programs are generally avoided.

In terms of the calculation of energies and geometries for drug molecules, the greatest advantage of the semi-empirical methods over their molecular mechanics counterparts is the explicit treatment of electron delocalization. This makes them particularly well-suited to the study of the highly conjugated heteroaromatic ring systems characteristic of many important drug molecules. The basic CNDO method was later extended to give INDO (intermediate neglect of differential overlap) (Pople et al., 1967). Dewar and coworkers also developed a semi-empirical approach known as MINDO (modified intermediate neglect of differential overlap) (Baird and Dewar, 1969; Dewar and Haselbach, 1970; Bingham et al., 1975) that is incorporated in the MOPAC (Molecular Orbital Package; Stewart and Dewar, 1983) program together with the more recently developed method, MNDO (modified neglect of diatomic overlap) (Dewar and Thiel, 1977). Both MINDO and MNDO were parameterized to reproduce experimental enthalpies of formation.

In general, programs at this level give good agreement with experimental values for molecular geometry (bond lengths and bond angles), but poorer agreement for relative conformational energies, even to the extent of favoring an experimentally minor conformation. Similarly, the association of partial charges with specific atoms, generally accomplished using Mulliken population analysis (Mulliken, 1962), frequently produces partial charge distributions for which there is little direct experimental support, such as the well-known prediction of alternation of induced charge by CNDO/2 in saturated molecules (Pople and Gordon, 1967; Stolow et al., 1981).

Although providing significant advantages over molecular mechanics-based approaches, semi-empirical programs take much longer to run and are effectively limited, primarily in terms of computer time, with respect to the size of molecule that may be studied. Even for relatively small molecules, geometry optimizations frequently require many hours of computer time to accomplish on medium-sized computers.

The ability to display and examine the optimized molecular geometries produced by theoretical calculations is an important feature of any graphical display system. Figure 1 shows the MNDO optimized structure of the basic tricyclic ring system contained in clozapine 1 with the central amine nitrogen atom replaced by carbon. This display, together with the others used in this chapter, was generated by Enigma, an interactive graphical display system developed by the Structural Chemistry group at Stuart Pharmaceuticals for direct use by medicinal chemists. The feature of interest in this particular study was the manner in which the angle between the planes containing the two benzene rings varied with the nature of the substituent as the central NH group was replaced by a range of structural alternatives. The ability to computationally model the structures of these molecules meant that the effects of a wide range of alternative bridging functions could be carefully evaluated in terms of their molecular geometries prior to the synthesis of the compounds.

Despite the general usefulness of semi-empirical approaches in determining molecular geometries, their weakness in accurately determining subtle energetic phenomena such as relative conformational stabilities or hydrogen bonding interactions has already been mentioned above. The most rigorous method of theoretically studying such interactions involves the use of *ab initio* quantum mechanical approaches.

Ab initio programs, such as Gaussian-80 (Binkley et al., 1981), Gaussian-82 (Binkley et al., 1983), HONDO (Dupius et al., 1976), and GAMESS (Generalized Atomic and Molecular Electronic Structure System) (Dupuis et al., 1980), represent the most sophisticated of the currently available calculations, but are also the most time consuming. In fact, the size of the calculation means that geometry optimizations for all but the smallest of drug molecules, performed using minimal basis sets, rapidly become prohibitive with the computational time increasing as the fourth power of the number of orbitals involved.

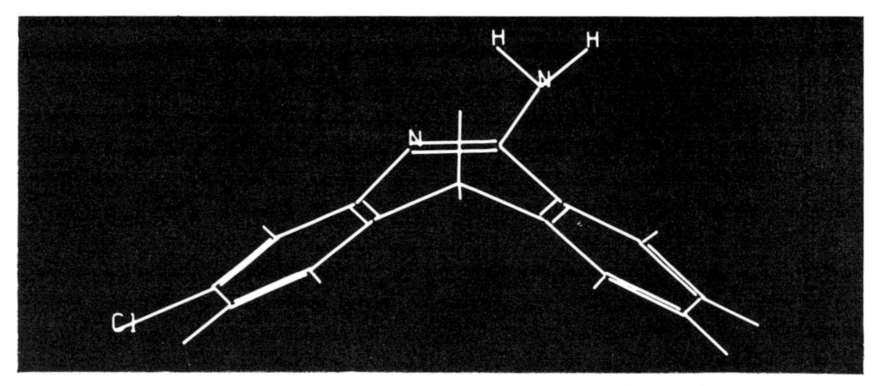

Fig. 1. MNDO-optimized structure of the central tricyclic ring system in clozapine, with central NH group replaced by CH_2.

3. Graphical Displays of Molecular Geometry and Steric Volume

In attempting the rational design of new drug compounds, a chemist is generally working from some form of initial lead. In a few fortunate cases, there may be a known enzyme as the target for inhibition for which X-ray crystallographic data of the target enzyme or a closely related analog is available. In the majority of cases, however, there is simply a lead compound upon which the chemist is trying to improve. Whichever situation applies, all attempts at rational drug design revolve around the generation of a series of hypotheses for the chemical basis of the observed biological behavior and the synthesis and testing of compounds chosen to exemplify them. On the basis of the observed biological behavior, the hypotheses are refined or abandoned and the process repeated in an iterative manner. The true power of computer modeling lies in its ability to assist in the formation of clear and rational hypotheses of the chemical basis of biological activity and in the objective evaluation of the extent to which selected synthetic targets accurately exemplify them.

Traditionally, one of the primary tools employed by the synthetic chemist in this endeavor has been the use of molecular models. Computer modeling significantly enhances this capability by providing precise, detailed displays of molecular structure, with geometries and conformational preferences optimized using theoretical calculations or those derived from experimental studies. This enables the chemist to work with much more precise structures than those formed from mechanical models and to interactively query the computer concerning the values of special bond lengths, bond angles, dihedral angles, and interatomic distances.

In addition to framework molecular models, such as Dreiding models, chemists have also used models that attempt to represent the spacial volume occupied by a molecule, the most familiar of these being "space filling" or CPK (Corey, Pauling, Koltun) models. Computer graphics offers a number of ways of displaying the steric requirements for a structure. Figure 2 shows the van der Waals surface for the pyrazolopyridine lactam structure 2, which has been investigated in the antianxiety program at Stuart.

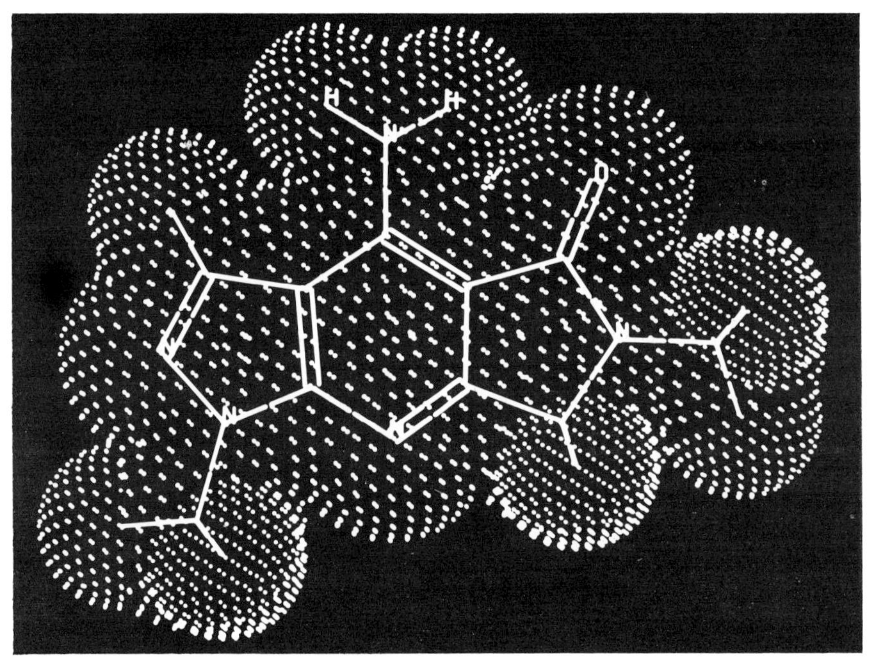

Fig. 2. Van der Waals surface display for an MNDO optimized pyrazolopyridine lactam structure.

As an alternative to the van der Waals surface, the Connolly (1983a,b) or probe surface for a structure is sometimes displayed. This is the surface obtained by "rolling" a spherical water molecule over the van der Waals surface and mapping out the area of contact. This has the effect of smoothing out indentations in the van der Waals surface and is sometimes referred to as the "solvent accessible surface."

Generally more important than the ability to display geometrically accurate structural representations is the ability to superimpose three-dimensional drug structures enabling detailed comparisons of their similarities and differences to be made. A simplified illustration of the potential usefulness of this type of approach is given in Fig. 3, which shows a proposed superposition of phenylethylamine 3 onto diphenylmethylidine 4 taken from the work of Salama et al. (1971). Displaying the van der Waals surfaces for the superimposed structures enables a researcher to study not only the detailed ability of one molecule to superimpose over another, but also to investigate both the common and differentiated spacial requirements of the two structures in their superimposed conformations.

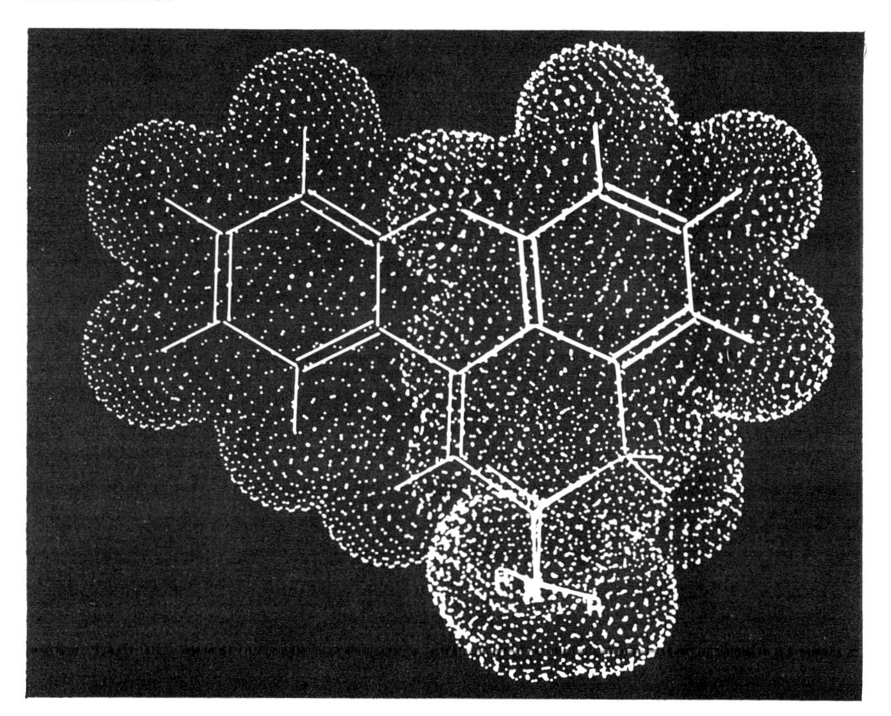

Fig. 3. Superposition of phenylethylamine and diphenylmethylidine matching the respective amine and phenyl groups. The van der Waals surfaces for the two structures are also displayed.

In the example chosen, the conformational flexibility of the two molecules is fairly limited; but for more flexible structures, for which the potential superpositioning can be performed in a variety of different ways, it becomes more difficult to determine the optimal superposition. When more than two structures are involved, the problem again becomes increasingly more complex and it is here that computational approaches afford a significant advantage. By way of illustration, the Enigma system developed at Stuart Pharmaceuticals is capable of performing rapid superpositions of up to four separate structures with the atom positions to be superimposed and the bonds about which conformational rotation is to be permitted being specified by the user. The program then performs an optimized superpositioning of the structures in accordance with the specified criteria and graphically displays the result, leaving the user free to examine the "reasonableness" of the proposed conformations.

4. Display of Electronic Properties

From the perspective of physical chemistry, the chemical properties of a drug molecule, and hence ultimately its biological activity, are directly determined by its steric and electronic requirements. In designing drug compounds, chemists have long been sensitive to the importance of the role of steric interactions in determining drug activity. This is caused in no small part by the widespread use of mechanical models as a means of examining and predicting steric interactions. Unfortunately, there is no correspondingly simple method of describing the electronic properties of a structure; and it is in the detailed study of such properties that computer modeling and graphical display techniques can play a major scientific role.

4.1. Limitations of Partical Charges in Modeling Electronic Interactions

Many of the theoretical calculations used to study molecular energetics provide information on the electronic properties of a molecule. Generally, this is provided in the form of excess electron densities or partial charges assigned to individual atoms. In the case of molecular mechanics programs, these are usually assigned on an empirical basis or are derived by performing an initial single point semi-empirical calculation. For most quantum mechanics calculations, partial charge information is derived from a Mulliken (1962) population analysis, which partitions the electron density in the molecular orbitals between individual atoms. The magnitude of the partial charge on an atom is then used as a direct measurement of its polar interactions.

Generally, the largest partial charges, typically those on highly electronegative atoms or on hydrogen atoms directly bonded to such atoms, are used to study the polar properties of the molecule, particularly in terms of its hydrogen bonding ability. There are, however, a number of fundamental limitations to the usefulness of this type of approach. Even when the results have been obtained from quantum mechanics calculations, the validity of assigning individual partial charges on the basis of a Mulliken population analysis has been subject to question (Wiberg, 1979). The use of such an approach also inherently assumes that the electron density around an individual atom is radially symmetric so that effects from the localization of lone pair electrons are lost.

Additionally, although the polar interaction between two highly charged atoms in close proximity to each other may be reasonably considered an isolated interaction, electrostatic interactions decrease only linearly with distance so that the presence of significant partial charges on nearby atoms may exert a significant perturbational effect.

4.2. Electrostatic Potential Displays

Explicit account of the influence of neighboring charges can be taken by defining polar interactions in terms of the interaction between an atom carrying a partial charge and an electrostatic field. The electrostatic field of a molecule at any given point in space is defined as the force acting upon a unit positive charge placed at that point and is obtained by summing the contributions made by all atoms in the molecule. The most common application of this approach involves the calculation of the electrostatic potential for an array of points distributed on the van der Waals or Connolly surface of a structure. Figure 4 shows the electrostatic potential for the pyrazolopyridine lactam 2, projected onto the Connolly surface for the molecule. To avoid the problems inherent in a Mulliken population analysis, the values of the electrostatic potential were calculated directly from the wave function describing the molecule. This retains asymmetries in the distribution of electron density about an atom so that lone pair directionality is preserved. In order to simplify the display, points on the surface with potentials in the range of −15 to +15 kcal/mol are omitted.

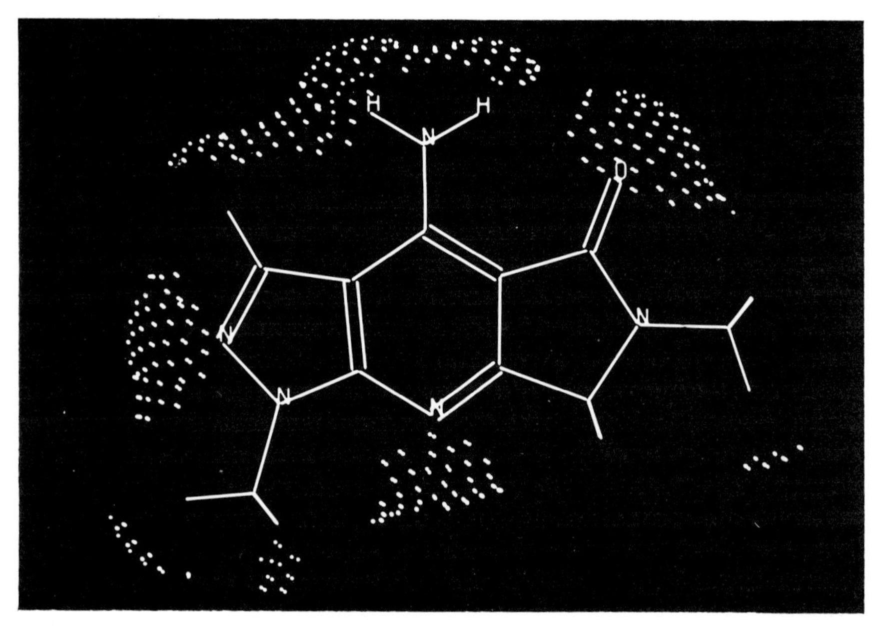

Fig. 4. Electrostatic Potential display generated using an STO-3G basis set projected onto the Connolly surface of a pyrazolopyridine lactam structure.

4.3. Qualitative Nature of Electrostatic Potential Displays

The display shown in Fig. 4 corresponds to the electrostatic potential for an isolated molecule in the gaseous phase. Bringing a water, or other polar, molecule close to this structure would significantly perturb the potential around it, since not only would the water molecule itself contribute to the value of the potential, but its approach, particularly in a hydrogen bonding orientation, would cause a polarization of the electron density distribution in the pyrazolopyridine lactam 2.

Consequently, displays of this nature are not intended to provide quantitative descriptions of polar interactions or hydrogen bonding ability. What they do provide, however, are qualitative images of the electronic properties of a molecule as a whole, indicating the major regions where strong polar interactions and hydrogen bond formation are most likely to occur and enabling the direct visual comparison of the electronic properties of different structures.

4.4. Isopotential Surface Displays

Since electrostatic interactions are attenuated only linearly with distance, it does not follow that the optimum position for a polar interaction occurs at the van der Waals contact distance. Consequently, instead of projecting electronic properties arbitrarily onto the steric surface of a structure, it is possible to define the actual potential surfaces around a structure corresponding to a given energy level. For any absolute value of the energy, this will give two possible surfaces, one corresponding to the positive and the other to the negative value of the potential. Figure 5 shows the 30 kcal/mol isopotential surfaces for the pyrazolopyridine lactam 2.

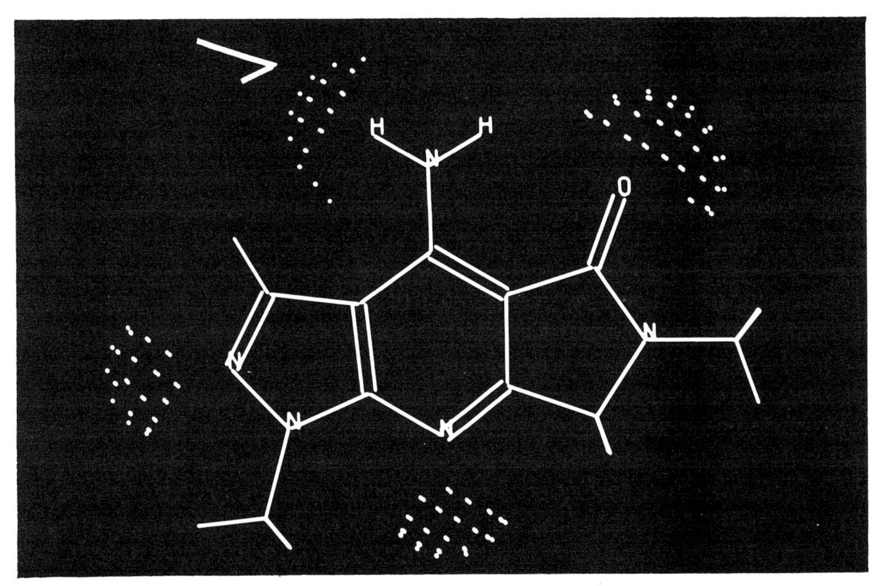

Fig. 5. 30 kcal/mol Isopotential surfaces for a pyrazolopyridine lactam structure calculated directly from the STO-3G wave function.

The display of the surface has been suppressed where it penetrates the van der Waals surface so that only the sterically accessible region of the isopotential surface is actually displayed. As with the electrostatic potential surface described above, this surface was also calculated directly from the molecular wave function so that the effects of lone pair directionality are preserved. As can be seen, the negative potentials corresponding to lone pairs on the carbonyl oxygen and two of the nitrogen atoms are clearly displayed, whereas no potential surface, at this energy level, is associated with the three remaining nitrogen atoms. Partial charges for the individual atoms in the molecule were obtained by performing a Mulliken population analysis of the *ab initio* results derived using an STO-3G (Slater Type Orbital 3-Gaussians) basis set. The basis set generally reflects the degree of sophistication of the mathematical treatment of the atomic orbitals in the molecule and STO-3G represents a widely used minimal basis set for general purpose calculations. The partial charges obtained in this manner are shown in Fig. 6. Note that neither of the two most negatively charged

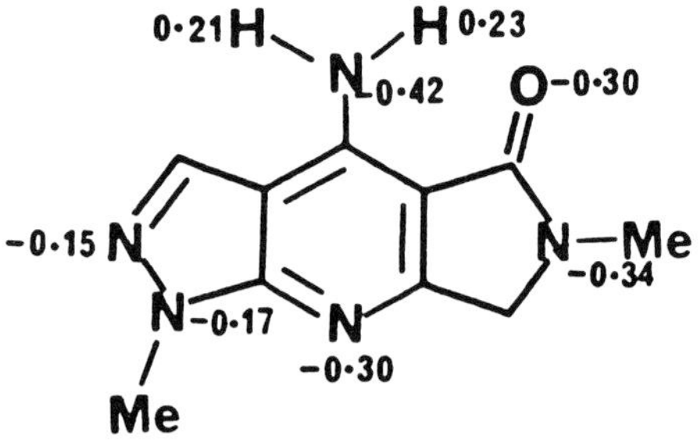

Fig. 6. Partial charges (displayed in positron units) derived from an STO-3G wave function.

nitrogens, the aromatic amine with −0.42 positrons and the lactam nitrogen with −0.34 positrons, gives rise to a negative potential surface. This is because both of these nitrogens are sp^2, hybridized giving rise to a more symmetrical distribution of the lone pair electrons, and they each have strongly positive neighbors that reduce the net value of the potential close to the atom. A similar effect is observed for the two amine hydrogens. Although carrying virtually identical partial charges, the potential caused by the hydrogen facing the carbonyl group is reduced by the negative contributions from the nearby nitrogen and oxygen atoms, whereas the hydrogen atom facing away from the carbonyl group gives rise to a significant positive potential. This display clearly indicates the danger of using isolated partial charge information to determine polar interactions, since it is the overall potential, not simply that contributed by a single atom, that plays the dominant role in determining such interactions.

Also included in this display is the position of one of the three water molecules found to be present in the crystal structure by X-ray analysis. As can be seen, the oxygen atom of the water molecule sits very close to the center of the positive potential surface surrounding the hydrogen atom. The remaining water molecules are also aligned close to the isopotential surface, although for visual simplicity, they have been omitted from this display.

It has long been known that the presence of substituents on a conjugation pathway can have significant chemical consequences. Electrostatic and isopotential displays provide a powerful method of studying the influence of substituents in such systems. By way of illustration, the 15 kcal/mol isopotential display for amiloride and its deschloro analog is shown in Fig. 7. As can be seen, the removal of the chlorine atom not only modifies the nature of the potential surface in the region of substitution, but significantly enhances the size of the negative potential surfaces for the nitrogen and oxygen atoms on the lower right-hand side of the structure so that the negative potential above the amine nitrogen atom, which would have been *para* to the chlorine, becomes clearly visible in the deschloro analog.

Used with appropriate care and discretion, isopotential surface displays provide a powerful method of qualitatively comparing the electronic properties of structurally disparate molecules and of systematically examining the influence of the nature and positioning of substituents in delocalized systems. Given the importance of electrostatic interactions for hydrogen bond formation, this approach also provides a sensitive method for studying the effects of substitution and molecular conformation on hydrogen bonding ability.

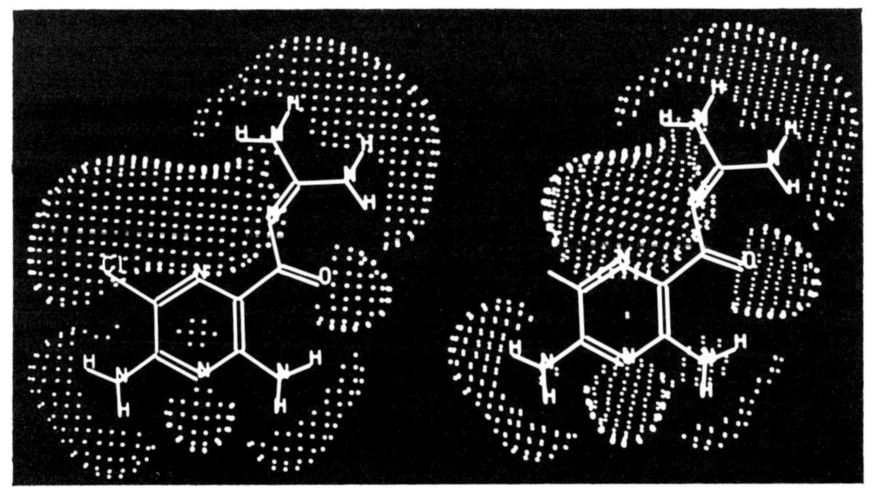

Fig. 7. 15 kcal/mol Isopotential surfaces for Amiloride and deschloro-Amiloride calculated directly from the STO-3G wave function.

5. Macromolecular Systems

In addition to their role in the direct study and comparison of drug molecules, computer modeling techniques may also be applied to the study of enzyme structure and enzyme substrate interactions. The theoretical calculation of the secondary and tertiary structure for even the smallest of enzyme systems, starting from the experimentally determined peptide sequence, is a formidable task that is well beyond the capabilities of most modern approaches. Hence, most theoretical studies of enzyme systems start from an experimentally determined X-ray crystal structure of either the required enzyme or a closely related analog.

Computer graphics can play a powerful role in this area by supporting a wide variety of visual displays of crystallographic data. Figure 8 shows the X-ray crystal structure of tosyl-porcine pancreatic elastase (Watson et al., 1970; Shotton and Watson, 1970). In order to simplify the display, the general convention of showing only

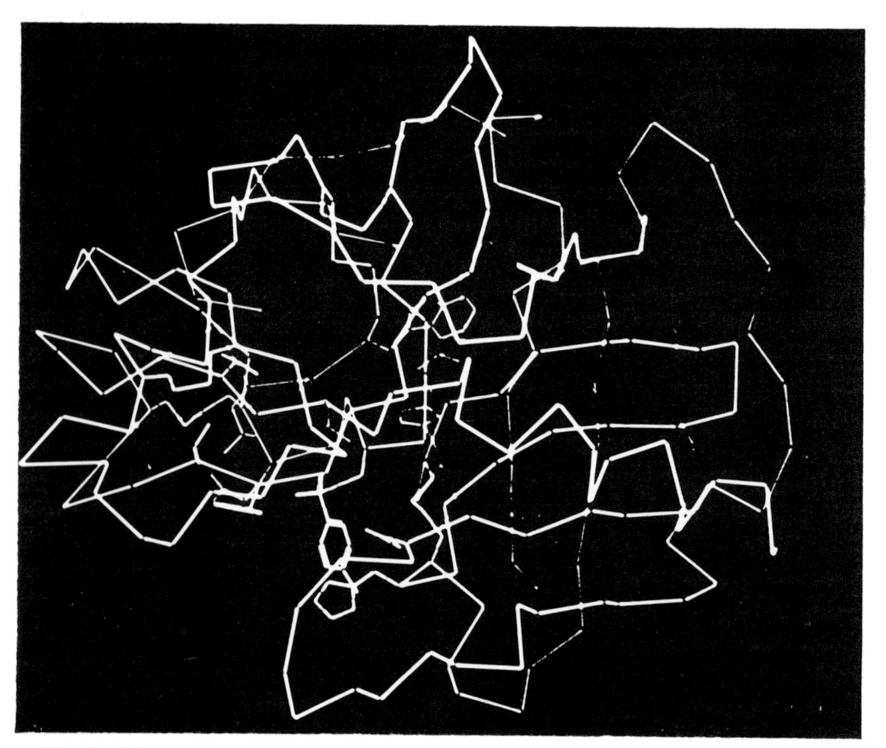

Fig. 8. X-ray crystal structure of tosyl-porcine pancreatic elastase.

the alpha carbon backbone is adopted. However, individual amino acid side chains are shown for residues close to the active site of the enzyme. Although displays of this type provide detailed structural information on the nature of the active site and its steric and electronic requirements, the application of calculational techniques to structures of this size is still a formidable obstacle even when the associated crystallographic data is available. The large number of atoms involved, even if only those residues defining the active site are considered, means that the application of quantum mechanics techniques is almost always precluded. A number of important approaches, based on the application of molecular mechanics techniques, have been specifically developed for use with macromolecular biological systems, including the ECEPP (Empirical Conformational Energy Program for Peptides) program of Momany et al. (1975), and the CHARMM (Chemistry at Harvard Macromolecular Mechanics) and AMBER (Assisted Modelbuilding with Energy Refinement) programs, developed by Brooks et al. (1983) and Weiner and Kollman (1981), respectively, which may be used both to study enzyme substrate interactions and to geometry optimize structures obtained by modifying the peptide sequence or conformation of an experimentally known structure.

From the perspective of new drug discovery, one of the major applications in this area is the study of enzyme–substrate interactions as models for inhibitor design. The theoretical study of such systems is, however, a complex process involving the difference in free energy between the enzyme and substrate free in solution and bound together in the complex. In the simplest case this is a binary process, but when additional substances, such as an enzyme cofactor, are involved, cooperative binding may occur so that the energetics of binding of both the substrate and the cofactor may need to be considered.

Another important factor that further complicates this process is that the active site of the enzyme is not necessarily empty prior to binding the substrate. Frequently, the active site is occupied by solvent molecules that are displaced on binding the substrate. Similarly, the substrate molecule may also have associated solvation that must be lost or reduced before it is able to bind to the enzyme. Such processes are difficult to treat theoretically and most applications of computer modeling to the study of enzyme substrate interactions have been predominantly qualitative in nature.

One of the most widely studied enzyme systems utilizing these techniques is dihydrofolate reductase (DHFR), which catalyzes the reduction of dihydrofolate 5 to tetrahydrofolate 6 by means of hydride transfer from NADPH. Tetrahydrofolate derivatives are utilized in a variety of one-carbon transfer reactions coupled with the action of thymidylate synthetase (Hitchings and Roth, 1980), and the inhibition of DHFR has been employed in both the treatment of cancer and the development of antibacterial agents. Methotrexate 7 is a potent DHFR inhibitor with K_i values of up to $10^{-11}M$ (Colwell et al., 1979). Figure 9 displays the X-ray structure of methotrexate and NADPH bound to DHFR derived from

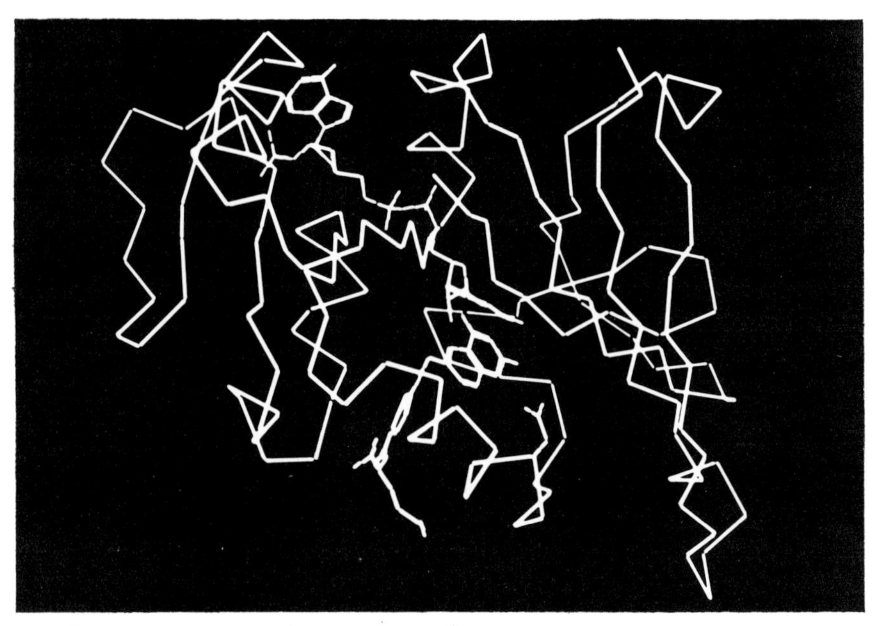

Fig. 9. X-ray crystal structure of methotrexate and NADPH bound to dihydrofolate reductase obtained from *L. casei*.

Lacto-bacillus casei (Matthews et al., 1978) showing the alpha carbon back-bone for the enzyme together with methotrexate and NADPH molecules. DHFR shows considerable species variation, and Hansch et al. (1982) have used computer graphics to study the binding of methotrexate 7 and trimethoprim 8, using the available crystallographic data for DHFR derived from both *L. casei* (Matthews et al., 1978) and *Escherichia coli* (Matthews et al., 1979; Baker et al., 1981). The X-ray data clearly shows that the active site of DHFR derived from *E. coli* is considerably more spacious than that derived from *L. casei*. This information was used to explain differences in the observed binding energies of inhibitors to enzyme derived from these two sources in terms of probable steric and polar interactions. It was also proposed that the larger site associated with the *E. coli* enzyme may permit the partial retention of inhibitor solvation on binding relative to the *L. casei* enzyme, a factor that would also contribute to the overall binding energy. In particular, computer graphics were used to clearly illustrate the tight steric fit of trimethoprim to the *L. case* enzyme and its much looser fit to the *E. coli* enzyme.

Despite the very strong binding of methotrexate 7 to DHFR, Andrews et al. (1984) have predicted that it binds in a high-energy conformation, a result that is also supported by potential energy calculations (Spark et al., 1982). Matthews et al. (1978) have also proposed that methotexate 7 binds in an orientation that is rotated by 180° from that expected by direct comparison with the natural substrate, an argument for which there is considerable indirect support (Fontecilla-Camps et al., 1979; Armarego et al., 1980).

Blaney et al. (1982) have also used crystallographic data (Blake and Oatley, 1977; Blake et al., 1978) to study the interaction of thyroxine with prealbumin as a model for the interaction of thyroid hormones with the nuclear thyroid hormone receptor (Eberhardt et al., 1979). By carefully examining the steric fit of a series of thyroid hormone analogs, an additional unused binding pocket was observed. Accordingly, four new analogs were designed to take advantage of this additional binding site and shown to bind to prealbumin in a manner corresponding with their ability to bind effectively at this site.

The mechanism of the zinc containing endopeptidase thermolysin has also been studied using a combination of crystallographic data (Monzingo and Matthews, 1984) and computer graphics techniques (Hangaur et al., 1984; Bush, 1984). By visually "docking" substrates with the enzyme and examining the steric and polar interactions incurred, the authors proposed a model in which the formation of the Michaelis complex involved the coordination of the substrate via a water molecule rather than directly to the zinc atom in the enzyme. It was also proposed that the mechanism of cleavage involved the formation of an intermediate tetrahedral complex in which the substrate coordinated to the zinc atom in a bidentate manner, causing the zinc to be pentacoordinate. The proposed model was also shown to be consistent with experimentally determined hydrolysis data for the enzyme (Morihara and Tsuzuki, 1970).

In each of the studies outlined above, the enzyme system was essentially treated as a rigid molecule. Even in such cases, it is clear that a great deal of valuable information may be obtained. The key to the future growth of such studies clearly lies in the availability of the necessary crystallographic data, and several major pharmaceutical companies have already established their own macromolecular X-ray crystallography groups.

Almost all of the larger pharmaceutical companies now have significant computer modeling capabilities, and it is clear that theoretical and graphical techniques will play an increasingly important role in the development of a more rational approach to the complex problem of drug design. Used appropriately, computational techniques can provide an important insight into the chemical basis of biological activity. It is important, however, to link theoretical predictions with experimental verification; and the most successful applications are likely to be those based upon an integrated multidisciplinary approach combining theoretical predictions with X-ray crystallographic data for structures in the solid state and other appropriate techniques, such as high resolution NMR data for structures in solution.

Acknowledgment

The authors would like to express their appreciation to Dr. C. L. Lerman of Stuart Pharmaceuticals for assistance in preparing the two enzyme structure displays shown in Figs. 8 and 9.

References

Abraham, R. J. and Hudson, B. (1984) Approaches to charge calculations in molecular mechanics. 2. Resonance effects in conjugated systems. *J. Comp. Chem.* **5**, 562–570.

Abraham, R. J. and Parry, K. (1970) Rotational isomerism. VIII. A calculation of the rotational barriers and rotamer energies of some halogenated compounds. *J. Chem. Soc.* **B**, 539–545.

Allinger, N. L. (1976) Calculation of molecular structure and energy by force-field methods. *Adv. Phys. Org. Chem.* **13**, 1–82.

Andrews, P. R., Craik, D. J., and Martin, J. L. (1984) Functional group contributions to drug–receptor interactions. *J. Med. Chem.* **27**, 1648–1657.

Armarego, W. L. F., Waring, P., and Williams, J. W. (1980) Absolute configuration of 6-methyl-5,6,7,8-tetrahydropterin produced by enzymatic reduction (dihydrofolate reductase and NADPH) of 6-methyl-7,8-dihydropterin. *Chem. Comm.* 334–336.

Baird, N. C. and Dewar, M. J. S. (1969) Ground state of sigma-bonded molecules. IV. M.I.N.D.O. method and its application to hydrocarbons. *J. Chem. Phys.* **50**, 1262–1274.

Baker, D. J., Beddell, C. R., Champness, J. N., Goodford, P. J., Norrington, F. E. A., Smith, D. R., and Stammer, D. K. (1981) The binding of trimethoprim to bacterial dihydrofolate reductase. *FEBS Lett.* **126**, 49–52.

Bingham, R. C., Dewar, M. J. S., and Lo, D. H. (1975) Ground states of molecules. XXV. MINDO/3: An improved version of the MINDO semi-empirical SCF-MO method. *J. Am. Chem. Soc.* **97**, 1285–1293.

Binkley, J. S., Frisch, M. J., DeFrees, D. J., Raghavachari, K., Whiteside, R. A., Schlegel, H. B., Fluder, E. M., and Pople, J. A. (1983) *GAUSSIAN 82* Carnegie-Mellon University, Pittsburgh, Pennsylvania.

Binkley, J. S., Whiteside, R. A., Krishman, R., Seeger, R., DeFrees, D. J., Schlegel, H. B., Topiol, S., Kahn, L. R., and Pople, J. A. (1981) QCPE program 406. University of Indiana, Bloomington, Indiana.

Blake, C. C. F. and Oatley, S. J. (1977) Protein–DNA and protein–hormone interactions in prealbumin: A model of the thyroid hormone nuclear receptor. *Nature* **268**, 115–120.

Blake, C. C. F., Geisow, M. J., Oatley, S. J., and Rerat, C. J. (1978) Structure of prealbumin: Secondary, tertiary and quaternary interactions determined by Fourier refinement at 1.8 angstroms. *J. Mol. Biol.* **121**, 339–356.

Blaney, J. M., Jorgenson, E. C., Connolly, M. L., Ferrin, T. E., Langridge, R., Oatley, S. J., Burridge, J. M., and Blake, C. C. F. (1982) Computer graphics in drug design: Molecular modeling of thyroid hormone-prealbumin interactions. *J. Med. Chem.* **25**, 785–790.

Brooks, B. R., Bruccoleri, R. E., Olafson, B. D., States, D. J., Swaminathan, S., and Karplus, M. (1983) CHARMM: A program for macromolecular energy, minimization and dynamics calculations. *J. Comp. Chem.* **4**, 187 217.

Bush, B. L. (1984) Interactive modeling of enzyme–inhibitor complexes at Merck Macromolecular Modeling graphics facility. *Comp. Chem.* **8**, 1–11.

Colwell, W. T., Brown, V. H., Degraw, J. T., and Morrison, N. E. (1979) Inhibition of mycobacterial dihydrofolate reductase by 2,4-diamino-6-alkylpteridines and deazapteridines. *Dev. Biochem.* 215–218.

Connolly, M. L. (1983a) Solvent-accessible surfaces of proteins and nucleic acids. *Science* **221**, 709–813.

Connolly, M. L. (1983b) Analytical molecular surface calculation. *J. Appl. Crystallogr.* **16**, 548–558.

Del Re, G. (1958) A simple M.O. L.C.A.O. method for calculating the charge distribution in saturated organic molecules. *J. Chem. Soc.* 4031–4040.

Dewar, M. J. S. and Haselbach, E.)1970) Ground states of sigma-bonded molecules. IX. The MINDO/2 method. *J. Am. Chem. Soc.* **92**, 590–598.

Dewar, M. J. S. and Thiel, W. (1977) Ground states of molecules. 38. The MNDO method. Approximations and parameters. *J. Am. Chem. Soc.* **99**, 4899–4907.

Dupius, M., Rys, J., and King, H. F. (1976) Evaluation of molecular integrals over Gaussian basis functions. *J. Chem. Phys.* **65**, 111–116.

Dupius, M., Spangler, D., and Wendoloski, J. J. (1980) NRCC software catalog 1, program QG01, Lawrence Berkeley Laboratory, University of California, Berkeley, California.

Eberhardt, N. L., Ring, J. C., Latham, K. R., and Baxter, J. D. (1979) Thyroid hormone receptors. Alterations of hormone binding specificity. *J. Biol. Chem.* **254**, 8534–8539.

Fontecilla-Camps, J. C., Bugg, C. E., Temple Jr., C., Rose, J. D., Montgomery, J. A., and Kisliuk, R. L. (1979) Absolute configuration of biological tetrahydrofolates. A. crystallographic determination. *J. Am. Chem. Soc.* **101**, 6114–6115.

Hangaur, D. G., Monzingo, A. F., and Matthews, B. W. (1984) An interactive computer graphics study of thermolysin-catalyzed peptide cleavage and inhibition by *N*-carboxymethyl dipeptides. *Biochemistry* **23**, 5730–5741.

Hansch, C., Li, R., Blaney, J. M., and Langridge, R. (1982) Comparison of the inhibition of *Escherichia coli* and *Lactobacillus casei* dihydrofolate reductase by 2,4-diamino-5-(substituted benzyl) pyrimidines: Quantitative structure–activity relationships, X-ray crystallography and computer graphics in structure–activity analysis. *J. Med. Chem.* **25**, 777–784.

Hendrickson, J. B. (1961) Molecular Geometry. I. Machine computation of the common rings. *J. Am. Chem. Soc.* **83**, 4537–4547.

Hitchings, G. H. and Roth B. (1980) Dihydrofolate Reductases as Targets for Selective Inhibitors, in *Enzyme Inhibitors as Drugs* (Sandler, M., ed.) Macmillan, London.

Lifson, S. and Warshel, A. (1968) Consistent force field calculations of conformations, vibrational spectra, and enthalpies of cycloalkane and *n*-alkane molecules. *J. Chem. Phys.* **49**, 5116–5129.

Matthews, D. A., Alden, R. A., Bolin, J. T., Filman, D. J., Freer, S. T., Suong, R., and Kraut, J. (1978) Dihydrofolate reductase from *Lactobacillus casei*. X-ray structure of the enzyme-methotrexate-NADPH complex. *J. Biol. Chem.* **253**, 6946–6954.

Matthews, D. A., Alden, R. A., Freer, S. T., Xuong, N. H., and Kraut, J. (1979) Dihydrofolate reductase from *Lactobacillus casei.* Stereochemistry of NADPH binding. *J. Biol. Chem.* **254**, 4144–4151.

Momany, F. A. (1978) Determination of partial atomic charges from ab-initio molecular electrostatic potentials. Application to formamide, methanol and formic acid. *J. Phys. Chem.* **82**, 592–601.

Momany, F. A., McGuire, R. F., Burgess, A. W., and Scheraga, H. A. (1975) Energy parameters in polypeptides. VII. Geometric parameters, partial atomic charges, nonbonded interactions and intrinsic torsional potentials for the naturally occurring amino acids. *J. Phys. Chem.* **79**, 2361–2381.

Monzingo, A. F. and Matthews, B. W. (1984) Binding of *N*-carboxymethyl dipeptide inhibitors to thermolysin determined by X-ray crystallography: A novel class of transition state analogues for zinc peptidases. *Biochemistry* **23**, 5724–5729.

Morihara, K. and Tsuzuki, H. (1970) Thermolysin: Kinetic study with oligopeptides. *Eur. J. Biochem.* **15**, 374–380.

Mulliken, R. S. (1962) Criteria for the construction of good self-consistent field molecular orbital wave functions, and the significance of L.C.A.O. M.O. population analysis. *J. Chem. Phys.* **36**, 3428–3439.

Pople, J. A. and Gordon, M. (1967) Molecular orbital theory of the electronic structure of organic compounds. I. Substituent effects and dipole moments. *J. Am. Chem. Soc.* **89**, 4253–4261.

Pople, J. A. and Segal, G. A. (1966) Approximate self-consistent molecular orbital theory. III. CNDO results for AB_2 and AB_3 systems. *J. Chem. Phys.* **44**, 3289–3296.

Pople, J. A., Beveridge, D. L., and Dobosh, P. A. (1967) Approximate self-consistent orbital theory. V. Intermediate neglect of differential overlap. *J. Chem. Phys.* **47**, 2026–2033.

Salama, A. I., Insalaco, J. R., and Maxwell, R. A. (1971) Concerning the molecular requirements for the inhibition of uptake of racemic ^{3}H-norepinephrine into rat cerebral cortex slices by tricyclic antidepressants and related compounds. *J. Pharmacol. Exp. Ther.* **178**, 474–481.

Shotton, D. M. and Watson, H. C. (1970) Three-dimensional structure of tosyl-elastase. *Nature* **225**, 811–816.

Spark, M. J., Winkler, D. A., and Andrews, P. R. (1982) Conformational analysis of folates and folate analogues. *Int. J. Quantum Chem., Quantum Biol. Symp.* **9**, 321–333.

Stewart, J. J. P. and Dewar, M. J. S. (1983) QCPE program 455, University of Indiana, Bloomington, Indiana.

Stolow, R. D., Samal, P. W., and Giants, T. W. (1981) On CNDO/2—predicted charge alternation. *J. Am. Chem. Soc.* **103**, 197–199.

Watson, H. C., Shotton, D. M., Cox, J. M., and Muirhead, H. (1970) Three-dimensional Fourier synthesis of tosyl-elastase at 3.5 angstrom resolution. *Nature* **225**, 806–811.

Weiner, P. K. and Kollman, P. A. (1981) AMBER: Assisted model building with energy refinement. A general program for modeling molecules and their interactions. *J. Comp. Chem.* **2**, 287–303.

Wiberg, K. B. (1965) A scheme for strain energy minimization. Application to the cycloalkanes. *J. Am. Chem. Soc.* **87**, 1070–1078.

Wiberg, K. B. (1979) Infrared intensities. The methyl halides. Effect of substituents on charge distributions. *J. Am. Chem. Soc.* **101**, 1718–1722.

Use of Intact Tissue Preparations in the Drug Discovery Process

FRANK R. GOODMAN

1. Introduction

In discussing the usefulness of isolated tissues in the drug discovery process, as with any model utilized in the characterization of a potential new drug, the question of relevance must be ascertained. The biologist usually interprets the effects of a drug on isolated tissues or organs as an action at the receptor level. However, the question does not have to be targeted toward a particular cellular or mechanistic parameter; it can also be more global in nature, e.g., clinical significance in comparison to known active agents. Usually, the more that is known about a chemical entity or clinical disorder, the more relevant are the in vitro tissue screens (*see* Williams and Malick, this volume).

In general, drugs are chemical entities that when they interact with biological systems, result in alterations in cell or tissue function. Because of the complex interrelationships of physiological and biochemical processes, as well as the fact that drugs often appear to have more than one action (therapeutic and side effects), studies directed toward obtaining experimental data on the basic aspects of drug actions are frequently targeted by investigation of isolated tissue systems. However, relating the results from in vitro studies to therapeutic actions in vivo can be misleading. For example, drugs can be metabolized to the active product and, in some cases, the metabolite is more active than the parent compound or may have a different pharmacological profile. Nevertheless, there are several advantages to utilizing the isolated tissue preparation versus in vivo studies. With an isolated tissue preparation, one eliminates (1) indirectly mediated in vivo responses, (2) feedback mechanisms that could mask the direct effect of the drug on a tissue or organ, and (3) the variability of drug distribution often encountered in vivo. In addition quantitative measurements of drug actions can usually be obtained with minimal effort and the resulting information utilized to further elucidate structure–activity relationships.

The present chapter describes some simple pharmacological tests useful in detecting and delineating mechanisms of action, as well as predicting potential therapeutic applications. In developing or identifying new therapeutic agents for the treatment of symptoms seen in disease states, isolated tissue assays have been widely employed in industry. Although the random screening process can be an effective way for finding a novel drug (Maxwell, 1984), selectivity and activity that reside in a given tissue may not necessarily exist in all tissues. For example, vascular smooth muscle contains a wide variety of drug receptors (Altura and Altura, 1970) and different densities of receptor types, and it is generally recognized that large vessels respond differently from small vessels to some drugs. Thus, elucidation of the particular mechanism(s), as well as the therapeutic target, becomes an important process in defining which in vitro assay is most appropriate for discovering a new drug. Although there are many examples of how isolated tissues can be utilized in drug discovery, only those required to make a particular point are described in the present chapter.

The intact tissue preparation is a level of complexity a step above the in vitro receptor binding assay (Enna, this volume), and a level below that of the whole animal. As indicated in the first chapter in this volume, the three approaches to the drug discovery process are of equal importance, and when evaluating a new chemical entity, although definitive data can be obtained from each of these approaches, such data should be compared using an increasingly vigorous approach to show that there is a congruency in assigning a specific mechanism to a given tissue or animal response. For instance, whereas activation of the adenosine receptor in guinea pig trachea can relax the contracted tissue, many agents without interactions at this receptor can produce identical effects. Thus an *a priori* approach is necessary to ascribe a particular mechanism in more complex systems.

2. General Concepts

2.1. Receptors

For a drug to induce an action it must react with some cellular constituent—the receptor (Ehrlich, 1913; Williams and Enna, 1986)

—and even though their chemical nature has remained, to a large extent, unknown, receptors have functioned as useful conceptual tools in pharmacology. Drug receptors and drug-receptor theory are important concepts for the understanding and consistent interpretations of dose–response data, and knowledge of receptor theory is mandatory for understanding the current research efforts directed at both the chemical and structural identification of receptors. For practical purposes receptors are defined as macromolecule components contained within or on the organism (e. g., smooth muscle) that interact with a specific group of agonists or antagonists. The strength of interaction between a drug and its receptor is influenced by the chemical structure of the drug. It should be noted, however, that the result of the interaction between a drug and its receptor is not always predictable. In smooth muscle, the interaction of the agonist with its receptor triggers a series of events that results in either a contraction or relaxation (Bolton, 1979). However, many biochemical events are interposed between the drug–receptor interaction and the physiologic/pharmacologic response. The ideas and equations utilized to analyze drug receptor interactions as well as the events beyond the initial reaction are referred to as drug receptor theory. This topic as well as comparison of the various models have been extensively covered in numerous reviews (Ariens and Simonis, 1962, 1964a,b; Furchgott, 1966; Waud, 1968; Ruffolo, 1982; Limbird, 1986; Williams and Enna, 1986) to which the reader is referred.

One of the many issues that confronts the pharmacologist in the ''drug discovery process'' is to explain what it is that makes some members of a drug series agonists and others antagonists. Of the various theoretical models proposed, there are two major theories. The ''classical'' occupation theory is attributed to A. J. Clark (1933), who is responsible for much of the present thinking about receptors. This theory assumes that the occupation of a receptor by a drug results in an effect that is proportional to the number of receptors occupied by the drug. The second theory, which is an extension of the occupation theory, was based on the assumption that drug efficacy results from the product of the rate of drug–receptor interaction (Paton, 1961). Paton's theory has been utilized to explain the lack of agonist activity by antagonists since antagonist–receptor complexes appear to turn over much slower than agonist–receptor complexes. However, it is important to realize that many of the equations utilized in receptor theory are based on idealized situations, and although certain assumptions about theoretical relationships may be true, exceptions are not uncommon. For example, the most straight-forward relationship between activation of the receptor and the resulting biological response is a linear one. However, there are numerous examples in the literature (Furchgott, 1955; Nickerson, 1956; Ariens et al., 1960; Venter, 1979) in which the relationship between these two events is non-linear; this observation has led to the postulation of ''spare receptors'' (Nickerson, 1956; Stephenson, 1956). This term, however, is only associated with a specific agonist and tissue since the number of ''spare receptors'' for any two agonists in a given tissue may be different.

Receptors can and have been studied using isolated tissues, organs, or enzymes. Even though an isolated test preparation may be considered superior to an in vivo experimental design for evaluating drug activity at the receptor level, experimental conditions and/or assumptions may vary from the ''ideal situation'' and could lead to erroneous or unjustified conclusions (Gillette, 1968). Desensitization of receptors, which results in decreased responsiveness to an agonist, agonist acting on more than one type of receptor, antagonist altering the response of an agonist by an action other than occupying the same receptor, concentration of free drug in the region of the receptor decreasing over time, and lack of equilibrium between medium and tissue concentration are a few of the commonly encountered variations that should be taken into account (Furchgott, 1968).

In addition to the drug–receptor interaction, agents can affect responsiveness of smooth muscle by acting on (1) the membrane, (2) nerves innervating the tissue, (3) the intermediate steps of excitation–contraction coupling, and (4) the contractile apparatus. Thus, it is obvious that a strip of smooth muscle is a complex system involving several factors (e.g., enzymes, diffusion processes) that must be taken into account when evaluating the effects of a drug on an isolated system.

Although utilization of isolated tissues to evaluate drug–receptor interaction at the receptor level is frequently fraught with difficulties, and experimental conditions may vary from idealized receptor theory, the concept that employment of these preparations are superior for collecting data on drug efficacy and affinity is still valid.

2.2. Dose–Response

As previously discussed, the pharmacological effect of a drug is generally the result of the interaction between the drug and its specific receptors. This interaction is best studied in terms of dose–response curves. The relationship between the amount of compound administered and the magnitude of the response is an example of a dose–response relation and is utilized as an initial step in understanding the mechanism of drug action. Frequently these relationships are of a simple hyperbolic form, and they are usually represented by a sigmoid-shaped curve. This curve represents the simplest model for drug–receptor interactions and assumes that the response is proportional to the number of receptors occupied, and that when all are occupied, a maximal response is obtained (Fig. 1A). A point on the curve that has particular significance is that point that corresponds to the half-maximal (50%) response. The height of the plateau is the maximal effect produced by the drug and is a measure of the drugs efficacy. Maximal efficacy of a drug is clearly a major characteristic. The location of its dose–response curve, indicating the magnitude of the effect of a given dose, is a measure of the potency of a drug. Thus, drugs that produce a certain effect may differ in either efficacy or potency or both (Fig. 1B). Whereas potency is influenced by absorption, distribution, elimination, and so on, efficacy is related to the fundamental action of the drug.

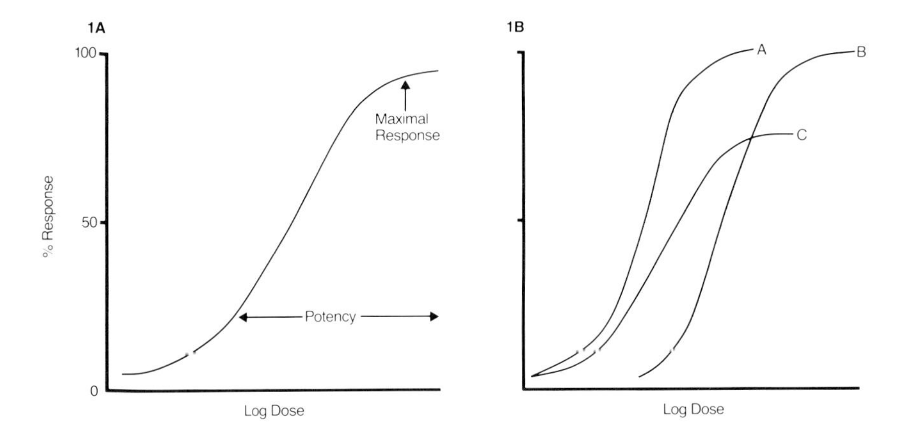

Fig. 1. (A) A representative sigmoidal dose–response curve. (B) Relationship between two agonists with similar efficacy but different potency (A more potent than B), and similar potency but different efficacy (A is a full agonist; C is a partial agonist). The curves also illustrate an agonist in the presence (B) or absence (A) of a competitive antagonist or an agonist in the presence (C) or absence (A) of a noncompetitive antagonist.

Utilizing isolated tissue preparations, it is possible to obtain dose–response data for agonists (an agent that induces an effect when interacting with its receptor). An advantage of this procedure is that the data can be obtained both before and after treatment of the preparation with other drugs (usually antagonists or chemical entities with unknown biological effects). Drug antagonisms can result from different mechanisms, but they have been generally classified into two categories. Antagonism in which both the agonist and antagonist combine with the same receptor (pharmacological) or different receptors or mechanisms (physiological). If the inhibition can be overcome completely by increasing the concentration of the agonist, then it is competitive (Furchgott, 1964; Ariens and Simonis, 1964a). Such is the case with acetylcholine and atropine. In contrast some pharmacological antagonists induce changes that are not overcome by the addition of the agonist. This is known as noncompetitive antagonism (Furchgott, 1964; Ariens and Simonis, 1964b). When evaluating drug action at the receptor level, it is important to utilize both agonists and antagonists. Pharmacological characterization of drug receptors requires the use of specific competitive antagonists. Under these experimental conditions, the dissociation constants (pA_2) can be obtained.

Since drugs are not usually limited to just one effect, dose–response curves can be obtained for each effect. However, these effects are not usually the same and therefore yield different values for potency, efficacy, and so on. Thus, formulating an hypothesis about the mechanism of action of a drug solely on the basis of the dose–response curve could be misleading. There are numerous other caveats that must be considered when interpreting dose–response data and the more complicated the drug effect, the more difficult it is to utilize the data for identifying the mechanism.

Although dose–response data can be obtained in the whole animal or any component (e.g., isolated tissue, enzyme), much of the current knowledge about the pharmacology of drugs has been obtained from experiments conducted on isolated tissue preparations. Use of isolated preparations allows for the observation of the direct actions of drug on the tissue of interest without being modified (unpredictably) by other factors. There are several factors that the investigator must consider when determining the appropriate isolated preparation to employ (Kenakin, 1984). For example, the existence of appropriate receptors on the smooth muscle is necessary for a pharmacological correspondence between the effects observed in the tissue bath and in humans. Receptor heterogeneity as well as the presence of different types of smooth muscle (e.g., longitudinal and circular) can markedly affect the magnitude of the contractile response, and in some instances (uterine tissue) the hormonal state of the animal is also important. The effects of many drugs vary from tissue to tissue and it is difficult to describe which is the best approach in generalized terms. Therefore, the choice of isolated preparations and experimental conditions needs to be evaluated on an individual basis. There are numerous reviews on vascular (Furchgott, 1955; Somylo and Somlyo, 1970; Bohr, 1973) and other preparations (Prosser, 1974; Bolton, 1979; Tallerida and Jacob, 1979) that summarize the experimental work on these various preparations, as well as techniques that have been employed to avoid the pitfalls of isolated tissues (Gaddum, 1953; Furchgott, 1968; Skaug and Detar, 1980).

3. Relevance to Drug Development

3.1. Calcium Antagonists

Calcium ions are known to play an important role in a variety of biologic processes. In the heart, calcium is involved in the generation of the cardiac action potential, and it is generally accepted that in vascular as well as other smooth muscle that increased tension is directly related to elevated levels of free intracellular calcium (Weiss, 1977). From these and other examples it is clear that calcium-dependent processes provide a variety of targets for pharmacological intervention. However, relative tissue specificity, selective site inhibition, or both are necessary in order to prevent undesirable in vivo side effects. Although agents that selectively inhibited the influx of calcium across the cell membrane have been known for some time, as a group, the calcium channel blockers have only recently emerged as agents that are effective in the treatment of a variety of cardiovascular disorders, including certain cardiac arrhythmias, hypertension, and angina (Ellrodt et al., 1980; Stone et al., 1980; Singh, 1982; Gould, this volume). Fleckenstein (1971) was the first to coin the term ''calcium antagonist'' for drugs that interfere with the contractile effect of calcium on cardiac and vascular smooth muscle. Utilizing the guinea pig papillary muscle preparation, Fleckenstein demonstrated that verapamil blocked the slow inward calcium current, abolished contraction, and had little or no effect on resting potential (Fleckenstein, 1977). In isolated smooth muscle, the importance of the slow inward calcium current with respect to excitation–contraction coupling is not as well defined as it is in cardiac tissue. It has been demonstrated, however, that addition of calcium to the bathing solution containing isolated smooth muscle preparations depolarized with high potassium results in a concentration-related tension response. This response is a function of calcium entering through voltage-sensitive channels (Bolton, 1979) and is inhibited by calcium antagonists. In contrast, contractile responses induced by agents that initiate the influx of calcium through receptor-operated channels (e.g., norepinephrine) are not inhibited by calcium antagonists (Karaki and Weiss, 1984).

In this context, several pharmacological screening tests were developed to identify and characterize potential calcium antagonists (Spedding and Cavero, 1984). Currently, calcium antagonists are being profiled by examining their activity on radioactive dihydropyridine binding, electrophysiological parameters, tension responses, and finally, selected in vivo preparations.

Binding studies using ^{3}H-dihydropyridine have indicated that although the receptor is specific for this class of compound, nondihydropyridine calcium antagonists can interact allosterically with this site (Glossman et al., 1982). With the advent of binding studies, a problem has emerged. Even though a fair correlation between ligand studies and pharmacological effects appear to exist (Janis et al., 1982), there are many examples in which discrepancies exist (Triggle, 1982; Murphy et al., 1983). Thus, it is possible that this receptor site may not be involved in the physiological regulation of these systems. The binding characteristics for several membrane preparations have been described, with varying results (Janis and Scriabine, 1983). Presumably these differences can be attributed to disruption of the physiological environment of the receptor, removal of essential cofactors, or alteration of receptor-properties caused by the isolation procedure. Regardless of these discrepancies, binding studies have provided some insight about how this chemical group, with many dissimilar chemical structures, might act by a similar mechanism. They all modulate the influx of calcium through the voltage-sensitive, but not the receptor-operated, channel. Unfortunately, binding studies do not delineate whether the chemical entity being examined is an agonist or antagonist or how the drug–receptor interaction results in an alteration of the voltage-operated channel, which, in turn, alters the pharmacological responsiveness.

Since the design of this group of agents can not be restricted to a specific chemical configuration and because the importance of the influx of calcium through the voltage-operated channel varies among tissues, numerous laboratories are utilizing smooth muscle preparations contracted by potassium depolarization as a primary screen to identify potential calcium antagonists and therapeutic usefulness. Figure 2 is an example of a flow chart depicting the initial pharmacological procedures useful for identifying calcium antagonist and tissue specificity. For more information on this topic and the use of isolated vascular smooth muscle as well as other in vitro assays to evaluate calcium antagonists, the reader is referred to the reviews of Spedding and Cavero (1984) and Cavero and Spedding (1983).

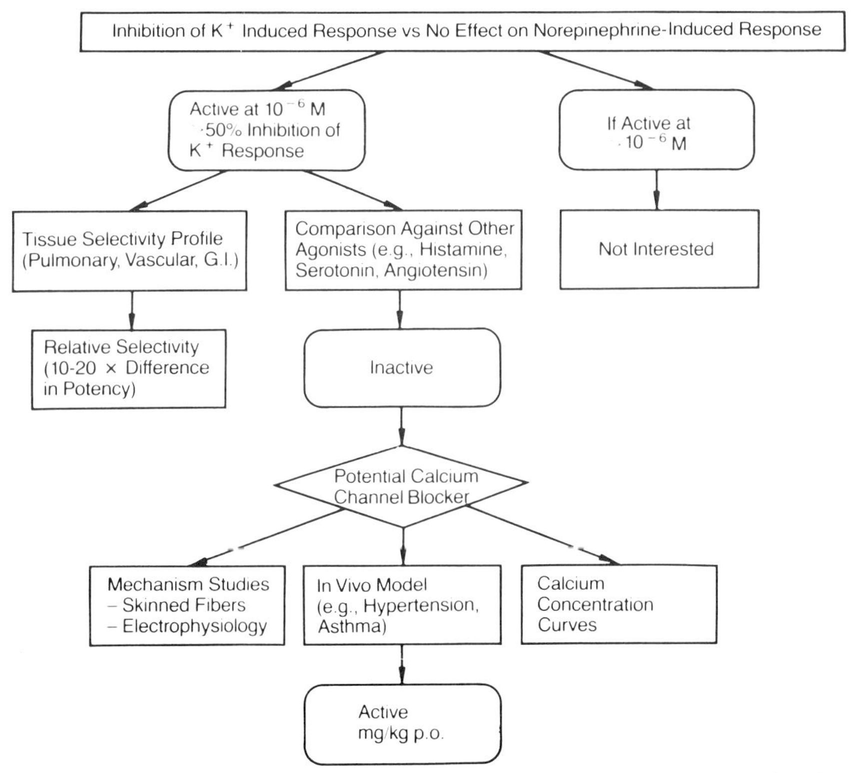

Fig. 2. Flow chart of initial pharmacological procedures utilized to identify and profile a calcium antagonist.

3.2. H_2 Antagonists

Another example of drug development, based on the identification of a receptor by utilizing different isolated tissue preparations is the H_2 receptor antagonists. The early characterization of H_1 receptors was made by Ash and Schild (1966), who proposed the terminology H_1 receptor based on experimental results obtained with histamine and typical antihistamines in isolated selected vascular beds. Furthermore, these investigators recognized that certain actions of histamine, such as gastric acid secretion, inhibition of rat uterine contractions, and stimulation of cardiac contractility, were not antagonized by typical antihistamines. Initially, it was not known whether these other effects of histamine were mediated by a common receptor since no specific antagonist existed.

Based on the analogous concept with beta-receptors and antagonists, Black and associates (Black et al., 1972; Gannellin and Brimblecombe, this volume) identified the first specific antagonists of the non-H_1 histamine responses and designated the receptors as H_2 receptors. The first specific H_2 antagonist described was burimamide. The specificity of this and other agents was demonstrated by their inability to inhibit histamine responses in guinea pig ileum. This discovery of selective antagonists of the H_2 receptor led to the characterization of histamine receptors in a variety of tissues as well as the development of more selective and orally active histamine H_2 antagonists. Although, initially this was primarly of research interest, clinical interest was aroused with the observation that these agents inhibited gastric acid secretion. If the in vitro data were correct, the hypothesis that histamine stimulated gastric acid secretion by activating H_2 receptors could be validated. The results of these clinical investigations resulted in the development of cimetidine. Thus, by systematic use of specific agonists and antagonists, Black and colleagues (1972, 1973) defined, characterized, and identified specific H_2 receptors and specific H_2 receptor antagonists, which, in turn, resulted in a new therapeutic class of agents.

3.3. Leukotriene Antagonists

Feldberg and Kellaway (1938) were the first investigators to describe slow reacting substance. These investigators were examining the smooth muscle contractile effects of substances released following the perfusion of cobra venom into guinea pig lungs. Because the contractile effect on isolated smooth muscle was slow to develop, they called the entity, "slow reacting smooth-muscle-stimulating substance" (SRS). A few years later, Kellaway and Trethewie (1940) demonstrated that the perfusion of guinea pig sensitized lungs with antigen resulted in the release of a substance with properties similar to SRS. Initially, the effects were not readily differentiated from those of histamine because of the lack of specific receptor antagonists and the fact that SRS was believed to be released with other mediators. With the development of the H_1 antihistamines, Brocklehurst (1960) demonstrated that this substance (SRS) induced a slow, prolonged contraction of the guinea pig ileum, and it was not inhibited by antihistamines. The term slow reacting substance of anaphylaxis (SRS-A) was chosen by Brocklehurst to differentiate the substance produced by lungs upon immunological challenge by specific antigens (SRS-A) from those generated upon nonimmunological challenge (SRS).

Characterization of SRS-A was slow because of limited availability of compound and lack of structural identity and suitable techniques to purify the substance. Under these circumstances an effective competitive antagonist was difficult to identify. Consequently, SRS-A activity was expressed in units (which differed among various laboratories), and bioassays were performed with impure standards. The usual bioassay of unpurified SRS-A employed the isolated guinea pig ileum. To achieve the desired selectivity, it was also necessary to have an antihistamine and an anticholinergic agent in the muscle bath (Orange and Austen, 1976). With the development of FPL 55712, a selective end-organ antagonist of SRS-A (Augstein et al., 1973), it was possible to inhibit the contractile response induced by the unpurified SRS-A. Both the development of FPL 55712 and the structure–activity relationships of other analogs were based on screening for inhibition of contractions induced by unpurified SRS-A on guinea pig ileal preparations. The early work utilizing FPL 55712 has been summarized by Chand (1979).

In addition to its activity on guinea pig ileum, SRS was shown to contract guinea pig and human tracheal smooth muscle in vitro. Thus, even though its chemical structure was unknown and the experiments were done with crude SRS, much information about this substances' biological and chemical properties was obtained. In fact, much of the information that is currently known about both SRS and SRS-A is based on the studies performed with these substances prior to their structural identification. The availability of synthetic material has permitted both the reaffirmation and more precise biological evaluation of these compounds in asthma and other disease states (Borgeat and Sirois, 1981; Bach, 1983).

Numerous investigators were involved in the elucidation of the structure (Bach et al., 1977; Borgeat and Samuelsson, 1979a,b; Radmark et al., 1980), synthesis of pure substance (Hammarstrom et al., 1979; Corey et al., 1980a,b) and characterization of the properties in isolated smooth muscle (Piper and Samhoun, 1981; Sirois et al., 1981). The term leukotriene was introduced to describe SRS and other compounds derived from the lipoxygenase pathway and containing a conjugated triene (Samuelsson et al., 1979). In addition, it was determined that SRS-A from a variety of sources contained several leukotrienes with SRS-A activity (e.g., LTC and LTD).

As a result of this new knowledge and the availability of synthetic leukotrienes, synthesis of receptor antagonists and inhibitors of the lipoxygenase pathway for therapeutic purposes has been facilitated. Although numerous attempts have been made to alter the actions of leukotrienes, and several compounds are currently undergoing clinical evaluation, their therapeutic usefulness remains to be determined. Experiments on the effects of leukotrienes in different biological systems will also continue to improve our understanding of the biological properties of the leukotrienes.

4. Summary

In conclusion, some of the factors affecting the use of isolated tissues to obtain information about drugs and their receptors have been reviewed. The effects of different drugs varies from animal to animal and from tissue to tissue. Thus, these effects should be interpreted on the basis of the validity of the model and the contribution of environmental factors to the experimental conditions. Whereas conclusions about drug–receptor interactions in vivo are influenced by both physical and humoral factors, in vitro measurements are not. More importantly, use of isolated tissues to obtain quantitative data defining drug affinity and efficacy is of initial use in the design of better therapeutic agents. The chances of a chemical entity becoming a useful drug are remote, but the delineation of a biological effect is dependent upon how broadly the compound is tested and the relevance of the laboratory test to the clinical situation. Although there are some major advantages to utilizing isolated tissues, it is necessary to perform both in vivo and in vitro experiments before delineating the mechanism of action of a new compound, because each experimental approach builds on the other.

References

Altura, B. M. and Altura, B. T. (1970) Heterogeneity of drug receptors in different segments of rabbit thoracic aorta. *Eur. J. Pharmacol.* **12**, 44–52.

Ariens, E. J. and Simonis, A. M. (1962) Drug–receptor interaction. *Acta Physiol. Pharmacol.* **11**, 151–172.

Ariens, E. J. and Simonis, A. M. (1964a) A molecular basis for drug action. *J. Pharm. Pharmacol.* **16**, 137–157.

Ariens, E. J. and Simonis, A. M. (1964b) A molecular basis for drug action. The interaction of one or more drugs with different receptors. *J. Pharm. Pharmacol.* **16**, 289–312.

Ariens, E. J., Van Rossum, J. M., and Koopman, P. C. (1960) Receptor reserve and threshold phenomena. *Arch. Int. Pharmacodyn. Ther.* **127**, 459–478.

Ash, A. S. F. and Schild, H. O. (1966) Receptors mediating some actions of histamine. *Br. J. Pharmacol.* **27**, 427–439.

Augstein, J., Farmer, J. B., Lee, T. B., Sheard, P., and Tattersall, M. L. (1973) Selective inhibitor of slow reacting substance of anaphylaxis. *Nature* (New Biol.) **245**, 215–217.

Bach, M. K. (1983) *The Leukotrienes: Their Structure, Actions, and Role in Diseases. Current Concepts.* Upjohn, Kalamazoo, Michigan.

Bach, M. K., Brashler, J. R., and Gorman, R. R. (1977) On the structure of slow reacting substance of anaphylaxis: Evidence of biosynthesis from arachidonic acid. *Prostaglandins* **14**, 21–38.

Black, J. W., Duncan, W. A. M., Durant, C. J., Ganellin, C. R., and Parsons, E. M. (1972) Definition and antagonism of histamine H_2-receptors. *Nature* **236**, 385–390.

Black, J. W., Duncan, W. A. M., Emmett, J. C., Ganellin, C. R., Hesselbro, T., Parsons, M. E., and Wyllie, J. H. (1973) Metiamide, an orally active histamine H_2 receptor antagonist. *Agents Action* **3**, 133–137.

Bohr, D. F. (1973) Vascular smooth muscle updated. *Circ. Res.* **32**, 665–672.

Bolton, T. B. (1979) Mechanisms of action of transmitters and other substances on smooth muscle. *Physiol. Rev.* **59**, 606–718.

Borgeat, P. and Samuelsson, B. (1979a) Transformation of arachidonic acid by rabbit polymorphonuclear leukocytes. Formation of a novel dihydroxyeicosatetraenoic acid. *J. Biol. Chem.* **254**, 2643–2646.

Borgeat, P. and Samuelsson, B. (1979b) Metabolism of arachidonic acid in polymorphonuclear leukocytes. Structural analysis of novel hydroxylated compounds. *J. Biol. Chem.* **254**, 7865–7869.

Borgeat, P. and Sirois, P. (1981) Leukotrienes: A major step in the understanding of immediate hypersensitivity reactions. *J. Med. Chem.* **24**, 121–126.

Brocklehurst, W. E. (1960) The release of histamine and formation of slow reacting substance (SRS-A) during anaphylactic shock. *J. Physiol.* (Lond.) **151**, 416–435.

Cavero, I. and Spedding, M. (1983) ''Calcium antagonists:'' A class of drugs with a bright future. I. Cellular calcium homeostais and calcium as a coupling messenger. *Life Sci.* **33**, 2571–2581.

Chand, N. (1979) FPL-55712—An antagonist of slow reacting substance of anaphylaxis (SRS-A): A review. *Agents Actions* **9**, 133–140.

Clark, A. J. (1933) *The Mode of Action of Drugs on Cells* Arnold, London.

Corey, E. J., Clark, D. A., Goto, G., Marfat, A., Mioskowski, C., Samuelsson, B., and Hammarstrom, S. (1980a) Stereospecific total synthesis of a slow reacting substance of anaphylaxis, leukotriene C-1. *J. Am. Chem. Soc.* **102**, 1436–1439.

Corey, E. J., Marfat, A., Goto, G., and Brian, F. (1980b) Leukotriene B: Total synthesis and assignment of stereochemistry. *J. Am. Chem. Soc.* **102**, 7984–7985.

Ehrlich, P. (1913) Chemotherapeutics—scientific principles, methods and results. *Lancet* **11**, 445.

Ellrodt, G., Chew, C. Y. C., and Singh, B. N. (1980) Therapeutic implications of slow-channel blockade in cardiocirculatory disorders. *Circulation* **62**, 669–679.

Feldberg, W. and Kellaway, C. H. (1938) Liberation of histamine and formation of lysocithin-like substances by cobra venom. *J. Physiol.* (Lond.) **94**, 187–226.

Fleckenstein, A. (1971) Specific Inhibitors and Promoters of Calcium Action in the Excitation–Contraction Coupling of Heart Muscle, in *Calcium and the Heart* (Opie, L. and Harris, P., eds.) Academic, London.

Fleckenstein, A. (1977) Specific pharmacology of calcium in myocardium, cardiac pacemakers, and vascular smooth muscle. *Ann. Rev. Pharmacol. Toxicol.* **17**, 149–166.

Furchgott, R. F. (1955) The pharmacology of vascular smooth muscle. *Pharmacol. Rev.* **7**, 183–265.

Furchgott, R. F. (1964) Receptor mechanisms. *Ann. Rev. Pharmacol.* **4**, 21–50.

Furchgott, R. F. (1966) The pharmacologial differentiation of adrenergic receptors. *Ann. NY Acad. Sci.* **139**, 553–570.

Furchgott, R. S. (1968) A Critical Appraisal of the Use of Isolated Organ Systems for the Assessment of Drug Action at the Receptor Level, in *Importance of Fundamental Principles in Drug Evaluation* (Tedeschi, D. H. and Tedeschi, R. E., eds.) Raven, New York.

Gaddum, J. H. (1953) Bioassays and mathematics. *Pharmacol. Rev.* **5**, 87–134.

Gillette, J. R. (1968) Problems Associated with the Extrapolation of Data from *In Vitro* Experiments to Experiments in Intact Animals, in *Importance of Fundamental Principles in Drug Evaluation* (Tedeschi, D. H. and Tedeschi, R. E., eds.) Raven, New York.

Glossman, H., Ferry, D. R., Lubbecke, F., Mewes, R., and Hofmann, F. (1982) Calcium channels: direct identification with radioligand binding studies. *Trends Pharmacol. Sci.* **3**, 431–437.

Hammarstrom, S., Murphy, R. C., Sammuelsson, B., Clark, D. A., Mioskowski, C., and Corey, E. J. (1979) Structure of leukotriene C: Identification of the amino acid part. *Biochem. Biophys. Res. Commun.* **91**, 1266–1272.

Janis, R. A. and Scriabine, A. (1983) Sites of action of Ca^{2+} channel inhibitors. *Biochem. Pharmacol.* **32**. 3499–3507.

Janis, R. A., Maurer, S. C., Sarmiento, J. G., Bolger, G. T., and Triggle, D. J. (1982) Binding of [^{3}H] nimodipine to cardiac smooth muscle membrane. *Eur. J. Pharmacol.* **82**, 191–194.

Karaki, H. and Weiss, G. B. (1984) Calcium channels in smooth muscle. *Gastroenterology* **87**, 960–970.

Kellaway, C. H. and Trethewie, E. R. (1940) Liberation of a slow reacting smooth muscle stimulating substance in anaphylaxis. *Q. J. Exp. Physiol.* **30**, 121–145.

Kenakin, T. P. (1984) The classification of drugs and drug receptors in isolated tissues. *Pharmacol. Rev.* **36**, 165–222.

Limbird, L. E. (1986) *Cell Surface Receptors: A Short Course On Theory and Methods* Nijhoff, Boston.

Maxwell, R. A. (1984) The state of the art of the science of drug discovery—an opinion. *Drug Dev. Res.* **4**, 375–389.

Murphy, K. M. M., Gould, R. J., Largent, B. L., and Snyder, S. H. (1983) A unitary mechanism of calcium antagonist drug action. *Proc. Natl. Acad. Sci. USA* **80**, 860–864.

Nickerson, M. (1956) Receptor occupancy and tissue response. *Nature* (Lond.) **178**, 697–698.

Orange, R. P. and Austen, K. F. (1976) The biological assay of slow reacting substances—SRS-A, bradykinin, prostaglandins. *Meth. Immunol. Immunochem.* **5**, 145–149.

Paton, W. D. M. (1961) A theory of drug action based on the rate of drug–receptor combination. *Proc. Roy Soc. London B Biol. Sci.* **154**, 21–69.

Piper, P. J. and Samhoun, M. N. (1981) The mechanism of action of leukotrienes C_4 and D_4 in guinea pig isolated perfused lung and parenchymal strips of guinea pig, rabbit and rat. *Prostaglandins* **21**, 793–803.

Prosser, C. L. (1974) Smooth muscle. *Ann. Rev. Physiol.* **36**, 503–535.

Radmark, O., Malmsten, C., Samuelsson, B., Clark, D. A., Goto, G., Marfat, A., and Corey, E. J. (1980) Leukotriene A: Stereochemistry and enzymatic conversion to leukotriene B. *Biochem. Biophys. Res. Commun.* **92**, 954–961.

Ruffulo, R. R., Jr. (1982) Review: Important concepts of receptor theory. *J. Auton. Pharmacol.* **2**, 277–295.

Samuelsson, B., Borgeat, P., Hammarstrom, S., and Murphy, R. C. (1979) Introduction of a nomenclature: Leukotrienes. *Prostaglandins* **17**, 785–787.

Singh, B. N. (1982) Pharmacological basis for the therapeutic applications of slow-channel blocking drugs. *Angiology* **33**, 492–515.

Sirois, P., Roy, S., Tetrault, J. P., Borgeat, P., Picard, S., and Corey, E. J. (1981) Pharmacological activity of leukotrienes A_4, B_4, C_4, D_4 on selected guinea pig, rat, rabbit and human smooth muscles. *Prostaglandins Med.* **7**, 327–340.

Skaug, N. and Detar, R. (1980) Indirect evidence for the importance of steady-state $[K]_0$ in determining the extent of the effects of hyperpolarizing electrogenesis on vascular smooth muscle reactivity. *Blood Vessels* **17**, 117–122.

Somlyo, A. P. and Somlyo, A. V. (1970) Vascular smooth muscle. II. Pharmacology of normal and hypertensive vessels. *Pharmacol. Rev.* **22**, 249–353.

Spedding, M. and Cavero, I. (1984) "Calcium antagonists:" A class of drugs with a bright future. II. Determination of basic pharmacological properties. *Life Sci.* **35**, 575–587.

Stephenson, R. P. (1956) A modification of receptor theory. *Br. J. Pharmacol.* **11**, 379–393.

Stone, P. H., Antman, E. M., Muller, J. E., and Braunwald, E. (1980) Calcium channel blocking agents in the treatment of cardiovascular disorders. II. Hemodynamic effects and clinical applications. *Ann. Intern. Med.* **93**, 886–904.

Tallarida, R. J. and Jacob, L. S. (1979) Isolated Preparations: Dose–Response Data, in *The Dose–Response Relation in Pharmacology* Springer-Verlag, New York.

Triggle, D. J. (1982) Counting Ca^{2+} channels? *Trends Pharmacol. Sci.* **3**, 465–466.

Venter, J. C. (1979) High Efficiency coupling between beta-adrenergic receptors and cardiac contractility: Direct evidence for "spare" beta adrenergic receptors. *Mol. Pharmacol.* **16**, 429–440.

Waud, D. R. (1968) Pharmacological receptors. *Pharmacol. Rev.* **20**, 49–88.

Weiss, G. B. (1977) Calcium and Contractility in Vascular Smooth Muscle, in *Advances in General and Cellular Pharmacology* (Narahashi, T. and Bianchi, C. P., eds.) Plenum, New York.

Williams, M. and Enna, S. J. (1986) The receptor: From concept to function. *Ann. Rep. Med. Chem.* **21**, 211–235.

Neuropsychopharmacological Drug Development

JEFFREY B. MALICK

1. Introduction

Behavioral pharmacology or psychopharmacology is the study of the effects of drugs on behavior, whereas neuropharmacology is concerned with the effects of drugs on nervous tissue. Thus, neuropsychopharmacology encompasses the whole spectrum, being concerned with the study of the effects of drugs on both nervous tissue in general and on behavior in particular.

Psychopharmacology entered its modern era with the discovery of chlorpromazine in France by Henri Laborit in 1952. The advent of the first antipsychotic drugs changed psychiatry dramatically, as demonstrated by the remarkable decline in the number of hospital beds occupied by chronically ill mental patients and the closing of many state hospitals after the introduction of these drugs. Prior to the discovery of the antipsychotics, physicians relied primarily upon barbiturates, stimulants, bromides, and classical soporific medications. It is difficult today to imagine a world in which no antidepressants, anxiolytics, or antipsychotics existed; these drugs are taken for granted today.

The first "psychopharmacology" laboratories in the pharmaceutical industry were created in the late 1950s, and they were headed and staffed primarily by psychologists. Since the pioneers in this field were chiefly psychologists, it was only natural that they were primarily concerned with the experimental analysis of behavior and utilized the neuropharmacological drugs primarily as variables that might facilitate behavioral research. In companies throughout the industry, chemists' shelves were covered with samples that were merely collecting dust in many instances; thus, these laboratories were created to fill a specific need—to evaluate the effects of these agents on behavior in the hopes of discovering other agents with profiles similar to those of the early psychotropics (i.e., chlorpromazine, imipramine). The early preoccupation with behavior has given way to a much more sophisticated multidisciplinary approach to the study of the effects of drugs on behavior and nervous tissue in general. Today's neuroscience teams are composed of experts from many different disciplines, among them neuropharmacologists, electrophysiologists, neurochemists, psychopharmacologists, neuroimmunologists, and so on. The complex term neuropsychopharmacology has evolved as a result of the marriage of these different disciplines with the common goal of discovering and developing new and novel treatments for the various psychiatric and neurological disorders.

It is clear that the initial discovery of the psychotropic medications (e.g., chlorpromazine, imipramine, lithium) was not achieved by the initiation of null hypothesis studies on specific agents. Rather, the first of the modern psychopharmaceutical agents were developed primarily via serendipity coupled with astute clinical observations. Even the recent discovery of the novel anxiolytic buspirone (*see* Eison et al., this volume) was the result of fortuitous clinical observations (Sathananthan et al., 1975; Goldberg and Finnerty, 1979), since the preclinical pharmacology had predicted that it would be an antipsychotic agent. The goal of neuropsychopharmacology today is to rationally design psychoactive agents utilizing laboratory animal models that mimic certain key features of the human mental disorder under investigation, coupled with neurochemical and neurophysiological studies designed to evaluate specific mechanistic approaches at specific neuronal/receptor sites within the central nervous system.

Although many of the drug developments in the mental disease area can be categorized as significant therapeutic advances, the psychoactive agents available today for the treatment of the classical mental diseases (depression, anxiety, psychosis) have many pitfalls. Although they are clearly effective, they all exhibit serious side effects or drawbacks. For example, although the antidepressants available are effective, there is still considerable room for therapeutic advances; newer agents will not only need to be more effective and have a more rapid onset of action, but they must exhibit fewer adverse side effects (e.g., anticholinergic, cardiovascular, weight gain) as well. Although the benzodiazepine anxiolytics are quite effective in the management of certain types of anxiety, newer agents must possess fewer side effects (e.g., sedation, ataxia, alcohol potentiation) and reduced abuse potential; in addition, agents that treat specific types of anxiety (e.g., panic states) must be discovered. Furthermore, with the antipsychotics, a more serious problem exists; not only do they cause adverse effects (e.g., extrapyramidal effects, hypotension), but they induce a serious disorder, tardive dyskinesia, which is irreversible in many instances.

Thus, there is still a great need for newer psychotropics, both safer and more effective, for the treatment of the classical mental disorders. In addition, there are many other psychiatric or neurological disorders (e.g., Alzheimer's dementia, impotence, stroke, and nerve repair) for which no good drugs presently exist. Thus, the goal of neuropsychopharmacology is to develop the next generation of psychoactive drugs for the treatment of all of the mental disorders.

The goal of this chapter will be to demonstrate the types of techniques that the neuropsychopharmacologist utilizes in the search for new and novel psychotherapeutics. No attempt will be made to present a comprehensive description of all of the tests that are currently being utilized in the discovery process for the myriad of mental disease areas. However, the types of techniques utilized for a single disease area, namely the search for new and novel antianxiety agents, will be used as examples of the types of approaches that are being used in the search for psychotropic drugs today. The types of procedures used to assess both therapeutic potential and side effect liabilities will be discussed.

2. Anxiolytic Drug Discovery and Development

In addition to their well-known use in the treatment of varying types of anxiety states, the anxiolytics, especially the benzodiazepines, also have important applications in anesthesia (Weidler and Hempelmann, 1983), the treatment of psychosomatic conditions, insomnia (Stern, 1979), alcohol withdrawal (Malick and Kubena, 1979), muscle spasms, and epilepsy (Overweg and Binnie, 1983). The types of methodologies utilized to detect the therapeutic potential of anxiolytics and to assess their side effect liabilities will be discussed in this section.

Although benzodiazepines are the most widely prescribed drugs for the treatment of anxiety and are effective in many neurotic patients, more selective compounds are required. For instance, the most common side effect with these agents is sedation and drowsiness, which limit their utility especially in the geriatric population and necessitate the warning concerning the operation of machinery and driving when taking benzodiazepines. Furthermore, although benzodiazepines are considered to be virtually suicide-proof when taken by themselves, they significantly potentiate the sedative properties of other CNS depressants (e.g., barbiturates, ethanol) (Lippa et al., 1979b; Palva and Linnoilia, 1978), and they can be extremely dangerous when taken in combination. In fact, the ethanol/benzodiazepine combination is one of the leading causes of drug overdose (Food and Drug Administration, 1980). In addition, the issues of drug dependence liability (Lader and Petursson, 1983) and amnesia following their administration have recently received a great deal of attention both in the lay press and scientific literature. Although drug dependence with the benzodiazepines is often more psychological than physiological, real withdrawal signs can be observed, especially if a patient who has been taking them for a long while abruptly terminates their use; however, the withdrawal signs are nowhere near as severe as those observed with opiates.

Thus, the goal of the neuropsychopharmacologist has been to search for new and novel anxiolytics that will be devoid of or exhibit markedly reduced side effects compared to the benzodiazepines. Most groups have concentrated on discovering a new agent from a nonbenzodiazepine chemical series, although newer types of benzodiazepines (e.g., partial agonists) that presumably have fewer side effects and perhaps a different spectrum of therapeutic activity, are still under investigation. In order to develop such compounds, a battery of tests is necessary to detect the anxiolytic activity and to determine side-effect liabilities; the following is a brief description of some of the more commonly used test procedures.

2.1. Antianxiety Activity

Many "approach–avoidance conflict" models have been developed in which "behavioral inhibition" or "anxiety" are induced in animals. These models have proved to be useful in predicting anxiolytic activity in man with both benzodiazepines and nonbenzodiazepines [e.g., buspirone (Eison et al., 1982), meprobamate, and barbiturates]. In most cases, rates of responding for reinforcement (e.g., food or water) are suppressed by the presentation of a mild punishment (e.g., electric shock, gust of air). All clinically effective anxiolytics will significantly increase responding that has been suppressed by punishment in these conflict models. The use of conflict models has been widely endorsed by most experts in the field (Geller and Seifter, 1960; Randall and Schallek, 1967; Vogel et al., 1971; Lippa et al., 1979b), and an excellent correlation exists between the oral minimum effective dose (MED) in rat anticonflict procedures and the average oral clinical dose for psychoneurotic anxiety (Cook and Sepinwall, 1975).

2.1.1. Shock-Induced Suppression of Drinking

One of the most useful primary screens for anxiolytic agents is the so-called "Vogel conflict procedure" (Vogel et al., 1971). This method has been used for many years in a large number of pharmaceutical houses to evaluate the antianxiety or disinhibitory properties of experimental agents because it is a simple and reliable conflict procedure. The Shock-Induced Suppression of Drinking test (SSD), the name given to the version of the test used at Stuart Pharmaceuticals, is a modification of the Vogel procedure (Patel and Malick, 1983). In this model, rats are water-deprived for 48 h and food-deprived for the last 24 h prior to testing. The test is initiated by gently placing a rat into a black plexiglass test chamber and permitting it to locate the drinking tube. When the animal completes 20 licks, the first electric shock is triggered. A 3-min timer is initiated at the termination of the first shock. During the 3-min test session, shocks are delivered following each 20th lick, and the number of shocks received during the session is recorded for each subject. Vehicle-treated control subjects only take a few (2–7) shocks, and their behavior is quickly inhibited, whereas animals treated with reference "antianxiety" agents (e.g., benzodiazepines) continue to drink and will accept a significantly greater number of shocks. Because there is a variability between both subjects and test days, a same-day positive control group (e.g., chlordiazepoxide- or diazepam-treated) is run in addition to a vehicle-treated group on each test day; eight subjects per dose are used.

In addition to the benzodiazepines and other classical disinhibitory agents (e.g., barbiturates, ethanol, meprobamate), several nonbenzodiazepines that exhibit neuropharmacological profiles predictive of potential anxiolytic activity [e.g., CL 218,872 (Lippa et al., 1979a), tracazolate (Patel and Malick, 1982; Malick et al., 1984), and CGS 9896 (Bernard et al., 1985)] have been shown to be active in this model (Table 1). The most noteworthy exception was that buspirone was inactive over a wide range of doses (0.5–50 mg/kg, po); although this agent has been shown to be effective in this pro-

Table 1
Effects of Anxiolytic Agents in the Shock-Induced Suppression of Drinking (SSD) Test in Rats

Drug	Minimal effective dose[a], mg/kg, po
Benzodiazepines	
Clonazepam	1
Flunitrazepam	5
Nitrazepam	5
Diazepam	5
Oxazepam	5
Bromazepam	5
Chlordiazepoxide	10
Medazepam	10
Chlorazepate	15
Flurazepam	15
Ripazepam	17
Halazepam	30
Desmethylchlordiazepoxide	30
Nonbenzodiazepines	
MK-801	1
Zopiclone	2
CL 218,872	5
CGS 9896	10
Phenobarbital	10
Fenobam	12.5
Tracazolate	20
Meprobamate	50
Ethanol	1200 ip
Buspirone	Inactive (50)[b]

[a]The lowest dose of drug producing a significant increase in the number of shocks received by at least 50% of the rats tested.

[b]Highest dose tested.

cedure in a couple of laboratories, the majority of investigators report it to be inactive in most anxiolytic screens in rodents (Sullivan et al., 1983). The validity of this test system is further strengthened by the fact that a wide range of psychotropic agents from other classes (e.g., analgesics, antipsychotics, stimulants, antidepressants) are totally inactive. This test is extremely useful as a primary screen because it is quick, reliable, has a high throughput, and does not require the use of highly trained subjects.

2.1.2. Geller-Seifter Conflict and Related Procedures

One of the oldest and most classic conflict models is the one developed by Geller and Seifter (1960). Briefly, food-deprived rats are trained in an operant chamber to bar press for food reinforcement on a multiple variable interval of 2 min/fixed ratio reinforcement (VI2/FR) schedule; the two schedule components alternate during the 60-min session. The session begins with a 13-min VI segment, followed by a 2-min FR segment. During the FR period, every lever press response is reinforced with food, but is accompanied by a mild electric foot shock; this segment is the "conflict period," and its occurrence is signaled by a light or tone cue. A trained rat will take very few shocks; however, subjects treated with anxiolytics will accept significantly more shocks than controls during this conflict period (Table 2). The Geller-Seifter test is remarkably specific for anxiolytic agents. Another major advantage of this conflict procedure is that since trained rats with stable baselines are used, each animal can serve as its own control, which markedly reduces variability. In addition, this procedure permits an assessment of both the therapeutic and the neurotoxic (sedative) effects of drugs in the same animal, since rates of responding in both conflict and nonconflict segments are measured. One major disadvantage to this test is that it requires considerable training (1 mo or more) to get a rat to perform well in this procedure.

Table 2
Effects of Selected Anxiolytics in Secondary Conflict Procedures in Rats

Drug	MED[a], mg/kg, po		
	Geller-Seifter[b]	Cook-Davidson[c]	Rat hold-down[d]
Diazepam	2.5 (ip)	0.63	1.25–2.5
Chlordiazepoxide	15 (ip)	2.2	9
Meprobamate	—	62.5	50
Oxazepam	10 (ip)	1.25	—

[a]MED, minimal effect dose; i.e., the lowest dose in which rats accepted significantly greater numbers of shocks compared to controls.
[b]MEDs were not established; values represent the lowest dose at which the mean number of shocks taken was at least 10 (Geller and Seifter, 1960).
[c]Data from Cook and Davidson (1973).
[d]Unpublished data, Patel et al.

Cook and Davidson (1973) have modified the Geller-Seifter conflict procedure. In their procedure, 5-min VI-30 s periods alternate with 2-min FR 10 segments in which every tenth response results in simultaneous delivery of a food pellet and a mild, brief electric footshock. As in the Geller-Seifter model, responding during the FR 10 segments is markedly suppressed, and anxiolytics selectively increase responding during the conflict (FR 10) as opposed to the nonconflict (VI-30) component of this multiple schedule (Table 2).

Another recent modification of this procedure, the rat hold-down test, was developed in this laboratory (Patel et al., 1984); this test is a rat version of a squirrel monkey conflict test developed by Patel and Migler (1982). A standard operant chamber with levers, stimulus light panels, food pellet dispenser, and so on, was used. Subjects, maintained at approximately 80% of their free-feeding body weight, were trained in the two-bar chamber to press for food reinforcement. The 1-h session consisted of a 40-min conflict period, followed by a 20-min nonconflict period. The onset of the conflict period was signaled by a yellow cue light over one bar; during this period, a 2-s depression (hold) of the lever was both rewarded with food and suppressed by a mild electric shock. The onset of the nonconflict segment was signaled by a cue light over the other lever, and a correct response (hold) was only food reinforced. Once trained, a rat will only respond during the nonconflict segment. As in the Geller-Seifter model, anxiolytics will significantly increase bar pressing in the conflict segment. Comparative minimal effective doses for selected agents in the three procedures are presented in Table 2.

Although there are marked differences in absolute potencies between the tests, the relative potencies remain constant (*see* Table 2); the choice of which test to use may be based more on investigator preference than on anything else. All three of these tests can serve as very useful secondary evaluation procedures that permit the investigator to more fully assess the anxiolytic profile of a new chemical entity.

2.1.3. Primate Conflict Models

It is essential to demonstrate anxiolytic activity in as many species as possible, especially in a nonhuman primate. Several monkey conflict procedures have been described (Hanson et al., 1967; Stein and Berger, 1971; Canon and Houser, 1978; Sepinwall et al., 1978). In general, the results observed with the reference anxiolytics in all of these tests are similar to what has been observed in the Geller-Seifter conflict model in rats.

The squirrel monkey hold-down procedure, developed by Patel and Migler (1982), is a particularly useful primate conflict model (Goldberg et al., 1983; Patel and Malick, 1983). Briefly, food-deprived squirrel monkeys were trained to depress (hold down) a lever for 5 s in order to obtain food reinforcement. Once the subject mastered the task, it was tested in a 6-h session that was divided into two 3-h portions: (a) a punished period, in which each correct response was both food-reinforced and suppressed by a brief, mild footshock, and (b) a nonpunished period during which each correct response was only food-reinforced. The number of shocks taken during the punished period was recorded and used as a measure of inhibition or anxiety. Untreated control subjects rarely take any shocks; in contrast, monkeys treated with anxiolytics will "accept" significant numbers of shocks in order to obtain food. Both benzodiazepine and nonbenzodiazepine anxiolytics dramatically increased the number of shocks taken during the punishment period; minimal effective doses (MEDs) are presented in Table 3. Buspirone and fenobam, as previously reported (Goldberg et al., 1983), were inactive up to the maximum dose that could be tested (Table 3). One of the important features of this test is that the 3-h

Table 3
Activity of Selected Reference Agents in the Squirrel Monkey Hold-Down Procedure[a]

Treatment	MED[b], mg/kg, po, for disinhibition
Diazepam	1.25
Chlordiazepoxide	2.5
Tracazolate	12.5
Zopiclone	12.5
CL 218,872	12.5
Phenobarbital	40
Meprobamate	50
Buspirone	Inactive (10)[c]
Fenobam	Inactive (25)[c]

[a]Data from Patel and Migler (1982) and Goldberg et al. (1983).
[b]MED, minimal effective dose; i.e., the lowest dose at which 50% or more of the monkeys took significant numbers of shocks compared to predrug control day.
[c]Highest dose tested.

conflict segment allows the investigator to obtain a measure of the time of onset, peak, and duration of the anxiolytic activity in a primate.

Sepinwall and coworkers (1978) used the model first developed by Cook and Davidson (1973) in rats to evolve a useful conflict model for squirrel monkeys. In this procedure, monkeys had a choice of working on two concurrent variable interval (VI) schedules; responses on the VI-1.5 min left lever schedule were suppressed by intermittent punishment on a variable ratio, 24-response schedule (VR 24), where responses on the VI-6 min right lever schedule were never punished. This procedure appears to be more sensitive to the effects of drugs since both chlordiazepoxide and diazepam were considerably more potent in this procedure than they were in the hold-down procedure; the MEDs were 0.62 and 0.31 mg/kg, po, for chlordiazepoxide and diazepam, respectively (Table 3). However, in both of these squirrel monkey procedures, diazepam was always twice as potent as chlordiazepoxide.

2.1.4. Neophobic Suppression of Drinking

A different type of conflict procedure that does not involve either foot-shock-induced inhibition of responding or operant conditioning is the neophobic suppression of feeding test first developed by Poschel (1971). This procedure capitalizes on the fact that rats will only consume very small quantities of any novel food source.

In the procedure used in our laboratory (Malick and Enna, 1979), a sweetened condensed milk solution (diluted 1:3 with tap water) was used as the novel food source. In spite of the fact that this solution will become a "preferred" food after several exposures, only very small quantities are consumed by rats upon initial exposure (Malick and Enna, 1979). Previous reports have demonstrated that benzodiazepine anxiolytics produced significant disinhibition in this procedure (Poschel, 1971; Poschel et al., 1974); that is, under the influence of benzodiazepines, rats consumed significantly greater quantities of the milk solution than did placebo-treated controls. Of all of the different types of psychoactive drugs that have been evaluated, only those that have been reported to have anxiolytic or disinhibitory activities have been found to be active in this procedure.

The results obtained with a range of agents that have been claimed to have anxiolytic activity are summarized in Table 4. The most active agents in this procedure are the benzodiazepine anxio-

Table 4
Effect of Selected Psychoactive Drugs in the Neophobic Suppression of Drinking Test in Rats

Drug	N[a]	MED[b], mg/kg, po, for significant disinhibition
Clonazepam	100	1
Flurazepam	45	3
Zopiclone	50	3
Diazepam	100	10
CL 218,872	25	10
Oxazepam	50	17
Chlordiazepoxide	155	20
Meprobamate	145	300
Phenobarbital	65	>177[c]
Buspirone	30	Inactive (100)[d]

[a]Number of rats tested.
[b]Minimal effective dose (i.e., the lowest dose producing a significant effect in 50% or more of the rats tested).
[c]Disinhibitory activity was evident, but a MED could not be determined.
[d]Highest dose tested.

lytics and the nonbenzodiazepines that presumably work through the same mechanism of action because they are potent displacers of ^{3}H-flunitrazepam binding from benzodiazepine receptors (e.g., zopiclone, CL 218,872). The anxiolytic or disinhibitory agents that are presumably acting via another mechanism of action (e.g., meprobamate, buspirone, phenobarbital), because they do not displace ^{3}H-flunitrazepam from binding sites in brain, are either much weaker disinhibitors in this procedure or are inactive (buspirone).

Thus, this procedure appears to be an effective means of detecting anxiolytic agents of the "benzodiazepine type," whereas it is much less useful as a model to detect nondisplacer (i.e., those that fail to displace ^{3}H-flunitrazepam) anxiolytics.

2.1.5. Anxiolytics: General Considerations

In addition to obtaining anxiolytic efficacy in several anxiolytic procedures in both rodents and primates, several other important features must be considered. For instance, it has been well established that benzodiazepine anxiolytics continue to exhibit efficacy after chronic administration in man; i.e., their antianxiety activities do not diminish or tolerate overtime. Thus, it is important to demonstrate that the antianxiety activity of a new drug candidate does not tolerate after at least 2 wk of continuous daily administration in rats and higher species (e.g., dog, monkey).

Furthermore, any new anxiolytic agent should demonstrate a reasonable duration of action. This can be assessed readily by dosing separate groups of animals with increasing absorption times (i.e., 1, 2, 4, 6 h) and then testing them in the appropriate procedure (e.g., shock-induced suppression of drinking, Geller-Seifter conflict).

At the present time, the detection of buspirone (Riblet et al., 1982) or buspirone-like anxiolytics [e.g., ipsapirone (TVX Q 7821; Tropon); SM-3997 (Sumitomo)] in classical conflict procedures remains a problem. The majority of the psychopharmacologists in the pharmaceutical industry report (personal communications; Sullivan et al., 1983) that it is very difficult, at best, to detect these types of agents in classical approach–avoidance conflict paradigms. This is not totally surprising since the history of buspirone is that it was taken to the clinic initially as an antipsychotic drug candidate, and its anxiolytic potential was first recognized in man. Although buspirone and related drugs have not been consistently reported to be active in conflict procedures, they have been shown to be effective in other behavioral models (e.g., footshock-induced aggression) (Traber et al., 1984; Eison et al., 1982) in which anxiolytics would also be detected. Thus, since buspirone is undoubtedly an effective anxiolytic in humans (it is marketed in Germany and was approved by the FDA in December, 1986), a real challenge for the industrial psychopharmacologist is to discover or create a laboratory model that will specifically measure the anxiolytic or disinhibitory potential of this type of agent.

Neuropharmacologists were puzzled about the anxiolytic mechanism of action of buspirone for many years. Since buspirone does not displace radiolabeled benzodiazepines from brain receptors, its mechanism of action did not appear to be the same as for the benzodiazepine anxiolytics. It was initially believed that the weak dopaminergic receptor blocking activity of buspirone was somehow responsible for its anxiolytic activity; however, subsequent disinhibitory compounds from the same chemical series (e.g., gepirone) (Eison et al., 1982) do not exhibit dopaminergic blocking properties. During the past year, a potential handle on the mechanism of action of buspirone and buspirone-like drugs (e.g., isapirone) has been uncovered. All of the buspirone-like anxiolytic agents (i.e., buspirone, gepirone, ipsapirone) exhibit activity as partial agonists at $5\text{-}HT_{1A}$ receptors in brain (Traber et al., 1984; Dompert et al., 1985; Peroutka, 1985). Thus it appears likely that a specific subtype of the $5\text{-}HT_1$ receptor may be important in the mediation of the anxiolytic activity of buspirone and related drugs. Although buspirone-like agents appear to act primarily at $5\text{-}HT_{1A}$ receptors, whereas benzodiazepine-like drugs appear to act directly upon the benzodiazepine/GABA/chloride ionophore complex, they may all have a final common pathway since all of these agents suppress the firing of dorsal raphe neurons (Sprouse and Aghajanian, 1985; Gehlbach and Vander-Maelen, 1985). The discovery of the apparent mechanism of action of the buspirone-type of anxiolytics is sure to provoke a flurry of activity in the pharmaceutical industry, the goal of which will be to rapidly identify new and better anxiolytics working via this new mechanism.

2.2. Anticonvulsant Activity

It is a well-established fact that the majority of the clinically effective anxiolytic agents (buspirone is a notable exception) exhibit varying degrees of activity as anticonvulsants in both animals (Niemegeers and Lewi, 1979) and humans (Overweg and Binnie, 1983). Clinically, these drugs (e.g., diazepam, clonazepam) are chiefly employed for the treatment of certain types of epilepsy, their most important use being the acute treatment of status epilepticus and the management of myoclonus.

The anxiolytics (e.g., benzodiazepines) and other disinhibiting agents (e.g., barbiturates) have been shown to be effective antagonists of convulsions induced by electroshock (Niemegeers and Lewi, 1979) and sound (Malick, 1985), as well as by a number of chemical substances, e.g., isoniazid (Costa et al., 1975a), strychnine (Zbinden and Randall, 1967), picrotoxin (Costa et al., 1975a), pentylenetetrazol (Lippa et al., 1979b), and bicuculline (Stone and Larid, 1978; Patel et al., 1980). Three techniques (i.e., antagonism of convulsions induced by pentylenetetrazol, bicuculline, and sound) that are particularly useful for the evaluation of the anticonvulsant activity of potential anxiolytics will be discussed.

2.2.1. Antagonism of Pentylenetetrazol-Induced Convulsions

In addition to predicting the anticonvulsant potential of anxiolytics, the ability of an agent to prevent pentylenetetrazol (metrazol)-induced convulsions in mice has been used by several laboratories as a primary screen for anxiolytic potential (Lippa et al., 1979b). In fact, the anxiolytics have been claimed to be "selective" antagonists of pentylenetetrazol convulsions since they antagonize seizures at doses that are below the doses at which they produce neuromuscular impairment or ataxia (Lippa et al., 1979b), and with a few exceptions (e.g., oxazepam), there is a good correlation between their relative potencies as antimetrazol agents and their effective anxiolytic doses in humans (Irwin, 1967; Lippa et al., 1979b).

Briefly, according to the procedure used in this laboratory, convulsive episodes are induced in male mice with pentylenetetrazol (82 mg/kg, sc); historically, this dose produced seizures in approximately 99% of vehicle-treated control mice. Groups of six mice were treated orally with various doses of test agents and then injected with pentylenetetrazol 45 min later. All mice were then placed into separate plexiglass cubicles and observed for signs of convulsive episodes for the next 30 min. The percent protection was calculated for each group, and the MED (minimal effective dose, i.e., the lowest dose producing 50% or greater antagonism of seizures) was determined for each drug.

Results obtained with selected anxiolytic drugs are presented in Table 5. All of the benzodiazepines tested significantly prevented pentylenetetrazol-induced convulsions in a dose-related manner. In an earlier report, Malick and Enna (1979) reported that a significant correlation [Spearman rank correlation coefficient (r_s) = 0.81; $p < 0.01$] existed between displacement of ^{3}H-diazepam binding and the potency of these benzodiazepines as inhibitors of pentylenetetrazol-induced seizures. Meprobamate and phenobarbital, representative anxiolytics from the carbamate and barbiturate series, respectively, also significantly antagonized pentylenetetrazol-induced convulsions in mice (*see* Table 5).

A similar procedure has been developed for use in rats (Niemegeers and Lewi, 1979; Goldberg et al., 1983), and it has been found that anxiolytics will also antagonize metrazol-induced convulsions in this species. For comparative purposes, ED_{50} values are presented for several clinically effective anxiolytics, as well as for several preclinical agents that are reported to exhibit potential anxiolytic activity in animals (Table 6). Performing an antimetrazol study in rats offers the advantage of determining anticonvulsant activity in the same species used to determine anxiolytic potential (e.g., SSD test, Geller conflict).

Table 5
Antagonism of Pentylenetetrazol-Induced Convulsions in Mice by Anxiolytic Drugs

Drug	MED, mg/kg, po, for antagonism of pentylenetetrazol-induced convulsions
Flunitrazepam	0.03
Clonazepam	0.25
Bromazepam	0.25
Diazepam	0.8
Chlorazepate	1.0
Nitrazepam	1.0
Oxazepam	1.3
Flurazepam	2.5
Chlordiazepoxide	2.5
Medazepam	4.0
Halazepam	8.0
Ripazepam	15.0
Phenobarbital	20.0
Meprobamate	200.0

2.2.2. Antagonism of Bicuculline-Induced Convulsions in Mice

Bicuculline, a GABA antagonist, is a potent convulsant agent. The convulsant activity produced by bicuculline is believed to be mediated via a blockade of postsynaptic GABA receptors located in the central nervous system (Curtis et al., 1970).

According to the procedure used in our laboratory (Patel et al., 1980), bicuculline (3 mg/kg, sc) was administered 30 min after the ip administration of test substances. The incidence of clonic and tonic seizures and/or death at each dose of drug was recorded for 30 min following bicuculline, and an ED_{50} (the dose expected to protect 50% of the mice) was calculated for each drug.

The results obtained with selected anxiolytics are presented in Table 7. All of the benzodiazepines evaluated were effective antagonists of bicuculline-induced seizures, and their ED_{50} values were below their respective values for producing significant impairment of rotorod performance (Patel et al., 1980). Similarly, both meprobamate and phenobarbital produced a dose-related antagonism of seizure activity, although they were considerably less potent than

Table 6
Effects of Selected Anxiolytics in the Pentylenetetrazol-Induced Convulsion Test in Rats

Drug	ED_{50}, mg/kg, po, for antagonism of seizures
Nitrazepam	0.25[a]
Oxazepam	0.7[a]
CGS 9896	2.0[b]
CL 218,872	2.5[b]
Zopiclone	2.6[b]
Chlorazepate	3.5[a]
Diazepam	4.0[b]
Sodium Phenobarbital	17.1[a]
Meprobamate	36.9[a]
Buspirone	Inactive[b] (50)[c]

[a]Data from Lippa et al. (1979b); median effective doses for protection.
[b]Data from Goldberg et al. (1983).
[c]Highest dose tested.

the benzodiazepines (Table 7). It has been hypothesized that anxiolytics antagonize seizures induced by bicuculline via GABAmimetic activity, and both benzodiazepines and barbiturates have been reported to significantly enhance ^{3}H-GABA binding to rat brain synaptosomal fractions (Costa and Guidotti, 1979; Willow and Johnston, 1980). Since the mechanism of action of benzodiazepine anxiolytics is believed to be, at least in part, caused by GABA-mimetic activity (Costa et al., 1975b; Goldstein et al., 1983), this may be a particularly appropriate model in which to assess the anticonvulsant potential of anxiolytics.

2.2.3. Antagonism of Sound-Induced Seizures in Mice

A great deal of evidence indicates that genetic factors control seizure susceptibility and the probability of developing epilepsy in humans. However, even if a genetically determined seizure-prone state exists, generally something must trigger or initiate the seizure; the initiators of seizures can be anything from environmental stimuli (e.g., sound in audiogenic mice) to neurochemical or hormonal imbalances or a foci produced by physical damage to nervous tissue. Therefore, the occurrence of epilepsy depends upon a combination of factors—a genetically determined propensity, and one or more endogenous or exogenous triggers.

Table 7
Effects of Anxiolytic Drugs on Bicuculline-Induced Seizures in Mice

Drug	ED_{50}, mg/kg, ip, for antagonism of seizures	95% Confidence limits
Diazepam	1.4	2.9–6.6
Oxazepam	1.8	1.1–3.8
Flurazepam	2.0	1.0–2.9
Chlordiazepoxide	4.7	2.9–6.6
Sodium Phenobarbital	34.3	19.0–64.8
Meprobamate	260.5	205.5–398.7

Unfortunately, until recently, much of the experimental research on epilepsy was done in rodent models in which seizures were induced by chemical or electrical means. However, several genetically linked animal models (e.g., audiogenic seizure-susceptible mice and rats, photosensitive baboons, epileptic dogs) have been developed that appear to more closely resemble human epilepsy. The audiogenic seizure-susceptible mouse model (Collins, 1972) appears to be particularly analogous to epilepsy in humans because it is genetically determined, and seizure episodes are precipitated by an environmental stimulus (i.e., sound); furthermore, all of the major therapies used to treat seizure disorders in humans (e.g., phenytoin, trimethadione, benzodiazepines, barbiturates, sodium valproate) prevented sound-induced seizures (Malick, 1985).

In this laboratory, DBA-2J mice were utilized between 20 and 22 d of age, a time at which sound-induced seizure activity was maximal. During the test, mice were evaluated for seizure activity in a sound-attenuated chamber, and an electric bell was used to generate a noise level between 116 and 118 dB. Historically, 100% of vehicle-treated control mice exhibited wild running, clonus, and tonus. In drug antagonism studies, all test agents were administered ip 30 min prior to exposure to the audiogenic stimulus and were observed for 60 s; any animal that failed to exhibit clonic seizures was considered to be significantly affected (blocked), and minimal effective doses (MED; i.e., the lowest dose of drug producing an inhibition of sound-induced seizures in 50% or more of the subjects tested) were established for each drug.

The results of studies with selected anxiolytics and disinhibitory substances are summarized on Table 8. All of the clinically effective anxiolytics, as well as those preclinical agents that are predicted

Table 8
Anticonvulsant Action of Benzodiazepine and Nonbenzodiazepine Anxiolytics in Audiogenic Seizure Susceptible Mice[a]

Drug	MED[b], mg/kg, ip, for antagonism of seizures
Alprazolam	0.06
Clonazepam	0.1
Halazepam	0.3
CGS 9896	0.3
Diazepam	1
Chlordiazepoxide	3
CL 218,872	5
Zopiclone	5
Sodium Phenobarbital	10
Tracazolate	20
Doxepin	20
Meprobamate	60
Buspirone	Inactive (0.1–40)[c]

[a]Data from Malick, 1985.
[b]MED, Minimal effective dose (i.e., the lowest dose inhibiting seizures in 50% or greater of the mice tested).
[c]Dose range evaluated.

to be anxiolytics on the basis of their pharmacological profiles (e.g., CL 218,872, CGS 9896, tracazolate), exhibited dose-related activity in this procedure. Buspirone was the one exception, and it also has failed to demonstrate anticonvulsant activity in any other laboratory model. Therefore, the audiogenic seizure-susceptible mouse model appears to be useful for evaluating the anticonvulsant potential of anxiolytics as well as other types of anticonvulsant agents.

Although all existing anxiolytics, with the exception of buspirone, are anticonvulsants, and the anticonvulsant tests are useful in characterizing the pharmacological profile of a potential anxiolytic, these procedures lack face validity as anxiolytic models. Although some investigators still rely heavily on the anticonvulsant (especially pentylenetetrazol) models as predictors of anxiolytic potential, these should not be used as a primary screen for such activity; rather, it is preferable to use the conflict procedures as anxiolytic primary screens and to relegate anticonvulsant tests to secondary evaluation of their activity profiles.

2.3. Muscle Relaxant Potential

In addition to their anxiolytic activity, most antianxiety agents, especially the benzodiazepines, exhibit significant muscle relaxant activity. Although it is debatable whether the muscle relaxant activity contributes significantly to the anxiolytic actions of these agents, since buspirone is an effective anxiolytic that is devoid of muscle relaxant activity, agents such as diazepam are widely used as muscle relaxants for the treatment of a variety of syndromes in which spasm is a prominent symptom (e.g., lower back problems). In addition, the muscle relaxant property may contribute to the utility of benzodiazepines as preanesthetic medications. Several types of animal models can be used to assess muscle relaxant potential.

2.3.1. Inhibition of the Linguomandibular Reflex in Cats

The ability of an agent to inhibit the polysynaptic linguomandibular reflex without concomitant inhibition of the monosynaptic patellar reflex in the cat has been considered indicative of central muscle relaxant activity (King and Unna, 1954). A detailed description of the methodology used in this preparation can be obtained in the recent manuscript by Novak and Zwolshen (1983). All of the muscle relaxants tested, including the benzodiazepine anxiolytics (i.e., diazepam, chlordiazepoxide), were inhibitors of the linguomandibular reflex. Therefore, this model could also be used to determine the muscle relaxant potential of anxiolytic drug candidates.

2.3.2. Antagonism of Etonitazene-Induced Rigidity

Another test that is predictive of central muscle relaxant activity is the etonitazene-induced rigidity model in rats (Barnett et al., 1975). Etonitazene is a potent narcotic analgesic drug that induces what has been described as "lead-pipe rigidity" of the trunk and limb musculature in rats. The degree of muscular rigidity induced by etonitazene is much greater than that produced by morphine; morphine-induced rigidity (Wand et al., 1974) has also been used as a screen for muscle relaxants. Etonitazene-induced rigidity was antagonized by centrally acting muscle relaxants (e.g., methocarbamol, zoxazolamine), including diazepam, which was the most potent compound evaluated by Barnett and coworkers (1975).

2.3.3. Morphine-Induced Straub Tail Test

The antagonism of morphine-induced Straub tail procedure (Ellis and Carpenter, 1974) has become one of the most frequently used and reliable tests for predicting muscle relaxant activity. Novak (1982) has recently reassessed this model. According to his procedure, mice were pretreated ip with either vehicle or test agent 15 min prior to administration of morphine sulfate (60 mg/kg, sc); this dose of morphine elicited Straub tail (90° from the horizontal elevation of the tail) in all vehicle-treated subjects. The presence or absence of Straub tail was observed 15 min postmorphine. The results obtained with selected muscle relaxants are summarized in Table 9. Both the anxiolytic benzodiazepines and the representative

Table 9
Effects of Selected Anxiolytics and Other Muscle Relaxants in the Morphine-Induced Straub Tail Test in Mice[a]

Treatment	ED_{50}, mg/kg, ip, for antagonism of Straub tail
Diazepam	1.3
Midazolam	1.4
Baclofen	3.6
Chlordiazepoxide	9.9
Dantrolene	16.8

[a]Data from Novak (1982).

muscle relaxants that were not anxiolytics (e.g., dantrolene, baclofen) were potent antagonists of Straub (rigid) tail in mice. In a comparison of the Straub tail procedure with other tests predictive of muscle relaxant efficacy (e.g., morphine-induced rigidity, decerebrate rigidity, polysynaptic reflexes), only the potency ranking in the morphine-induced Straub tail procedure was significantly correlated with the potency ranking for muscle relaxant activity in man (Novak and Zwolshen, 1983). Thus this simple procedure appears to be the most reliable predictor of muscle relaxant potential and can be used to quickly and accurately predict the muscle relaxant activity of potential anxiolytic agents.

2.4. Assessment of Sedation and Neuromuscular Impairment

One of the major side-effect liabilities with anxiolytics is that they produce sedation (drowsiness) and impair the ability to perform tasks requiring fine motor movements or neuromuscular coordination. Therefore, it is important to assess the potential for producing these effects with any new anxiolytic that is developed, and several of the techniques that have been widely used for this purpose will be discussed.

2.4.1. Rotorod Performance

The rotorod procedure used in this laboratory to assess neuromuscular coordination is a modification of the method described by Kinnard and Carr (1957). Briefly, rats are trained to maintain themselves for at least 1 min on a rotating wooden rod (rotorod) moving at a constant speed of 6 rpm. Groups of rats that have demonstrated the ability to negotiate the rod for at least 1 min were treated with either vehicle or test agent and retested to determine their ability to maintain rotorod performance at 30, 60, 90, and 120 min postdrug administration. Any subject that failed to stay on the rotorod for the 1-min test duration was considered to be neuromuscularly impaired (ataxic), and ED_{50} values were calculated by the method of Litchfield and Wilcoxon (1949).

Since absolute values (i.e., mg/kg doses) for producing impaired performance can be somewhat misleading, a neuromuscular impairment liability (NIL) index, which is calculated by dividing the ED_{50} for rotorod impairment by the MED value for therapeutic efficacy (e.g., MED in the SSD test), is used in this laboratory to compare the activity of a test agent to that of selected reference standards. For example, diazepam exhibited an ED_{50} for rotorod impairment of 10.2 mg/kg, po, and had a MED in the primary anxiolytic screen (SSD) of 5.0 mg/kg, po; therefore, the NIL index for diazepam was 2.0. On the basis of a NIL index of 2, it would be predicted that diazepam would exhibit little, if any, margin of safety between the anxiolytic and neuromuscular impairing doses in humans, i.e., it would be expected, as is the case, that neuromuscular impairment would be evident at the anxiolytic dose range in humans. Naturally, it is hoped that any new anxiolytic would exhibit a high NIL index so that it could be considered "anxioselective" (i.e., produce anxiolytic activity at doses that are markedly below those producing ataxia and sedation).

2.4.2. Inclined Screen

Another simple procedure that has been widely used to assess neuromuscular impairment in rodents is the inclined screen. The apparatus used in this method is very readily made and is inexpensive; it consists of a piece of heavy-gage wire screening mounted onto a wooden frame and positioned such that it is on a 45° angle (Malick, 1976). Normal rats or mice can readily traverse the inclined surface. However, the test is sensitive to the effects of any agent that causes neuromuscular impairment. Any rat that slips when walking on the screen or cannot negotiate the screen for any reason is considered to be neuromuscularly impaired or ataxic. One big advantage to this procedure is that it can readily be used to assess neuromuscular impairment at any time since the subjects do not require any previous training. For instance, after a subject has been tested for any behavioral measure (e.g., for anxiolytic activity), it can then be placed upon the screen and evaluated for neuromuscular impairment.

2.4.3. Wire Maneuver Test

Another simple procedure that does not require elaborate apparatus is the wire maneuver test (File, 1982). The apparatus for rats consists of a horizontally placed wire (40 cm long and 2 mm in diameter) that is suspended 46 cm above the surface. When a normal rat is suspended from the wire by its forepaws, it will readily heave one or both paws onto the wire. The test duration is 120 s, and any rat that fails to lift one or both hindpaws onto the wire is considered to be neuromuscularly impaired or ataxic. This procedure can also be used for mice. Naturally, a smaller-gage wire should be used for mice. A mouse that is suspended by its forepaws can easily climb onto the wire and balance itself for a few seconds, in contrast to rats that are too heavy to climb onto the wire. Separate groups of rodents are used at several doses of the test compound, and ED_{50} values are calculated by the method of Litchfield and Wilcoxon (1949).

2.5. Evaluation of Drug-Interaction Liabilities

It is well recognized that anxiolytics, benzodiazepines in particular, potentiate the CNS depressant properties of a variety of psychoactive compounds (e.g., barbiturates, alcohol). Since this adverse drug-interaction represents a serious side-effect liability with existing anxiolytics, the methods that are utilized to assess the potential for such interactions will be discussed.

2.5.1. Potentiation of Barbiturates

A wide variety of classes of agents, including anxiolytics, that produce sedation have been shown to potentiate the sedative properties of barbiturates. Thus, the potentiation of barbiturate-induced sleeptime procedure has been used both to predict potential for interaction with barbiturates and to give another measure of sedative liability.

The potentiation of sodium barbital-induced sleeptime test in rats is a simple and reliable procedure for measuring the interaction between various agents and barbiturates. Sodium barbital is used because it is not extensively metabolized, thereby minimizing the possibility that an agent could be potentiating sleeptime by inhibition of metabolizing enzymes. In this procedure, test drug or placebo is administered orally 30 min prior to ip administration of sodium barbital (100 mg/kg); this dose of barbital was chosen since it does not produce sleep and yet can be readily potentiated by CNS depressants. Rats are observed after barbital for a loss of the righting reflex, and the length of time (min) spent sleeping (sleeptime) is recorded. Diazepam has been shown to significantly potentiate barbital-induced sleeptime at a dose as low as 5 mg/kg, po (unpublished observations); thus, diazepam will significantly potentiate the sedative properties of barbital at the same dose as its minimal effective dose (MED) for anxiolytic activity in the SSD conflict test. Although it is preferable to perform barbiturate interaction studies in rats, since the most reliable conflict procedures are also performed in this species, this study can also be performed readily in mice (Patel and Malick, 1982).

2.5.2. Interaction with Alcohol

In addition to possessing anxiolytic efficacy, it is also important that any new and novel antianxiety agent exhibit a reduced liability to adversely interact with alcohol compared to the benzodiazepines. Therefore, it is critical that the alcohol-potentiation liability of an experimental agent be established in the same species and by the same route of administration that are used to determine antianxiety activity.

In our laboratory, the potentiation of the sedative properties of alcohol are measured on the rotorod; a detailed description of the rotorod procedure is given earlier in this report (section 2.4.1). In this procedure, trained rats are pretreated with vehicle or test drug 15 min prior to the administration of a subthreshold dose of alcohol (0.8 g/kg, ip). All rats are then retested for rotorod performance at 15, 30, and 60 min postalcohol. If a dose-related potentiation of alcohol is obtained, an ED_{50} value can be obtained. For example, the ED_{50} value for diazepam was found to be 2.5 mg/kg, po. Therefore, in this laboratory, diazepam significantly potentiated alcohol-induced impairment of rotorod performance at a dose that was one half its anxiolytic dose (5 mg/kg, po) in the shock-induced suppression of drinking test in rats. This test is a very sensitive measure of the alcohol potentiation liability of anxiolytic agents.

The potentiation of alcohol-induced lethality test in mice is another procedure that can be used to measure this interaction liability (Patel and Malick, 1980a). However, although the latter procedure appears to be reliable, it is a less sensitive measure of alcohol/anxiolytic interactions.

2.6. Dependence Liability

Although physical addiction to benzodiazepines is not nearly as significant a problem as with some other psychoactive drugs (e.g., narcotics), psychological dependence on benzodiazepines has been reported, and withdrawal symptoms have been observed following abrupt cessation after long-term treatment with moderate to high doses in humans (Lader and Petursson, 1983). For example, following 5 mo treatment with typical clinical anxiolytic doses of chlordiazepoxide, subjects experienced a mild withdrawal reaction that included tension, dizziness, loss of appetite, and tremors (Covi et al., 1973). One of the reasons signs of benzodiazepine dependence may be missed or misdiagnosed is that, unlike withdrawal from opiates or alcohol, anxiety is the cardinal symptom of the benzodiazepine withdrawal syndrome.

In rodents, withdrawal signs (e.g., hyperexcitability) were observed after cessation of chronic treatment with low doses of diazepam (Turnbull et al., 1981). Furthermore, mice that had been chronically treated with diazepam (30 mg/kg, po, for 4 wk) exhibited long-lasting (15 d) increases in pentylenetetrazol-induced seizures following sudden drug withdrawal (Boast et al., 1985); in contrast, CGS 9896, a pyrazoloquinoline that behaves like an agonist/antagonist in a battery of behavioral tests (Malick et al., 1986; Patel et al., 1986) and that may be a partial agonist at brain benzodiazepine receptors, failed to exhibit withdrawal signs after comparable treatment (Boast et al., 1985). A good review of the literature on the methods used to assess abuse liability with benzodiazepines in animals and humans was prepared by Griffiths and coworkers (1985).

The judicious use (e.g., short-term therapy, caution with patients having a history of alcohol or drug abuse) of anxiolytics (especially benzodiazepines) can minimize their abuse liabilities (Ananth, 1982).

2.7. Behavioral Assessment of Benzodiazepine Antagonist Activity

In recent years, several agents have been discovered that have been classified as benzodiazepine antagonists because they exhibit the following profile: although they potently displace radiolabeled benzodiazepines (e.g., ^{3}H-flunitrazepam) from their binding sites in brain, they lack the anticonflict and anticonvulsant activities characteristic of benzodiazepine agonists (e.g., diazepam). Furthermore, these agents significantly antagonize many of the properties of benzodiazepines (e.g., anticonflict, anticonvulsant, sedative, muscle relaxant, and so on) and even exhibit anxiogenic activity under certain circumstances (File et al., 1982). Several structurally dissimilar antagonists have been identified, such as the imidazobenzodiazepine, Ro 15-1788 (Hunkeler et al., 1981), CGS 8216, a pyrazoloquinoline (Bernard et al., 1981), and ethyl-β-carboline-3-carboxylate (β-CCE) (Braestrup et al., 1980).

In searching for new benzodiazepine antagonists, any agent that is a potent displacer of ^{3}H-flunitrazepam (*see* Enna, this volume) and is inactive as an anticonflict agent is worthy of further consideration. If a compound exhibits the aforementioned profile, it is important to ascertain whether it is, in fact, active in vivo; i.e., are pharmacokinetic problems (e.g., absorption, distribution) responsible for its inactivity in conflict paradigms? One of the quickest ways of determining this is to perform an ex vivo ^{3}H-flunitrazepam binding assay in which animals are dosed with drug, sacrificed, and their brains assayed for ligand binding. If such a compound is still a potent displacer ex vivo, it is likely that it may be an antagonist. The experimental compound can next be tested to determine whether it exhibits benzodiazepine antagonist activity in vivo; for example, can it antagonize diazepam-induced neuromuscular impairment on the rotorod? Naturally the antagonism of other behavioral or neuropharmacological properties of benzodiazepines (e.g., anticonflict, anticonvulsant) should be used to fully characterize the antagonist profile of the agent.

In recent years, a new category of drug has been discovered, the benzodiazepine agonist/antagonist. CGS 9896 represents the prototypical agent in this class (Bernard et al., 1985). CGS 9896 exhibits the profile of a full benzodiazepine agonist in that it produces potent, dose-related anxiolytic activity in conflict procedures and exhibits anticonvulsant activity. However, CGS 9896 exhibits benzodiazepine antagonist activity since it antagonized the rotorod impairment produced by diazepam (Bernard et al., 1985); recently we have fully characterized the antagonist activity of this unique agent in a range of behavioral assays (Malick et al., 1986; Patel et al., 1986). Agonist/antagonist anxiolytics (e.g., CGS 9896) appear to offer the potential for reduced dependence liability compared to full agonists (e.g., diazepam) since preliminary studies indicate that CGS 9896-treated mice do not exhibit the characteristic withdrawal syndrome observed with typical benzodiazepines (Boast et al., 1985; Patel et al., unpublished observations).

2.8. Behavioral Differentiation of Benzodiazepine Displacer and Nondisplacer Anxiolytics

Very simplistically, anxiolytic agents can be divided into two types: (1) the displacers, i.e., the typical or benzodiazepine-like agents that displace ^{3}H-benzodiazepines from brain receptors (e.g., diazepam, zopiclone, CL 218,872, CGS 9896) and (2) the nondisplacers, i.e., atypical anxiolytics that do not displace ^{3}H-flunitrazepam and that presumably act via a different mechanism of action [e.g., meprobamate, barbiturates, buspirone, clonidine, 5-HT_2 antagonists (Patel and Malick, 1984), fenobam]. These two types of anxiolytics can be readily differentiated either neurochemically (i.e., ^{3}H-flunitrazepam binding assay) or behaviorally as follows. In a recent study, Patel and coworkers (1982) demonstrated that the displacers could be separated from the nondisplacers behaviorally by determining whether their anticonflict activity in the shock-induced suppression of drinking test could be antagonized by a benzodiazepine antagonist (e.g., CGS 8216, Ro 15-1788). The results obtained with a variety of anxiolytics are presented in Fig. 1. Whereas the anxiolytic activities of the displacer anxiolytics (e.g., chlordiazepoxide, CL 218,872, zopiclone, CGS 9896) were significantly antagonized by either CGS 8216 or Ro 15-1788, the disinhibitory activity of a variety of compounds that fail to displace ^{3}H-flunitrazepam (e.g., meprobamate, phenobarbital, fenobam, tracazolate) was not significantly altered (Fig. 2). Thus, this behavioral analysis, in which the anxiolytic activity of a compound is assessed both alone and in combination with a benzodiazepine antagonist, can predict whether an agent is a benzodiazepine-like agonist.

2.9. Potential False Positives

The prediction of therapeutic efficacy in humans, based upon the pharmacological activity of an agent in animals, is always

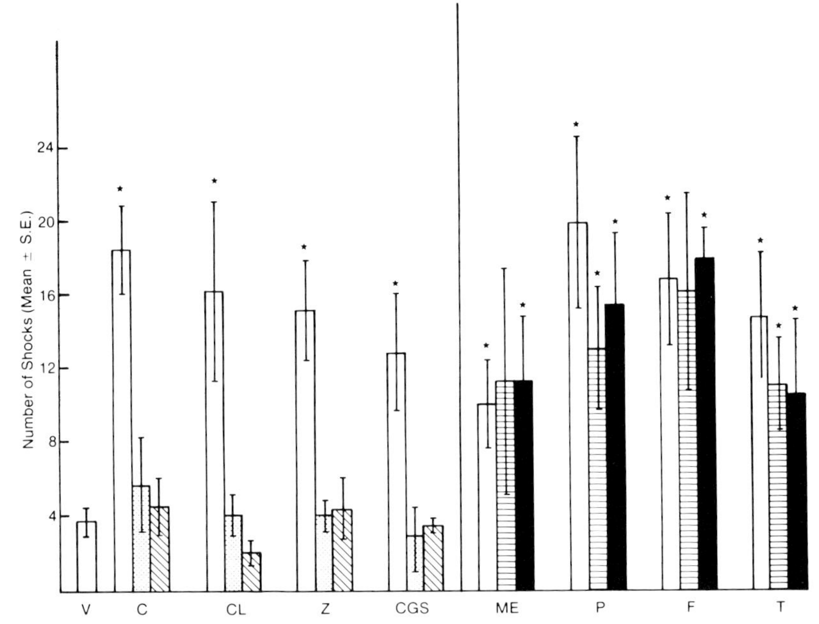

Fig. 1. Mean number of shocks taken in the SSD test following oral treatment (60 min pretest) of selected disinhibitory agents alone and in combination with antagonist (30 min pretest). Symbols and abbreviations: *, $p < 0.05$; V, vehicle; C, chlordiazepoxide, 18 mg/kg; CL, CL 218,872, 5 mg/kg; Z, zopiclone, 5 mg/kg; CGS, CGS 9896, 10 mg/kg; M, meprobamate, 50 mg/kg; P, phenobarbital, 20 mg/kg; F, fenobam, 12.5 mg/kg; T, tracazolate, 20 mg/kg; □, treatment alone; □, treatment + Ro 15-1788, 10 mg/kg, ip; ■, treatment + Ro 15-1788, 20 mg/kg, ip; □, treatment + CGS 8216, 10 mg/kg, ip; □, treatment + CGS 8216, 20 mg/kg, ip.

fraught with many difficulties. Predictability has been especially difficult in the area of psychopharmacology, an area in which many of the major breakthroughs (e.g., the discovery of the first antipsychotic) were made in the clinic based upon astute observations rather than upon predictions made as a result of animal experimentation.

In many areas of psychotropic drug research (e.g., depression, schizophrenia, Alzheimer's disease), the animal models are, for the most part, extremely theoretical or speculative. In the past, industrial neuropsychopharmacologists have been proficient at predicting that a "me-too" agent (i.e., one that closely mimics the profile of an existing agent, but may have better potency or reduced side-effect liabilities) will be effective in humans. However, it has been extremely difficult to predict that a truly new and novel agent, which presumably works via a new mechanism of action, will be effective. The most recent example of this is buspirone, which was predicted to be an antipsychotic based upon its pharmacological profile in animals, but which turned out to be an anxiolytic in humans. In recent years, great strides have been made in the understanding of brain mechanisms involved in behavior; hopefully, this knowledge will be reflected in better predictability from animals to humans.

In the area of anxiolytic drug research, it has been somewhat easier to predict activity in humans with new agents because the behavioral models appear to have at least partial face validity. For example, it has been shown that human subjects experience an increase in anxiety when performing an approach–avoidance paradigm similar to that used in animals; furthermore, diazepam pretreatment produced significant antianxiety or disinhibitory activity in these subjects, allowing them to take a greater number of shocks to obtain reinforcement (Beer and Migler, 1975).

Since predictability can be difficult, it is always critical to consider the possibility that a "false positive" (i.e., a compound that is active in a given procedure for some reason other than the fact that it possesses the desired therapeutic effect) could be detected in any test. For instance, the majority of conflict procedures utilize a mild electric footshock to induce behavioral suppression; thus, it is obvious that there is the distinct possibility that an analgesic agent could show up as a false positive in these tests. With very few exceptions, however, analgesics have been shown to be inactive in conflict models (Cook and Davidson, 1973; Lippa et al., 1979b). Occasionally, morphine-treated subjects will accept a greater number of shocks than controls, but this activity is not dose-related (i.e., it occurs only at a single dose level), and therefore morphine must be considered inactive in these procedures. Thus, it is unlikely that the anticonflict effects of anxiolytics are in any way related to analgesic or threshold-elevating actions of these agents. Furthermore, if an experimental agent exhibited activity in a conflict test because of analgesic activity, it would be a simple matter to discover this by performing a couple of classical analgesic tests. Potential anxiolytics should be evaluated for analgesic activity to be certain that antinociceptive activity is not responsible for the anticonflict action of the agent.

A few years ago, Patel and Malick (1980b) reasoned that any agent that had a direct action on the primary drives or motivational states (i.e., hunger, thirst) might show up as a ''false positive'' in conflict tests that depend upon food or water reinforcement. Isoproterenol, a beta-adrenergic agonist, was evaluated in the shock-induced suppression of drinking (SSD) test because it increases water consumption in rats (Schwob and Johnson, 1975) and because chlordiazepoxide has been reported to increase both water (Knowles and Ukena, 1973) and food (Niki, 1965) consumption. Isoproterenol exhibited significant dose-related anticonflict activity in the SSD procedure (Patel and Malick, 1980b); thus, β-agonists and other agents that enhance water intake could indeed be detected as ''false positives'' in a conflict procedure motivated by thirst. Furthermore, chlordiazepoxide significantly enhanced drinking at anxiolytic doses in nondeprived rats (Patel and Malick, 1980b). Thus, although it is clear that chlordiazepoxide possesses ''real'' anticonflict activity, it is also evident that its effects in procedures involving water reinforcement (i.e., the SSD test) might be related, at least in part, to its direct effects on neuronal systems controlling water intake. Thus, the magnitude of the anticonflict effect (i.e., the number of shocks taken) might be greater for an agent such as chlordiazepoxide, for which the anxiolytic and water intake effects appear to be additive, compared to an agent that is devoid of activity on thirst mechanisms. Thus, drugs that enhance water intake might show up as false positives in the SSD test, and similarly, drugs that enhance food intake might be false positives in conflict procedures utilizing food reinforcement (e.g., Geller-Seifter conflict).

The possibility of research findings being confounded by false positive activity is an important issue in drug development and one that must always be addressed thoroughly prior to recommending an agent for clinical evaluation.

3. Summary

In this chapter, the author has attempted to demonstrate the ways in which an industrial neuropsychopharmacologist would seek to discover new and novel psychoactive drugs for the treatment of psychiatric disorders. The basic approach currently used in many pharmaceutical laboratories to search for antianxiety agents was used as an example since the author felt that to skim the surface of many areas would not be as meaningful as to probe a single area in depth. This chapter contains a thorough discussion of the behavioral or neuropharmacological methodologies used to predict therapeutic efficacy, not only as an anxiolytic, but also for the additional types of indications for which classical benzodiazepines are utilized, such as muscle relaxant or anticonvulsant effects. In addition, the types of studies that would be required to assess the side-effect liabilities (e.g., sedation or neuromuscular impairment, alcohol and barbiturate potentiation, dependence and withdrawal) commonly associated with anxiolytics of the benzodiazepine type are thoroughly discussed.

References

Ananth, J. (1982) Benzodiazepines: Selective use to avoid addiction. *Postgraduate Med.* **72**, 271–276.

Barnett, A., Goldstein, J., Fiedler, E., and Taber, R. (1975) Etonitazene-induced rigidity and its antagonism by centrally acting muscle relaxants. *Eur. J. Pharmacol.* **30**, 23–28.

Beer, B. and Migler, B. (1975) Effects of Diazepam on Galvanic Skin Response and Conflict in Monkeys and Humans, in *Predictability in Psychopharmacology: Preclinical and Clinical Correlations* (Sudilovsky, A., Gershon, S., and Beer, B., eds.) Raven, New York.

Bernard, P., Bergen, K., Sobiski, R., and Robson, R. (1981) CGS 8216 (2-phenylpyrazolo) [4,3-c] quinolin-3(5H)-one), an orally effective benzodiazepine antagonist. *Pharmacologist* **23**, 150.

Bernard, P. S., Bennett, D. A., Pastor, G., Yokoyama, N., and Liebman, J. M. (1985) CGS 9896: Agonist–antagonist benzodiazepine receptor activity revealed by anxiolytic, anticonvulsant and muscle relaxation assessment in rodents. *J. Pharmacol. Exp. Ther.* **235**, 98–105.

Boast, C. A., Gerhardt, S. C., Gajary, Z. L., and Brown, W. N. (1985) Lack of withdrawal effects in mice after chronic administration of the non-sedating anxiolytic CGS 9896. *Soc. Neurosci. Abst.* **11**, 424.

Braestrup, C., Nielsen, M., and Olsen, C. E. (1980) Urinary and brain β-carboline-3-carboxylates as potent inhibitors of brain benzodiazepine receptors. *Proc. Natl. Acad. Sci. USA* **77**, 2288–2292.

Canon, J. and House, V. P. (1978) Squirrel monkey active conflict test. *Physiol. Psychol.* **6**, 215–222.

Collins, R. L. (1972) Audiogenic Seizures, in *Experimental Models of Epilepsy—A Manual for the Laboratory Worker* (Purpura, D. P., Penry, J. K., Tower, D., Woodbury, D. M., and Walter, R., eds.) Raven, New York.

Cook, L. and Davidson, A. B. (1973) Effects of Behaviorally Active Drugs in a Conflict-Punishment Procedure in Rats, in *The Benzodiazepines* (Garattini, S., Mussini, E., and Randall, L. O., eds.) Raven, New York.

Cook, L. and Sepinwall, J. (1975) Behavioral Analysis of the Effects and Mechanisms of Action of Benzodiazepines, in *Mechanism of Action of Benzodiazepines* (Costa, E. and Greengard, P., eds.) Raven, New York.

Costa, E. and Guidotti, A. (1979) Molecular mechanisms in the receptor actions of benzodiazepines. *Ann. Rev. Pharmacol. Toxicol.* **19**, 531–545.

Costa, E., Guidotti, A., and Mao, C. C. (1975a) Evidence for Involvement of GABA in the Action of Benzodiazepines: Studies on Rat Cerebellum, in *Mechanism of Action of Benzodiazepines* (Costa, E. and Greengard, P., eds.) Raven, New York.

Costa, E., Guidotti, A., Mau, C., and Suria, A. (1975b) New concepts in the mechanism of benzodiazepines. *Life Sci.* **17**, 167–186.

Covi, L., Lippman, R. S., and Pattison, J. H. (1973) Length of treatment with anxiolytic sedatives and response to their sudden withdrawal. *Acta Psychiatr. Scand.* **49**, 51–64.

Curtis, D. R., Duggan, A. W., Felix, D., and Johnston, G. A. R. (1970) GABA, bicuculline and central inhibition. *Nature* **226**, 1222–1224.

Dompert, W. U., Glaser, T., and Traber, T. (1985) ^{3}H-TVX Q 7821: Identification of 5-HT$_1$ binding sites as target for a novel putative anxiolytic. *Naunyn Schmiedebergs Arch. Pharmacol.* **38**, 467–470.

Eison, M. S., Taylor, D. P., Riblet, L. A., New, J. S., Temple, D. L., Jr., and Yevich, J. P. (1982) MJ 13805-1: A potential nonbenzodiazepine anxiolytic. *Soc. Neurosci. Abst.* **8**, 470.

Ellis, K. O. and Carpenter, J. F. (1974) A comparative study of dantrolene sodium and other skeletal muscle relaxants with the Straub tail mouse. *Neuropharmacology* **13**, 211–214.

File, S. E. (1982) Chlordiazepoxide-induced ataxia, muscle relaxation and sedation in the rat: Effects of muscimol, picrotoxin and naloxone. *Pharmacol. Biochem. Behav.* **17**, 1165–1170.

File, S. E., Lister, R. G., and Nutt, D. J. (1982) The anxiogenic action of benzodiazepine antagonists. *Neuropharmacology* **21**, 1033–1037.

Food and Drug Administration (1980) *FDA Consumer* **13**, 21–23.

Gehlbach, G. and VanderMaelen, C. P. (1985) The non-benzodiazepine antidepressant-anxiolytic candidate, gepirone, inhibits serotonergic dorsal raphe neurons in the rat brain slice. *Soc. Neurosci. Abst.* **11**, 186.

Geller, I. and Seifter, J. (1960) The effects of chlordiazepoxide and chlorpromazine on a punishment discrimination. *Psychopharmacologia* (Berl.) **3**, 374–385.

Goldberg, H. L. and Finnerty, R. J. (1979) The comparative efficacy of buspirone and diazepam in the treatment of anxiety. *Am. J. Psychiat.* **136**, 1184–1187.

Goldberg, M. E., Salama, A. I., Patel, J. B., and Malick, J. B. (1983) Novel nonbenzodiazepine anxiolytics. *Neuropharmacology* **22**, 1499–1504.

Goldstein, J. M., Knobloch, L. C., and Malick, J. B. (1983) GABAmimetic properties of anxiolytic drugs. *Life Sci.* **32**, 613–616.

Griffiths, R. R., Lamb, R. J., Ator, N. A., Roache, J. D., and Brady, J. V. (1985) Relative abuse liability of triazolam: Experimental assessment in animals and humans. *Neurosci. Biobehav. Rev.* **9**, 133–151.

Hanson, H. M., Witoslawaski, J. J., and Campbell, E. H. (1967) Drug effects in squirrel monkeys trained on a multiple schedule with a punishment contingency. *J. Exp. Anal. Behav.* **10**, 565–569.

Hunkeler, W., Mohler, H., Pieri, L., Polc, P., Bonetti, E. P., Cumin, R., Schaffner, R., and Haefely, W. (1981) Selective antagonists of benzodiazepines. *Nature* **290**, 514–516.

Irwin, S. (1967) Anti-neurotics: Practical Pharmacology of Sedative-Hypnotic and Minor Tranquilizers, in *Psychopharmacology: A Review of Progress 1957–1967* (Efron, D. H., ed.) US Public Health Service publication no. 1836, Bethesda, Maryland.

King, E. E. and Unna, K. R. (1954) The action of mephenesin and other interneuron depressants on the brain stem. *J. Pharmacol. Exp. Ther.* **111**, 293–301.

Kinnard, W. T. and Carr, C. T. (1957) A preliminary procedure for the evaluation of central nervous system depressants. *J. Pharmacol. Exp. Ther.* **121**, 354–361.

Knowles, W. and Ukena, T. (1973) The effects of chlorpromazine, pentobarbital, chlordiazepoxide and *d*-amphetamines on rates of licking in the rat. *J. Pharmacol. Exp. Ther.* **184**, 385–397.

Lader, M. and Petursson, H. (1983) Abuse Liability of Anxiolytics, in *Anxiolytics: Neurochemical, Behavioral and Clinical Perspectives* (Malick, J. B., Enna, S. J., and Yamamura, H. I., eds.) Raven, New York.

Lippa, A. S., Coupet, J., Greenblatt, E. N., Klepner, C. A., and Beer, B. (1979a) A synthetic non-benzodiazepine ligand for benzodiazepine receptors: A probe for investigating neuronal substrates of anxiety. *Pharamcol. Biochem. Behav.* **11**, 99–106.

Lippa, A. S., Nash, P. A., and Greenblatt, E. N. (1979b) Preclinical Neuropsychopharmacological Testing Procedures for Anxiolytic Drugs. in *Industrial Pharmacology, Vol. 3: Anxiolytics* (Fielding, S. and Lal, H., eds.) Futura, Mount Kisco, New York.

Litchfield, J. F., Jr. and Wilcoxon, F. (1949) A simplified method of evaluating dose–effect experiments. *J. Pharmacol. Exp. Ther.* **96**, 99–115.

Malick, J. B. (1976) Pharmacological antagonism of mouse-killing behavior in the olfactory bulb lesion-induced killer rat. *Aggressive Behav.* **2**, 123–130.

Malick, J. B. (1985) Audiogenic seizure susceptible (AGS) mice: Utility as a screen for anti-epileptic agents. *Fed. Proc.* **44**, 1106.

Malick, J. B. and Enna, S. J. (1979) Comparative effects of benzodiazepines and non-benzodiazepine anxiolytics on biochemical and behavioral tests predictive of anxiolytic activity. *Commun. Psychopharmacol.* **3**, 245–252.

Malick, J. B. and Kubena, R. K. (1979) Benzodiazepines: Treatment of Acute Alcohol Withdrawal and Long-Term Management of Alcoholics, in *Industrial Pharmacology* vol. 3 *Anxiolytics* (Fielding, S. and Lal, H., eds.) Futura, Mount Kisco, New York.

Malick, J. B., Patel, J. B., Salama, A. I., Meiners, B. A., Giles, R. E., and Goldberg, M. E. (1984) Tracazolate: A novel nonsedative anxiolytic. *Drug Develop. Res.* **4**, 61–73.

Malick, J. B., Patton, S. P., and Patel, J. B. (1986) Dissimilar characteristics of the benzodiazepine antagonist profiles of CGS 9895 and CGS 9896. *Pharmacologist* **28**, 112.

Niemegeers, C. J. E. and Lewi, P. J. (1979) The Anticonvulsant Properties of Anxiolytic Agents, in *Industrial Pharmacology* vol. 3 *Anxiolytics* (Fielding, S. and Lal, H., eds.) Futura, Mount Kisco, New York.

Niki, H. (1965) Chlordiazepoxide and food intake in the rat. *Jap. Psychol. Res.* **7**, 80–85.

Novak, G. D. (1982) Studies on the efficacy and depressant potential of muscle relaxants in mice. *Drug Develop. Res.* **2**, 383–386.

Novak, G. D. and Zwolshen, J. M. (1983) Predictive value of muscle relaxant models in rats and cats. *J. Pharmacol. Meth.* **10**, 175–183.

Overweg, J. and Binnie, C. D. (1983) Benzodiazepines in Neurological Disorders, in *The Benzodiazepines: From Molecular Biology to Clinical Practice* (Costa, E., ed.) Raven, New York.

Palva, E. S. and Linnoilia, M. (1978) Effect of active metabolites of chlordiazepoxide and diazepam, alone and in combination with alcohol, on psychomotor skills related to driving. *Eur. J. Pharmacol.* **13**, 345–350.

Patel, J. B. and Malick, J. B. (1980a) Lack of tolerance to the potentiation of ethanol-induced depression by benzodiazepines following subacute administration in mice. *Fed. Proc.* **39**, 318.

Patel, J. B. and Malick, J. B. (1980b) Effects of isoproterenol and chlordiazepoxide on drinking and conflict behavior in rats. *Pharmacol. Biochem. Behav.* **12**, 819–821.

Patel, J. B. and Malick, J. B. (1982) Pharmacological properties of tracazolate: A new non-benzodiazepine anxiolytic agent. *Eur. J. Pharmacol.* **78**, 323–333.

Patel, J. B. and Malick, J. B. (1983) Neuropharmacological Profile of an Anxiolytic, in *Anxiolytics: Neurochemical, Behavioral and Clinical Perspectives* (Malick, J. B., Enna, S. J., and Yamamura, H. I., eds.) Raven, New York.

Patel, J. B. and Malick, J. B. (1984) Evaluation of the anxiolytic potential of selected 5-HT receptor antagonists in laboratory conflict procedures. *Soc. Neurosci. Abst.* **10**, 1070.

Patel, J. B. and Migler, B. (1982) A sensitive and selective monkey conflict test. *Pharmacol. Biochem. Behav.* **17**, 645–649.

Patel, J. B., Nelson, L. R., and Malick, J. B. (1980) Effects of selected psychoactive agents on bicuculline-induced convulsions in mice. *Brain Res. Bull.* **5** (suppl. 2), 639–642.

Patel, J. B., Martin, C., and Malick, J. B. (1982) Differential antagonism of the anticonflict effects of typical and atypical anxiolytics. *Eur. J. Pharmacol.* **86**, 295–298.

Patel, J. B., Stengel, J., Malick, J. B., and Enna, S. J. (1984) Neurochemical characteristics of rats distinguished as benzodiazepine responders and non-responders in a new conflict test. *Life Sci.* **34**, 2647–2653.

Patel, J. B., Ross, L. E., and Malick, J. B. (1986) Detection of anxiolytics with partial agonist and mixed agonist/antagonist profiles in a battery of neuropharmacological tests. *Pharmacologist* **28**, 112.

Peroutka, S. J. (1985) Selective interaction of novel anxiolytics with 5-hydroxytryptamine$_{1A}$ receptors. *Biol. Psychiatry* **20**, 971–979.

Poschel, B. P. H. (1971) A simple and specific screen for benzodiazepine-like drugs. *Psychopharmacologia* (Berl.) **19**, 193–198.

Poschel, B. P. H., McCarthy, D. A., Chen, G., and Ensor, C. R. (1974) Pyrazopon (CI-683): A new antianxiety agent. *Psychopharmacologia* (Berl.) **35**, 257–271.

Randall, L. O. and Schallek, W. (1967) Pharmacological Activity of Certain Benzodiazepines, in *Psychopharmacology: A Review of Progress 1957–1967* (Efron, D. H., ed.) US Public Health Service publication no. 1836, Bethesda, Maryland.

Riblet, L. A., Taylor, D. P., Eison, M. S., and Stanton, H. P. (1982) Pharmacology and neurochemistry of buspirone. *J. Clin. Psychiatry* **43**, 11–16.

Sathananthan, G. L., Sanghvi, I., Phillips, N., and Gershon, S. (1975) MJ 9022: Correlation between neuroleptic potential and stereotypy. *Curr. Ther. Res.* **18**, 701–705.

Schwob, J. and Johnson, A. (1975) Evidence for involvement of renin-angiotensin system in isoproterenol dipsogenesis. *Soc. Neurosci. Abst.* **1**, 467.

Sepinwall, J., Grodsky, F. S., and Cook, L. (1978) Conflict behaviors in the squirrel monkey: Effects of chlordiazepoxide, diazepam and *N*-desmethyldiazepam. *J. Pharmacol. Exp. Ther.* **204**, 88–102.

Sprouse, J. S. and Aghajanian, G. K. (1985) Serotonergic dorsal raphe neurons: Electrophysiological responses in rats to 5-HT_{1A} and 5-HT_{1B} receptor subtype ligands. *Soc. Neurosci. Abst.* **11**, 47.

Stein, L. and Berger, B. (1971) Psychopharmacology of 7-chlor-5-(*O*-chlorophenyl)-1,3-dihydro-3-hydroxy-2H-1,4-benzodiazepine-2-one (lorazepam) in squirrel monkey and rat. *Arzneimittelforsch.* **21**, 1073–1078.

Stern, W. C. (1979) Sleep and Anxiolytics, in *Industrial Pharmacology* vol. 3 *Anxiolytics* (Fielding, S. and Lal, H., eds.) Futura, Mount Kisco, New York.

Stone, W. E. and Larid, M. J. (1978) Benzodiazepines and phenobarbital as antagonists of dissimilar chemical convulsants. *Epilepsia* **19**, 361–368.

Sullivan, J. W., Keim, K. L., and Sepinwall, J. (1983) A preclinical evaluation of buspirone in neuropharmacologic, EEG, and anticonflict test procedures. *Soc. Neurosci. Abst.* **9**, 434.

Traber, J., Davies, M. A., Dompert, W. U., Glaser, T., Schuurman, T., and Seidel, P. R. (1984) Brain serotonin receptors as a target for the putative anxiolytiic TVX Q 7821. *Brain Res. Bull.* **12**, 741–744.

Turnbull, M. J., Watkins, J. W., and Wheeler, H. (1981) Demonstration of withdrawal hyperexcitability following administration of benzodiazepines to rats and mice. *Br. J. Pharmacol.* **72**, 495.

Vogel, J., Beer, B., and Clody, D. (1971) A simple and reliable conflict procedure for testing anti-anxiety agents. *Psychopharmacologia* **21**, 1–7.

Wand, P., Kusinsky, K., and Sontag, K. H. (1973) Morphine-induced muscular rigidity in rats. *Eur. J. Pharmacol.* **24**, 189–193.

Weidler, B. and Hempelmann, G. (1983) Intravenous Use of Benzodiazepines, in *The Benzodiazepines: From Molecular Biology to Clinical Practice* (Costa, E., ed.) Raven, New York.

Willow, M. and Johnston, G. A. R. (1980) Enhancement of GABA binding by pentobarbitone. *Neurosci. Lett.* **18**, 323–327.

Zbinden, G. and Randall, L. O. (1967) Pharmacology of benzodiazepines: Laboratory and clinical considerations. *Adv. Pharmacol.* **5**, 213–291.

Biochemical Approaches for Evaluating Drug–Receptor Interactions

S. J. ENNA

1. Introduction

A productive drug screening program entails various levels of analysis. Although it is ultimately necessary to demonstrate a biological response in intact animals, in vivo tests are labor-intensive and require significant quantities of compound, limiting their utility as primary screens. More desirable are simple, inexpensive tests that allow for the rapid identification of inactive or less active compounds. This can be achieved in many instances by measuring an in vitro biochemical response related to the action of a particular drug class. Such tests require only small amounts of sample and can be used for analyzing dozens of compounds in a matter of hours. By providing the pharmacologist and chemist with information about potency, efficacy, and structure–activity characteristics, this methodology enables them to select a few lead agents that can then be subjected to more detailed examination in secondary and tertiary screens.

Many of the biochemical methods used for drug evaluation are based on the receptor theory of drug action. If, as the ancients believed, drugs act primarily by modifying humors, it would be impossible to study their effects in vitro, since pharmacological responses would be dependent upon the relative balance of bodily fluids. The modern concept of drug action evolved at the turn of this century when Ehrlich and Langley proposed that compounds interact with specific cellular constituents (receptors) (Ehrlich, 1913;

Langley, 1906). Support for this theory has accumulated over the intervening years to the point where it now appears that the vast majority of drugs operate in this manner. Receptor theory is based upon the observations that slight alterations in chemical structure modify pharmacological responses, and that drugs influence specific organs and tissues. For example, stereoisomers of the same compound often differ dramatically in regard to potency. Moreover, structural modifications can sometimes completely eliminate efficacy, yielding an agent that antagonizes the effect of the parent compound. Indeed, the existence of specific antagonists provides the most compelling evidence to support the receptor concept of drug action.

Receptors also account for the fact that drugs are organ- or tissue-selective. Atropine is a potent antagonist of the negative chronotropic effect of acetylcholine, but is relatively inactive in blocking cholinergic transmission at the myoneural junction. Conversely, curare inhibits the skeletal muscle response to acetylcholine, but has little effect on the action of this transmitter in cardiac tissue. Such findings are consistent with the notion that drug responses are mediated by a selective interaction with specialized components of the cell.

Any cellular constituent can serve as a drug receptor. This includes inorganic ions, enzymes, and structural elements. Inasmuch as some of the more useful drugs activate or inhibit hormone or neurotransmitter receptors, major efforts have been made to develop methods for examining these sites. The aim of this chapter is to describe two of the more common in vitro approaches for evaluating drug–receptor interactions. Both theoretical and technical aspects will be discussed and comments made with regard to the strengths and weaknesses of these assays as tools for drug evaluation. Readers desiring more detailed descriptions of individual topics are urged to consult other sources (Goldstein et al., 1974; Yamamura et al., 1985; Bylund, 1980; Enna and Strada, 1983; Shulster and Levitzki, 1980; Berridge and Irvine, 1984; Williams and Enna, 1986).

2. Neurotransmitter Receptor Binding

The first step in a receptor-mediated event is attachment of the ligand (e. g., neurotransmitter, hormone, or drug) to the receptor site. A ligand may be defined as any substance that forms a complex with another agent. The neurotransmitter norepinephrine is a ligand that attaches to adrenergic receptors on the surface of plasma membranes. This coupling initiates a series of events that culminate in a modification of cellular activity (Lefkowitz and Hoffman, 1980). Propranolol, an adrenergic antagonist, binds to the same receptor as norepinphrine, but is incapable of initiating a cellular response and, by its presence, interfers with the attachment of the catecholamine. That portion of the receptor to which the ligand attaches is referred to as the recognition site. This component imparts selectivity to the receptor by recognizing only a specific chemical structure or configuration. Assays have been developed for studying ligand attachment (binding) in vitro to assess the ability of a test compound to interact with a particular receptor.

The principle of the binding assay is that the number and kinetic properties of receptor recognition sites can be quantified by analyzing radioligand attachment to these components. Such assays are useful for drug evaluation since nonradioactive agonists or antagonists selectively interfer with radioligand binding to a receptor. Although the theoretical basis of these assays was known for years, the method has been generally applicable for only the past decade because of technical difficulties associated with labeling specific receptor sites. Most troublesome is the fact that receptor densities are quite low, and therefore only a small amount of radioligand adheres to these sites, making them difficult to quantify. Moreover, ligands nonselectively bind to a variety of membrane constituents, making it necessary to distinguish this nonselective attachment from that associated with the receptor of interest. These problems were overcome by developing radioligands with high specific activities (>10 Ci/mmol), which made it possible to accurately quantify the small quantity of compound bound to receptor recognition sites and to minimize nonspecific binding. A number of tritium-labeled substances are now available for use as radioligands, and therefore binding assays have been developed for a variety of neurotransmitter and hormone receptors (Bennett, 1978; Williams and U'Prichard, 1984).

2.1. Criteria for Identification of a Receptor Binding Site

In a typical neurotransmitter receptor binding assay, a membrane fraction is prepared from tissue known to possess the receptor of interest. For example, if β-adrenergic receptors are to be studied, brain, heart, or lung membranes could be utilized. Although the tissue preparation varies somewhat among different binding assays, the basic approach is quite similar. The tissue is homogenized in water or a nonosmotic buffer and the sample centrifuged to separate soluble constituents from the particulate fraction, which includes the membrane fragments. Resuspension and centrifugation of the membranes is usually repeated several times to rid the tissue of endogenous substances that may interfer with radioligand attachment. For some tissue, such as brain, a subcellular fractionation is sometimes performed initially to obtain a sample enriched in synaptosomal fragments. This fraction is then lysed and rinsed prior to assay. In other circumstances it may be necessary to treat the sample to eliminate unwanted cellular constituents such as connective tissue. In any case, the tissue sample used for analysis is always a thoroughly washed membrane suspension.

Portions of this suspension (0.25–1.0 mg protein) are added to a series of incubation tubes containing a measured concentration (~1–10 n*M*) of radioligand. Some of the assay tubes also contain a relatively high concentration (~0.1–100 μM) of an unlabeled ligand known to selectively interact with the receptor of interest. In a drug screening program there would be additional tubes containing radioligand, tissue, and known quantities of the test substance. The final volume of the incubation mixture seldom exceeds 2 mL. The reaction is initiated with the addition of tissue and the binding allowed to proceed to equilibrium at a predetermined temperature. The assay is normally terminated either by centrifuging the samples or by pouring them over glass fiber filters maintained under reduced pressure. These maneuvers separate the membrane fragments from unbound radioligand. The tissue is rinsed rapidly with cold buffer and the bound radioactivity extracted into an appropriate scintillation cocktail and quantified by liquid scintillation spectrometry. Total binding is defined as that which occurs when only tissue and radioligand are incubated together, whereas nonspecific binding is the radioactivity remaining when unlabeled ligand is also present. The difference between these values is referred to as the saturable or specific binding component, which should represent the amount of radioligand attached to the receptor recognition site. These designations are based upon the premise that the radioligand binds to a limited number of specific sites and that this attachment is competitively inhibited by the unlabeled species. Because the amount of nonspecific binding is virtually limitless, the unlabeled ligand does not compete for these sites. For some receptor binding assays (e. g., cholinergic muscarinic, benzodiazepine, and $GABA_A$), displaceable or specific binding can represent greater than 80% of the total binding under the proper incubation conditions. Displaceable binding, however, may be at best 10–20% of the total in other cases. The latter are generally of limited utility in drug screening programs, since the data generated are less precise.

Although displaceable binding is the *sine qua non* of ligand binding assays, it is insufficient to prove that the radioligand attaches to the receptor of interest (Burt, 1985; Hollenberg and Cuatrecasas, 1979; Laduron, 1984). To this end it must be ascertained whether the displaceable component meets certain requirements. The first of these is that the binding must be saturable within a physiologically or pharmacologically relevant range of radioligand concentrations (Fig. 1). As the concentration of radioligand is elevated, total binding rises and the nonspecific component increases linearally, reflecting the fact that the number of nonspecific ''sites'' is virtually limitless. Specific binding increases over a narrow concentration range, however, and then plateaus (saturates). Such data

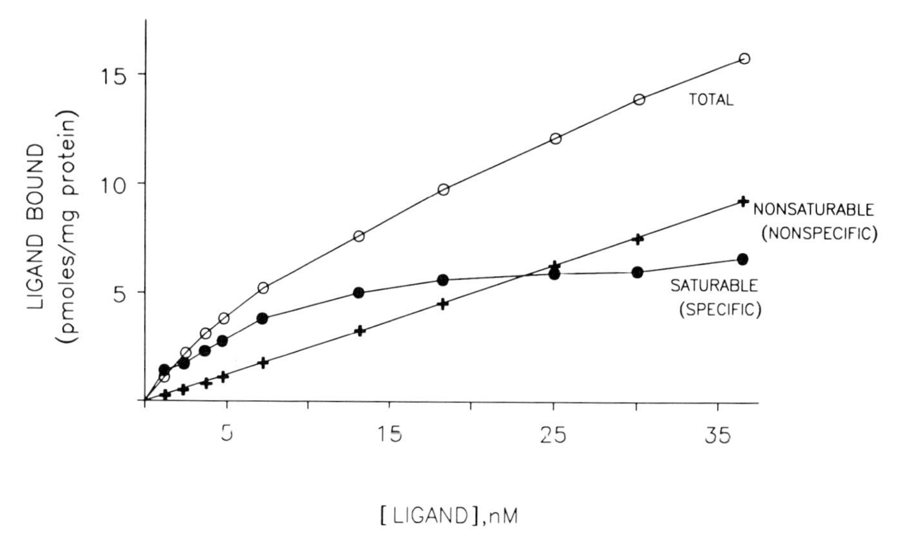

Fig. 1. Radioligand binding site saturation data. Total binding represents the amount of radioactive ligand bound in the absence of unlabeled ligand, whereas nonsaturable (nonspecific) binding is that remaining in the presence of the unlabeled species. Saturable (specific) binding is the calculated difference between the total and nonsaturable values.

indicate that the displaceable or specific binding component is saturable at low ligand concentrations, which is characteristic of attachment to a biologically relevant receptor.

It must also be demonstrated that displaceable binding is present only in those tissues or tissue fractions known to possess the receptor of interest (tissue specificity). For example, GABA receptor binding should be abundant in central nervous system tissue, but absent in heart and kidney—organs devoid of GABAergic neurons. Localization of the specific binding to tissues known to possess this particular receptor is crucial since saturable binding can sometimes be detected with inert materials (Cuatrecasas and Hollenberg, 1975). Specific binding should also be most enriched in those subcellular fractions known to contain the highest concentration of receptor sites. In addition, because neurotransmitter and hormone receptors are proteins, displaceable binding should be destroyed by heat denaturation.

Displaceable binding must also be substrate-specific (Table 1). If the binding represents attachment to a particular receptor, then only those substances known to be active at this site should be

Table 1
Substrate Selectivity of ^{3}H-GABA Binding to Rat Brain Membranes[a]

Compound	Inhibition of ^{3}H-GABA binding, IC_{50}, n*M*
GABA	20
Muscimol	3
trans-3-Aminocyclopentane-*1*-carboxylic acid	5
cis-3-Aminocyclopentane-*1*-carboxylic acid	3,000
(+)-Bicuculline	5,000
(−)-Bicuculline	200,000
2,4-Diaminobutyric acid	>1,000,000

[a]Adapted from Enna and Snyder (1977).

capable of inhibiting radioligand binding to the membrane fraction. As illustrated for $GABA_A$ receptor binding, GABA inhibits 50% of the specific binding (IC_{50}) at a concentration of 20 n*M*. The known $GABA_A$ receptor agonists muscimol and *trans*-3-aminocyclopentane-1-carboxylic acid have IC_{50} values of 3 and 5 n*M*, respectively, indicating that they may be more potent than GABA at this site. Importantly, the *cis*-isomer of 3-aminocyclopentane-1-carboxylic acid is almost 1000-fold less active than the *trans* form in the binding assay and in electrophysiological tests, demonstrating stereoselectivity for the site. Likewise, (+)-bicuculline, a GABA receptor antagonist, is substantially more potent than (−)-bicuculline in inhibiting ^{3}H-GABA binding in vitro and as a convulsant. 2,4-Diaminobutyric acid, a GABA analog known to be inactive in $GABA_A$ receptors, is inactive in the binding assay. Such data suggest that, under these incubation conditions, ^{3}H-GABA attaches to a physiologically and pharmacologically relevant receptor site for this neurotransmitter (Enna and Snyder, 1975; Enna and Snyder, 1977).

It is noteworthy that (+)-bicuculline is several hundredfold weaker than GABA in the binding assay, even though it is known to be a potent antagonist in vivo. In general, agonists as a group are more potent inhibitors of radiolabeled agonist attachment, whereas antagonists are more potent as a group when the radioligand is an antagonist. This suggests that although antagonists appear to competitively interact with the receptor recognition site, they may be attaching to a different portion or configuration of the receptor molecule than agonists (Snyder, 1975). This concept is important in a drug screening program, since a compound that may appear to be relatively weak in an agonist binding assay may be a potent antagonist. The converse is true if an antagonist ligand is used to label the receptor.

Although fulfillment of these criteria is not absolute proof that the binding component is a biologically relevant receptor, it is a minimum requirement for suggesting that a particular assay may be appropriate for studying this site. It should be borne in mind that the stability of the radioligand, incubation time, temperature, and buffer can all affect receptor site selectivity (Enna, 1984a). If for some reason these must be altered to accomplish a certain task, the investigator must reestablish the above criteria before utilizing the assay in a screening program.

2.2. Analysis of Binding Data

Ligand binding assays reveal two important characteristics of receptors; their density (B_{max}) and affinity (K_d). The latter is a measure of the degree of attraction between the receptor and the ligand. These properties are difficult to quantify when the data are plotted as in Fig. 1, since the relationship between the variables is nonlinear. A rough approximation can be made, however, from this type of display since the B_{max} is reflected by the asymptote of the specific binding curve, and the K_d is the concentration of radioligand necessary to achieve half-maximal occupancy of the receptor sites (Bylund, 1980). In this example the B_{max} is roughly 5–7 pmol/mg protein, and the K_d is around 5 nM. A more precise way to define these parameters is to transform the data using the method of Rosenthal (1967) or Scatchard (1949). For a Scatchard plot, the data are transformed according to a general equation that linearizes the results by comparing the ratio of bound radioligand over the concentration of free to the amount bound (Fig. 2). Because it is assumed that the concentration of free ligand remains unchanged over time, it is important that the assay be conducted under conditions in which less than 10% of the radioligand is bound at equilibrium (Bennett, 1978).

Scatchard analysis generally yields one of two kinds of binding curves (Fig. 2). If there is a linear relationship between all of the

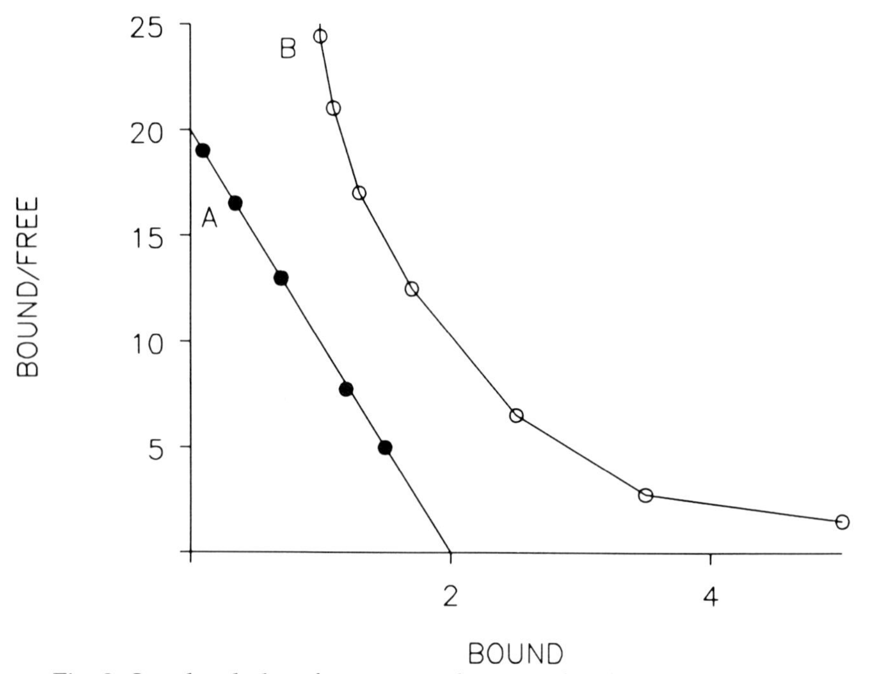

Fig. 2. Scatchard plot of saturation data. On the abscissa is represented the amount of specifically bound ligand at each concentration of free radioligand, whereas the ordinate is the ratio of the bound ligand over the concentration of free isotope. The affinity of the ligand for the receptor (K_d) is equal to the slope of the line, and the concentration of binding sites (B_{max}) is the x-intercept.

points, the results indicate a single class of noninteracting sites (Fig. 2A). On the other hand, biphasic plots suggest that the ligand is attaching to two or more sites, each of which has a different affinity for the radioligand (Fig. 2B). Although it is relatively simple to calculate a line of best fit when only a single component is present, it is more difficult to accurately calculate the values for the individual components of a more complex curve. Accordingly, these data are best calculated by computer-assisted analysis (Minneman et al., 1979; Munson and Rodbard, 1980; DeLeon et al., 1981).

The existence of multiple binding sites suggests there may be pharmacologically and functionally distinct receptors for the radioligand. This is an important principle in a drug screening program, since it may be found that a test compound interacts with only one population of sites. Such a finding would be suggested if the test agent inhibits only a fraction of the specifically bound radioligand. More typically a test compound yields a biphasic concentration–response curve indicating different affinities for the different binding constituents. In either case the results suggest that the test substance has a greater selectivity for a subpopulation of receptors. Thus, Scatchard analysis is crucial for defining the basic characteristics of a receptor recognition site and provides important information as to the possible existence of multiple binding components.

In many instances the rate at which a radioligand attaches to an available recognition site is unrelated to the number of receptors already occupied. In some cases, however, the rate of ligand binding increases with increasing receptor occupancy. This phenomenon is referred to as positive cooperativity, since receptor occupancy increases the affinity of unoccupied sites. Conversely, negative cooperativity occurs when receptor occupancy lowers the affinity of the remaining sites, thereby decreasing ligand attachment as receptor saturation is approached. Such interactions are determined graphically by use of a Hill (1910) plot (Fig. 3). If the slope of the line, or Hill coefficient (n_H), is approximately 1.0 (0.8–1.2), it may be concluded that there is no interaction between occupied and unoccupied receptors. In other words, a Hill coefficient of 1.0 suggests that the rate at which the ligand binds to an unoccupied receptor is unrelated to the extent of receptor occupancy. If n_H is >1.2 or <0.8, this would suggest positive or negative cooperativity, respectivity (Fig. 3). Hill plots can be constructed from data obtained when examining the effect of a test agent on radioligand binding. Although a Hill coefficient of <1.0 can result from technical artifacts (Cuatrecasas and Hollenberg, 1975), it may also indicate

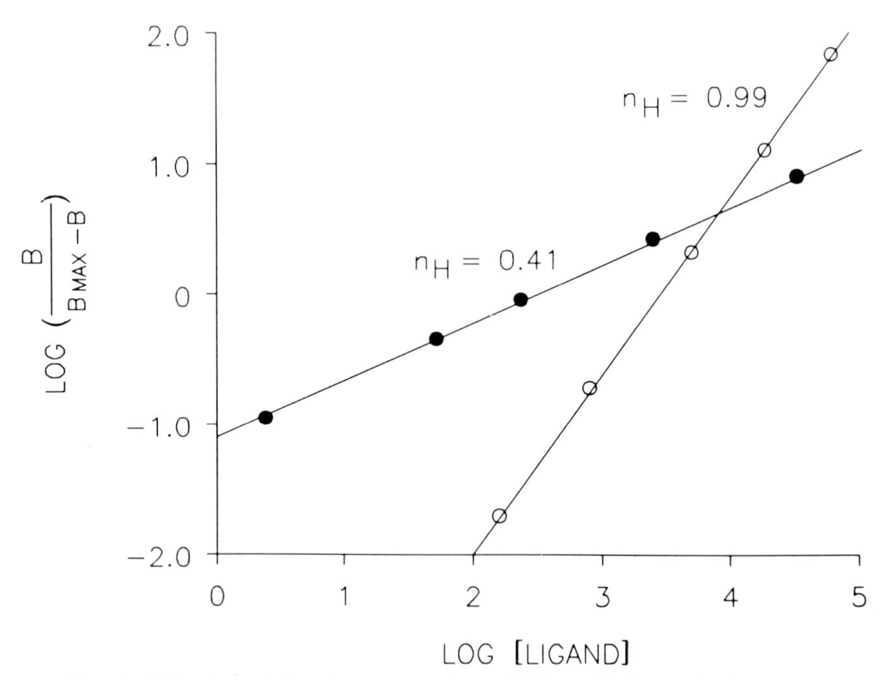

Fig. 3. Hill plot of displacement data. Hill coefficients (n_H) near unity indicate a single class of noninteracting sites, whereas an n_H greater or less than 1.0 suggests that the ligand is binding to sites having multiple affinities.

that the labeled ligand binds to multiple sites and that the test compound has a greater affinity for one site over the other. Such a finding may be particularly useful in a drug screening program. For example, it is conceivable that Scatchard analysis might suggest that a radioligand binds to a single class of noninteracting sites, although it actually binds to two or more sites with equal affinity. When the radioligand binding is challenged with a compound having different affinities for these sites, however, the inhibition data would yield a Hill coefficient of <1.0. Such findings may be taken as evidence that the test compound is more selective for a subgroup of receptors.

The most routine use of binding assays in a screening program is to estimate the relative potencies of a series of test compounds as inhibitors of ligand binding. For these studies tissue is incubated with the radioligand and one of various concentrations of the test agents to determine the percent displacement of specifically bound isotope. By plotting these data on logit paper (Rodbard and Frazer,

1975), it is possible to generate a straight line and estimate the concentration of test compound necessary to inhibit 50% of the specifically bound radioligand (IC_{50}) (Fig. 4). Although IC_{50} values are useful for comparing relative potencies within an experiment, they are less accurate when attempting to compare potencies between assays. The reason for this is that the IC_{50} is a function of radioligand concentration with respect to K_d. A more absolute determination of relative potencies is made by calculating the inhibition constant (K_i). The K_i value is calculated according to the following equation:

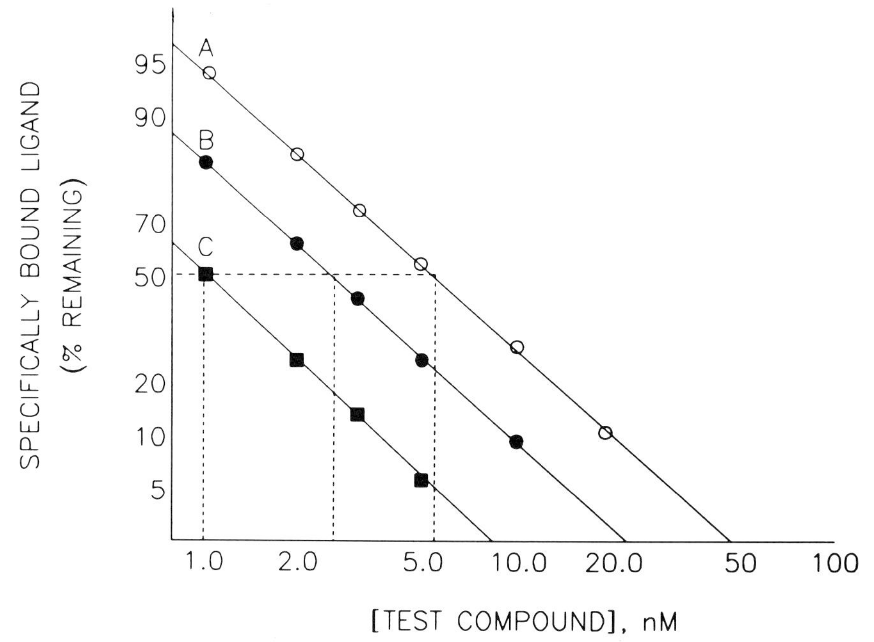

Fig. 4. The ability of compounds A, B, and C to inhibit specifically bound radioligand attachment was determined by examining various concentrations of each in the binding assay. The precent displacement is plotted against the concentration of each compound using logit paper. The concentration of the test compound that inhibited 50% of the specifically bound radioligand (IC_{50}) can be estimated directly from this type of display (stippled lines).

$$K_i = \frac{IC_{50}}{1 + F/K_d}$$

If the concentration of free ligand (F) is equal to the K_d, then the denominator is 2 and the K_i = 1/2 the IC_{50}.

Not only do inhibition studies reveal the IC_{50} or K_i of a compound, they also provide information as to the type of interaction that exists between the labeled and unlabeled substances (Fig. 4). In this theoretical display, compound A is the least potent of the three agents, with an IC_{50} of approximately 5.5 n*M*. Compound B is more active, having an IC_{50} of 2.7 n*M*, and compound C the most active with an IC_{50} of 1.0 n*M*. If the Hill coefficients derived from these data are 1.0, then it can be concluded that the compounds are interacting with the binding site in a competitive manner. A Hill coefficient that differs significantly from 1.0, however, would suggest either a noncompetitive interaction or a competitive interaction at multiple sites.

Concrete evidence that an interaction with a particular binding site may be related to the pharmacological activity of a compound can be provided by conducting correlation studies (Creese et al., 1976). This analysis is usually performed later in the drug development process, since it requires data with regard to the biological activity of the test compounds (Fig. 5). For this evaluation, a log–log plot is constructed comparing the dose of each drug necessary to elicit a half-maximal response (ED_{50}) against its K_i value in the receptor binding assay. If the line of best fit yields a slope of 1.0, it can be concluded that there is a correlation between the two variables. Because correlations do not prove causality, however, such data can never be taken as an absolute indicator that the drug–receptor interaction is responsible for a particular response. Rather, these comparisons should be used only as additional evidence that such a relationship may exist.

2.3. Drug–Binding Site Interactions

A drug can interact with a neurotransmitter or hormone receptor recognition site in several ways. A competitive agent binds to the same site as the radioligand and therefore directly competes for attachment to the receptor. Such an interaction is usually the most desirable for drug candidates since it indicates selectivity of action. There are a number of ways to determine whether a compound inhibits binding in a competitive manner. One approach is to perform a saturation experiment with the radioligand in the presence of a fixed (IC_{50}) concentration of test compound. A Scatchard plot yielding a K_d significantly greater (lower affinity) than that obtained in the absence of the test agent, but with the same B_{max}, would be indicative of a competitive interaction. This reflects

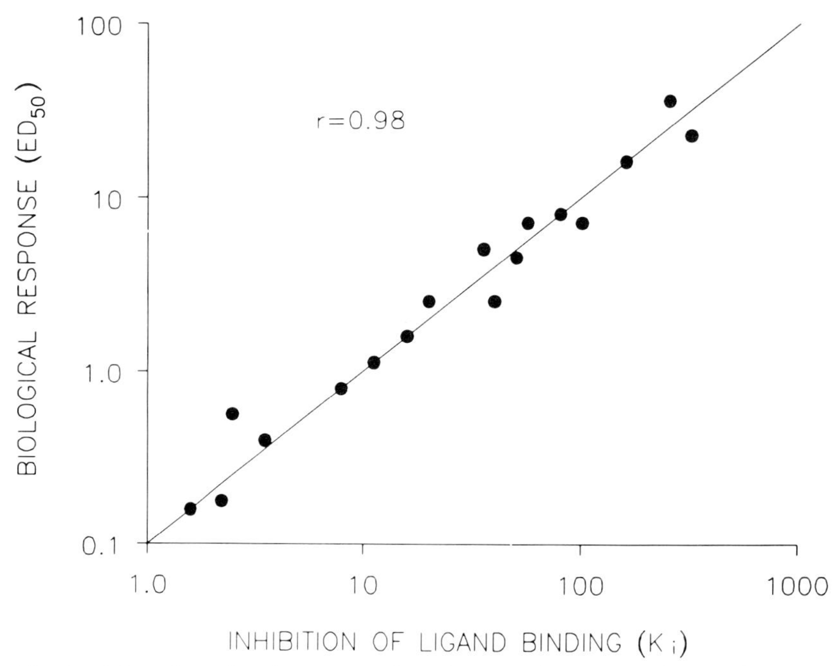

Fig. 5. A log–log plot for estimating the correlation coefficient between two varaibles. The points represent different drugs, each of which was tested in both a biological and receptor binding assay. The dose of the drug necessary to provoke a 50% response in the biological test (ED_{50}) was plotted against the concentration of drug necessary to inhibit 50% of the specifically bound radioligand (K_i). The correlation coefficient (r) is the slope of the line of best fit.

the fact that a competitive substance lowers the apparent affinity of the receptor for the radioligand since the radiolabeled species must vie with the unlabeled compound for the same binding site.

A noncompetitive interaction is normally less interesting from a pharmacological standpoint, since the drug probably does not interact specifically with this particular binding site. A noncompetitive agent normally has no effect on the apparent receptor affinity for the radioligand, but rather alters the number of binding sites (B_{max}). In general, substances that act in a noncompetitive manner are poor drug candidates, since nonselective interactions can lead to a host of unpredictable consequences.

A third possibility is a noncompetitive, but selective, interaction between the test compound and the receptor recognition site.

The interaction between GABA and benzodiazepine binding sites exemplifies this type of relationship (Fig. 6). Unlabeled muscimol, a $GABA_A$ receptor agonist, enhances ^{3}H-benzodiazepine binding to rat brain membranes (Gallagher et al., 1978). As the concentra-

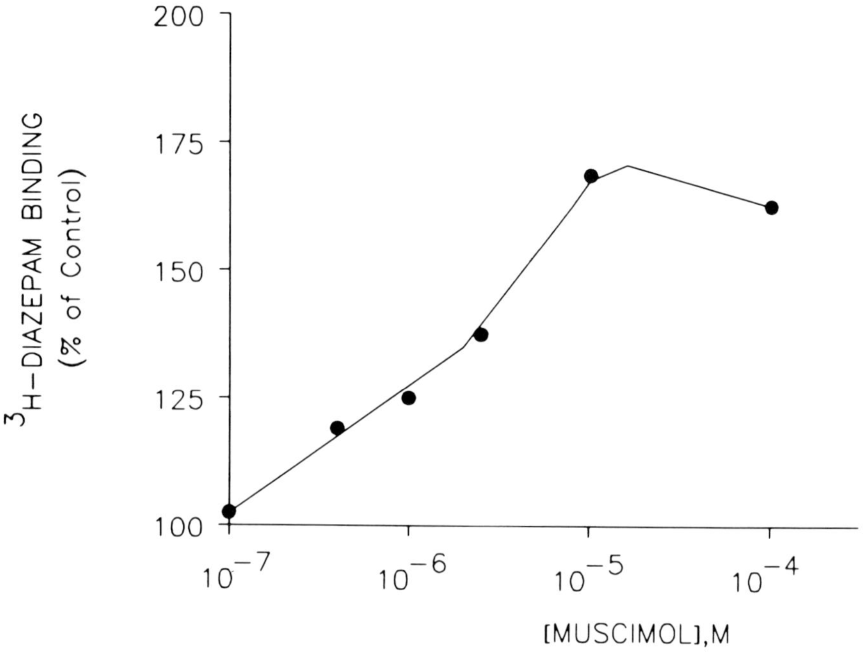

Fig. 6. Effect of muscimol on ^{3}H-diazepam binding to rat brain membranes. Each point represents the amount of specifically bound ^{3}H-diazepam in the presence of various concentrations of unlabeled muscimol (adapted from Braestrup et al., 1980).

tion of the GABA agonist is increased, ^{3}H-diazepam binding increases, almost doubling at saturating concentrations of muscimol. Scatchard analysis shows that the increase in diazepam binding is caused by an increase in the affinity of the benzodiazepine site, with no change in the number of these binding components. This effect is apparently mediated by the activation of $GABA_A$ receptors by muscimol, since the enhancement is blocked by $GABA_A$ receptor antagonists. Studies have shown that the GABA and benzodiazepine binding sites are distinct, but closely associated, membrane constituents. Indeed, under the proper conditions, benzodiazepines are capable of enhancing ^{3}H-GABA binding (Guidotti et al., 1978). These data indicate that the GABA and benzodiazepine binding sites are coupled such that activation of either causes an allosteric modification in the other, resulting in a change in binding site affinity. Accordingly, it is conceivable that a screening program may yield a compound that enhances rather than inhibits radioligand attachment. Because such a finding may be artifactual, further experimentation should be performed to rule out technical explanations. If it can be shown that the test compound selectively facilitates binding of the radioligand by attaching to a different receptor, it may be concluded that the drug has a novel mechanism of action and may perhaps display a unique clinical profile.

3. Receptor-Mediated Responses

Although ligand binding assays can provide information about the interaction of a compound with a particular receptor recognition site, they seldom reveal anything about the efficacy of a test agent. Thus, both agonists and antagonists compete for attachment to a binding site and displacement data alone will therefore not establish whether a compound will activate or inhibit the receptor. Although it is possible to differentiate agonists from antagonists in some binding assays (Creese, 1978), many are incapable of making this distinction. Rather, compounds must be examined using tests that measure receptor function in order to determine efficacy.

A variety of receptor-related biochemical responses have been characterized. With regard to neurotransmitters, receptor activation initiates two types of change, either of which can be studied in vitro to measure receptor function (Enna and Strada, 1983). In some cases receptor activation modifies membrane permeability for a particular inorganic ion. Accordingly, ion flux experiments can be used to assess the functional capacity of receptors and the intrinsic activity of a suspected ligand (Thampy and Barnes, 1984). For the most part such assays have not yet been developed to the point where they can be used as a drug screening device.

The second type of response to neurotransmitter receptor activation is stimulation of membrane-associated enzyme systems that catalize the production of a second messenger. Once formed, second messengers, such as cyclic nucleotides or inositol triphosphate, influence a number of cellular processes. Unlike changes in ion flux, second messenger production is a robust biochemical response that can be readily quantified in vitro. For this reason analysis of these substances has been used as the basis for in vitro tests aimed at measuring the intrinsic activity of a test compound.

The criteria for establishing the relationship between a given second messenger system and receptor function are similar to those described for ligand binding assays (Enna and Strada, 1983). This includes saturability as well as tissue and substrate specificity of the response. By measuring the effect of a compound on enzyme activity, or on the production of a second messenger, it is theoretically possible to determine its relative efficacy at this receptor. Conversely, if the test agent is a receptor antagonist, it will inhibit the response to an agonist substance.

Data derived from these assays can be evaluated in a manner similar to that described for ligand binding procedures. Thus, the potency of a ligand and the maximal response can be determined by performing saturation studies. EC_{50} and IC_{50} values can be determined by a logit analysis of activation or inhibition data as described for binding assays (Fig. 4). Data interpretation is less straightforward than for a ligand binding assay, however, given the number of factors associated with a receptor-coupled event. Nevertheless, functional assays are important in a drug screening program when used in conjunction with ligand binding assays since the results complement one another and provide significantly greater insights than either test alone.

3.1. Cyclic AMP

One of the most popular second messenger systems used for studying receptor function is that associated with the enzyme adenylate cyclase (Perkins, 1975; Drummond, 1984) (Fig. 7). Adenylate cyclase may be coupled to both neurotransmitter and hormone receptors, depending upon the tissue and animal species. Receptor activation facilitates the coupling between the recognition site and a membrane-bound regulatory protein (N protein). The attachment of N protein to the recognition site increases the binding of guanosine triphosphate (GTP) to the N protein. The N protein–GTP complex in turn activates adenylate cyclase (C). GTP binding to N protein also reduces the affinity of the recognition site for the agonist ligand. This latter effect facilitates the dissociation of the neurotransmitter or hormone from the binding site, terminating receptor activation. Adenylate cyclase catalyzes the conversion of ATP to cyclic AMP, the second messenger. Cyclic AMP is metabolized to 5′ AMP by the enzyme phosphodiesterase. A number of

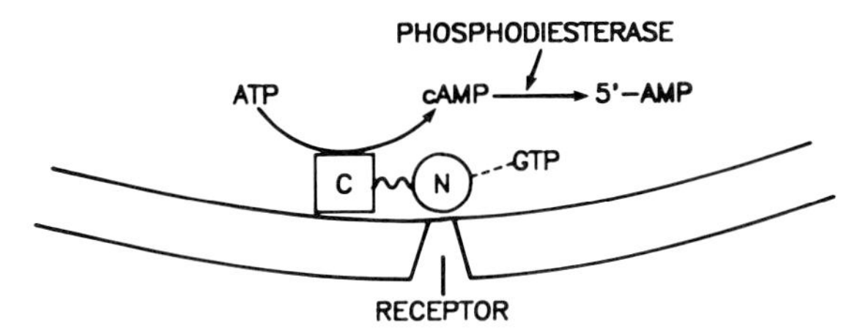

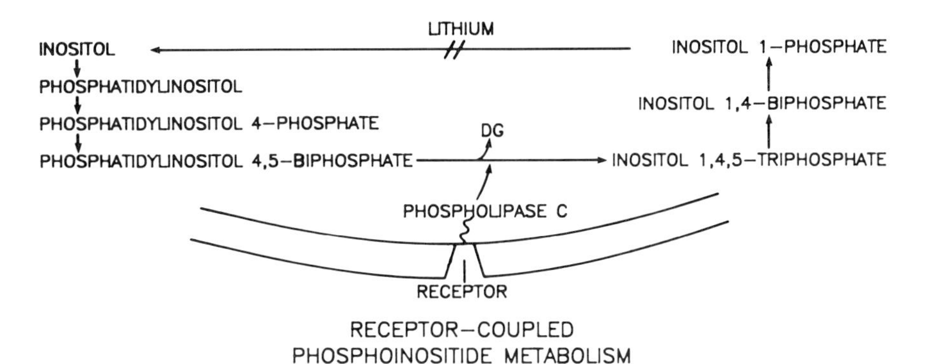

Fig. 7. Schematic representation of the cellular constituents and reactions associated with receptor-coupled stimulation of cyclic AMP and inositol phosphate production.

endogenous substances are known to stimulate adenylate cyclase activity through an interaction with a specific receptor site. This includes norepinephrine, serotonin, histamine, adenosine, vasoactive intestinal peptide, and insulin. These responses are tissue-specific and, in some cases, species-specific as well. Therefore it is possible to examine the functional activity of these receptors by measuring adenylate cyclase activity or cyclic AMP accumulation in vitro.

Cyclic AMP-associated receptor responses may be measured in two ways. In certain circumstances it is possible to quantify adenylate cyclase activity in membrane fractions (Salomon et al., 1974). With this approach, membrane fragments are prepared and incubated with radiolabeled ATP in the presence and absence of a receptor agonist. Activation of the receptor increases enzyme activity, which is reflected by an increase in the production of radiolabeled cyclic AMP. By measuring the amount of cyclic AMP formed per unit of time, it is possible to obtain a measure of enzyme activity. A test compound that acts as an agonist will stimulate cyclic AMP formation, whereas antagonists will reduce the stimulatory action of an agonist. Many receptors, however, cannot be easily analyzed by this procedure since with some tissues, in particular brain, the transmitter response is greatly reduced in membrane fragments. In these cases it is best to study receptor activity by measuring cyclic AMP formation in intact cells (Shimizu et al., 1974). With this assay the cyclic AMP is normally measured by either radioimmunoassay or by labeling the ATP pool with ^{3}H-adenine (prelabeling technique). Using the prelabeling technique, the amount of radiolabeled cyclic AMP formed in slices of rat brain cerebral cortex in the absence of a receptor agonist (basal) represents less than 0.1% of the total pool of labeled substances in the tissue (Fig. 8). Exposure of the sample to norepinephrine results in a concentration-dependent increase in cyclic AMP production, with saturating concentrations of the catecholamine causing an eight-

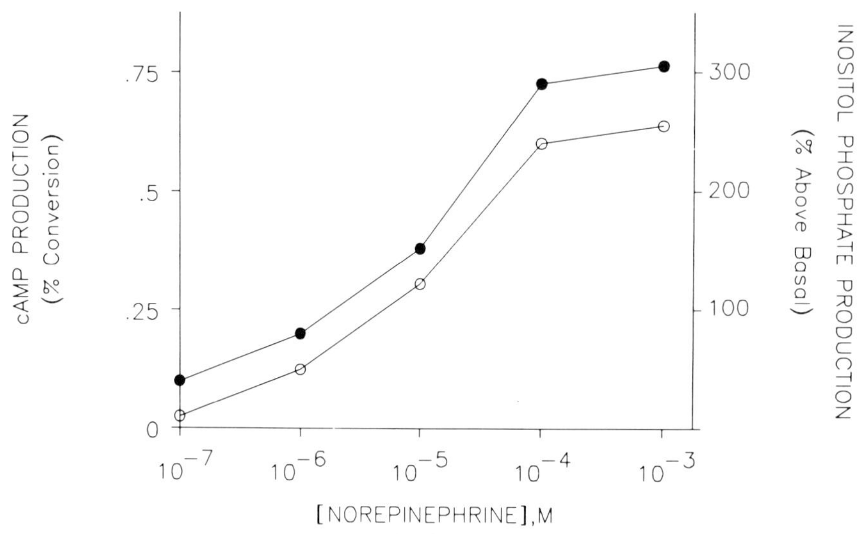

Fig. 8. The effect of norepinephrine on cyclic AMP (cAMP) and inositol phosphate production in slices of rat brain cerebral cortex. Each point represents the mean of 3–6 separate experiments. Cyclic AMP production is calculated by determining the percent of the total tritium pool represented by ^{3}H-cAMP (% conversion). Tritiated inositol phosphates are analyzed and the data expressed as the percent above that obtained in unstimulated tissue (% above basal). ●——●, cAMP production; ○··○, inositol phosphate production.

to ten-fold increase in the relative concentration of radiolabeled cyclic AMP. From these data it can be estimated that the concentration of norepinephrine necessary to induce a half-maximal stimulation (EC_{50}) of cyclic AMP production is about 10 μM. This response to norepinephrine is thought to be mediated primarily by activation of β-adrenergic receptors, although stimulation of α-receptors may contribute as well (Schwabe and Daly, 1977). The concentration–response curve for norepinephrine is shifted in a parallel manner to the right in the presence of a competitive antagonist. Noncompetitive antagonists reduce the maximal response to norepinephrine without influencing the potency of the agonist. Thus, this type of test is useful for establishing whether a test compound that is known to be active in an adrenergic receptor binding assay is an agonist or antagonist at this site. It can also be used to establish the relative potencies of the test agents.

If a compound is found to influence cyclic AMP production, it cannot automatically be concluded that it is acting through the receptor recognition site. Thus, cyclic AMP levels are determined by the interplay between several cellular constituents, including N protein, adenylate cyclase, and phosphodiesterase (Ross and Gilman, 1980). Accordingly, a test compound could increase cyclic AMP production by inhibiting phosphodiesterase or by directly stimulating adenylate cyclase. Conversely, an agent could activate phosphodiesterase, inhibit adenylate cyclase, or interfer with the coupling between the recognition site and N protein and thereby reduce the response to an agonist. Although such substances may yield data indicating noncompetitive antagonism, they could also influence the system in such a way as to indirectly modify recognition site affinity. Therefore, the results from these assays should be interpreted with care and are best evaluated in conjunction with ligand binding data. If a compound is found to inhibit hormone or neurotransmitter-stimulated cyclic AMP production, but is devoid of activity in the appropriate binding assay, it is possible that it is acting at a point beyond the receptor recognition site.

3.2. Phosphatidylinositol Metabolism

Another second messenger system used for examining neurotransmitter receptor responses is that associated with the phosphotidylinositol metabolism (Fig. 7). Inositol is converted to phosphotidylinositol, which is phosphorylated to phosphotidylinositol 4-phosphate and phosphotidylinositol 4,5-diphosphate by the action of kinases (Berridge and Irvine, 1984). Transmitter receptor stimulation leads to activation of phospholipase C, which catalyzes the conversion of phosphotidylinositol 4,5-diphosphate to inositol 1,4,5-triphosphate and diacylglycerol (DG). These latter substances are thought to act as second messengers, with inositol 1,4,5-triphosphate mobilizing intracellular calcium and diacylglycerol activating protein kinase C. The inositol phosphates are sequentially metabolized by phosphatases to inositol (Fig. 7). The conversion of inositol *1*-phosphate to inositol is inhibited by lithium ion (Hallcher and Sherman, 1980). This discovery made it possible to use this assay as a measure of receptor function, since the metabolism of this second messenger is normally too rapid to accurately analyze inositol phosphate levels. A number of putative neurotransmitters are known to stimulate phospholipase C through an interaction with plasma membrane receptors. This includes norepinephrine, acetylcholine, serotonin, vasopressin, and substance P (Berridge, 1984).

In a typical assay, tissue (intact cells) is incubated with ^{3}H-inositol, which is accumulated and converted to phosphotidyl-^{3}H-inositol 4,5-diphosphate. Exposure of the cells to an appropriate receptor agonist induces a concentration-dependent increase in the amount of ^{3}H-inositol phosphates that are extracted and quantified (Fig. 8). The EC_{50} value for norepinephrine to activate inositol phosphate production in rat brain slices is quite similar to that found for cyclic AMP production. Whereas norepinephrine-stimulated cyclic AMP accumulation appears to be related to β-adrenergic receptors, inositol phosphate production is associated with α-adrenergic sites. Therefore, if binding assays suggest that a test compound has some affinity for α_1- rather than β-adrenergic receptors, the agent should be tested as an activator or inhibitor of inositol phosphate production.

As for the cyclic AMP system, the finding that a test substance activates or inhibits inositol phosphate accumulation does not prove that it is interacting with a receptor recognition site. A drug could enhance the formation of phosphotidylinositol 4,5-diphosphate by activating the kinases that catalyze the phosphorylation reactions. On the other hand, the test compound could inhibit phospholipase C activity by modifying intracellular calcium levels, rather than through an interaction with the transmitter receptor site. Because of these and other possibilities, it is prudent to obtain receptor binding data to establish whether a test agent is acting through a plasma membrane receptor.

The phosphoinositide system may also be uniquely useful for assessing whether a compound may be effective for the treatment of manic–depressive illness. Lithium is currently the most common therapy for the treatment of this disorder (Baldessarini and Lipinski, 1975). Lithium administration is accompanied, however, by a number of side effects and toxicities, making it desirable to discover new therapeutic agents. Since it is possible that the action of lithium on inositol-*1*-phosphate metabolism may contribute to its therapeutic effect, this assay could be used to screen for compounds with similar activity. This illustrates how second messenger assays can also be useful for identifying drugs with a site of action other than a transmitter receptor.

4. Drug Screening

Because in vitro receptor assays are technically simple and require only small quantities of test compound, it is possible to analyze a large number of agents for their potential as receptor-active drugs. Both binding and functional assays have been developed for the majority of neurotransmitter and hormone receptors. However, these assays should not be considered as two different means to the same end, since the data complement one another. That is, ligand binding assays reveal little about efficacy, whereas functional analysis does not necessarily identify the site of action.

A drug screening program designed around these methods should entail an initial evaluation with the binding assays. If the aim is to develop new adrenergic drugs, the ability of test agents to inhibit ligand attachment to α_1-, α_2-, and β-adrenergic receptors should be assessed. Since most active compounds inhibit binding with a $K_i < 10\ \mu M$, it is prudent to perform the initial screen at a concentration of 100 μM. If no inhibition (or stimulation) of binding occurs at this concentration, then the substance is unlikely to influence the receptor system. If significant (>50%) inhibition of binding is observed in one or more of these assays, however, complete inhibition data should be obtained to accurately estimate the K_i. Assays can also be performed to determine whether the compound modifies the number of binding sites or binding site kinetics (K_d) to provide precise information about the nature of the receptor interaction. A Hill coefficient of 1.0 would indicate that the compound is attaching with a uniform affinity to the same site or sites as those labeled by the radioactive substance. A Hill coefficient significantly less than 1.0 might suggest that the compound differentiates between two binding components for the radioligand. Such information provides important clues as to which compound in a series should be examined in greater detail.

The next stage of analysis would be to examine the functional activity of those drugs that are most potent in the binding assays. Norepinephrine- or isoproterenol-stimulated cyclic AMP accumulation are useful measures of β-adrenergic receptor activity, whereas norepinephrine-stimulated inositol phosphate production can be used to measure α_1-adrenergic receptor function. The compounds would be tested in these assays by incubating them with the tissue alone or in combination with a known agonist. When taken together with the results of the binding assay, such data provide compelling evidence that the test compound is a biologically active adrenergic receptor agent.

At this point it would be advisable to examine the drug using an in vivo assay, since a compound demonstrating significant activity in vitro may be inactive when administered systemically. For example, if the in vivo analysis is aimed at measuring a central nervous system response, it is possible that the test compound may be inactive if it does not cross the blood–brain barrier. Other explanations for differences between in vivo and in vitro results include rapid metabolism or excretion. Accordingly, in vitro analysis provides no information about pharmacokinetic properties, which are of vital importance when considering pharmacological responses.

Ligand binding and second messenger assays can also be used to assess the effect of drug administration on receptor properties and function (Enna, 1986). For these studies a test compound is administered chronically at a dose known to cause a physiological response. The duration of drug treatment can vary anywhere from a few days to several weeks, with the mode of administration (ip, sc, implantation) dependent upon the known duration of action. At 24–48 h after the termination of drug treatment, receptor binding and second messenger production are analyzed in the target tissue to determine whether there is any change in receptor number, affinity, or function. For example, it may be found that chronic administration of a behaviorally active drug increases serotonin receptor binding and function in certain brain areas. Further analysis may indicate that the drug has no affinity for serotonin binding sites, nor does it influence serotonin-mediated second messenger responses. Such data would suggest that the test agent indirectly modifies serotonergic neuronal function in vivo. Since chronic administration increases the number and function of serotonin receptors, it could be speculated that the compound is decreasing serotonergic transmission, since a decline in receptor occupancy is known to induce an increase in receptor sensitivity (Enna, 1984b).

It is possible that a drug treatment could modify receptor binding without altering receptor function or vice versa. This may occur if there has been a change in the coupling between the recognition site and the second messenger generating system. For example, new recognition sites may not be properly coupled to N protein or, in cases where the number of recognition sites has diminished, the remaining sites may be more efficiently coupled to the regulatory subunit. Thus, a functional analysis is obligatory before deciding whether a drug-induced binding site alteration may have a physiological consequence.

Chronic studies should not be considered an obligatory part of a routine screening program, but rather should be performed after a compound has clearly demonstrated interesting pharmacological activity in vivo. In some cases, however, as for antidepressants, chronic studies may be the only means for estimating therapeutic efficacy, making it necessary to utilize this more cumbersome and expensive procedure earlier in a screening program (Enna, 1986).

The ability to study receptor interactions in vitro has been a boon for drug development programs. The ease of these assays has made it possible for the pharmacologist to accumulate important biological information in a timely and cost-effective manner. This allows for a more efficient utilization of resources, since it enables the staff to concentrate on the design, synthesis, and testing of only those substances most likely to possess pharmacological activity. Although data derived from in vitro tests are insufficient for drawing firm conclusions about therapeutic utility, they provide crucial information about whether a compound should be tested using more elaborate and costly procedures. As basic research continues on the molecular properties of receptors, it is likely that new methods will evolve that will lend themselves to drug development. This should further simplify the initial screening procedure and facilitate the design of drugs with novel mechanisms of action.

Acknowledgments

Preparation of this manuscript was made possible in part by the support of the National Institute of Mental Health, the National Science Foundation, and Bristol-Myers, Inc. I thank Mrs. Doris Thornton for her excellent secretarial assistance.

References

Baldessarini, R. J. and Lipinski, J. G. (1975) Lithium salts: 1970–1975. *Ann. Intern. Med.* **83**, 527–533.

Bennett, J. P. (1978) Methods in Binding Studies, in *Neurotransmitter Receptor Binding* 1st Ed. (Yamamura, H. I., Enna, S. J., and Kuhar, M. J., eds.) Raven, New York.

Berridge, M. J. (1984) Inositol triphosphate and diacylglycerol as second messengers. *Biochem. J.* **220**, 345–360.

Berridge, M. J. and Irvine, R. F. (1984) Inositol triphosphate, a novel second messenger in cellular signal transduction. *Nature* **312**, 315–321.

Braestrup, C., Nielsen, M., Krogsgaard-Larsen, P., and Falch, E. (1980) Two or More Conformations of Benzodiazepine Receptors Depending on GABA Receptors and Other Variables, in *Receptors for Neurotransmitters and Peptide Hormones* (Pepeu, G., Kuhar, M. J., and Enna, S. J., eds.) Raven, New York.

Burt, D. R. (1985) Criteria for Receptor Identification, in *Neurotransmitter Receptor Binding* 1st Ed. (Yamamura, H. I., Enna, S. J., and Kuhar, M. J., eds.) Raven, New York.

Bylund, D. B., ed. (1980) *Receptor Binding Techniques* Society for Neuroscience, Washington, DC.

Creese, I. (1978) Receptor Binding as a Primary Drug Screen, in *Neurotransmitter Receptor Binding* 1st Ed. (Yamamura, H. I., Enna, S. J., and Kuhar, M. J., eds.) Raven, New York.

Creese, I., Burt, D. R., and Synder, S. H. (1976) Dopamine receptor binding predicts clinical and pharmacological potencies of antischizophrenic drugs. *Science* **192**, 481–483.

Cuatrecasas, P., and Hollenberg, M. D. (1975) Binding of insulin and other hormones to nonreceptor materials; saturability, specificity and apparent "negative cooperativity." *Biochem. Biophys. Res. commun.* **62**, 31–41.

DeLeon, A., Hancock, A. A., and Lefkowitz, R. J. (1981) Validation and statistical analysis of a computer modeling method for quantitative analysis of radioligand binding data for mixtures of pharmacological receptor subtypes. *Mol. Pharmacol.* **21**, 5–16.

Drummond, G. I., ed. (1984) *Cyclic Nucleotides In the Nervous System* Raven, New York.

Ehrlich, P. (1913) Chemotherapeutics: Scientific principles, methods and results. *Lancet* **2**, 445–447.

Enna, S. J. (1984a) Principles of Receptor Binding Assays: The GABA Receptor, in *Investigations of Membrane-Located Receptors* (Reid, E., Cook, C. M. W., and Morre, D. J., eds.) Plenum, New York.

Enna, S. J. (1984b) Receptor Regulation, in *Handbook of Neurochemistry* vol. 6 (Lajtha, A., ed.) Plenum, New York.

Enna, S. J. (1986) *In Vivo* Receptor Modifications as a Measure of Antidepressant Potential, in *Receptor Binding in Drug Research* (O'Brien, R. A., ed.) Marcel Dekker, New York.

Enna. S. J. and Snyder, S. H. (1975) Properties of γ-aminobutyric acid (GABA) receptor binding in rat brain synaptic membrane fractions. *Brain Res.* **100**, 81–97.

Enna, S. J. and Snyder, S. H. (1977) Influences of ions, enzymes and detergents on GABA receptor binding in synaptic membranes of rat brain. *Mol. Pharmacol.* **13**, 442–453.

Enna, S. J. and Strada, S. J. (1983) Postsynaptic Receptors: Recognition Sites, Ion Channels and Second Messengers, in *Clinical Neurosciences* vol. V. (Rosenberg, R., Grossman, R., Schochet, S., Heinz, E. R., and Willis, W., eds.) Churchill Livingstone, New York.

Gallagher, D. W., Thomas, J. W., and Tallman, J. F. (1978) Effect of GABAergic drugs on benzodiazepine binding site sensitivity in rat cerebral cortex. *Biochem. Pharmacol.* **27**, 2745–2749.

Goldstein, A., Aronow, L., and Kalman, S. M. (1974) *Principles of Drug Action: The Basis of Pharmacology* 2nd Ed., pp. 1–117, Wiley, New York.

Guidotti, A., Toffano, G., and Costa, E. (1978) An endogenous protein modulates the affinity of GABA and benzodiazepine receptors in rat brain. *Nature* **275**, 553–555.

Hallcher, L. M. and Sherman, W. R. (1980) The effects of lithium ion and other agents on the activity of myo-inositol-*1*-phosphatase from bovine brain. *J. Biol. Chem.* **255**, 10896–10901.

Hill, A. W. (1910) The possible effects of the aggregation of the molecules of hemoglobin on its dissociation curves. *J. Physiol.* **40**, 4–7.

Hollenberg, M. D. and Cuatrecasas, P. (1979) Distinction of Receptor from Nonreceptor Interaction in Binding Studies, in *The Receptors, A Treatise* vol. 1 (O'Brien, R. D., ed.) Plenum Press, New York.

Laduron, P. M. (1984) Specificity Criteria for Receptor Sites in Binding Studies, in *Investigation of Membrane-Located Receptors* (Reid, E., Cook, G. M. W., and Morre, D. J., eds.) Plenum, New York.

Langley, J. N. (1906) On nerve endings and on special excitable substances in cells. *Proc. Roy. Soc.* **B78**, 170–178.

Lefkowitz, R. J. and Hoffman, B. B. (1980) Adrenergic receptors. *Adv. Cyclic Nucleotide Res.* **12**, 37–47.

Minneman, K. P., Hegstrand, L. R., and Molinoff, P. B. (1979) Simultaneous determination of beta-1 and beta-2-adrenergic receptors in tissues containing both receptor subtypes. *Mol. Pharmacol.* **16**, 34–46.

Munson, P. J. and Rodbard, D. (1980) Ligand: A versatile computerized approach for characterization of ligand-binding systems. *Anal. Biochem.* **107**, 229–231.

Perkins, J. P. (1975) Regulation of Adenylate Cyclase Activity by Neurotransmitters and Its Relation to Neural Functions, in *The Nervous System* vol. 1 (Brady, R. O., ed.) Raven, New York.

Rodbard, D. and Frazier, G. R. (1975) Statistical analysis of radioligand assay data. *Meth. Enzymol.* **37**, 3–22.

Rosenthal, H. E. (1967) Graphic method for the determination and presentation of binding parameters in a complex system. *Anal. Biochem.* **20**, 525–532.

Ross, E. M. and Gilman, A. G. (1980) Biochemical properties of hormone-sensitive adenylate cyclase. *Annu. Rev. Biochem.* **49**, 533–564.

Salomon, Y., Londos, C., and Rodbell, M. (1974) A highly sensitive adenylate cyclase assay. *Anal. Biochem.* **58**, 541–548.

Scatchard, G. (1949) The attraction of proteins for small molecules and ions. *Ann. NY Acad. Sci.* **51**, 660–672.

Schwabe, U. and Daly, J. W. (1977) The role of calcium ion in accumulation of cyclic adenosine monophosphate elicited by alpha and beta adrenergic agonists in rat brain slices. *J. Pharmacol. Exp. Ther.* **202**, 134–143.

Shimizu, H., Daly, J. W., and Creveling, C. R. (1974) A radioisotopic method for measuring the formation of adenosine 3′,5′-cyclic monophosphate in incubated slices of brain. *J. Neurochem.* **16**, 1609–1619.

Schulster, D. and Levitzki, A., eds. (1980) *Cellular Receptors for Hormones and Neurotransmitters* Wiley, New York.

Snyder, S. H. (1975) Neurotransmitter and drug receptors in the brain. *Biochem. Pharmacol.* **24**, 1371–1374.

Thampy, K. G. and Barnes, E. M. (1984) γ-Aminobutyric acid-gated chloride channels in cultured cerebral neurons. *J. Biol. Chem.* **259**, 1753–1757.

Williams, M. and Enna, S. J. (1986) The receptor: From concept to function. *Ann. Rep. Med. Chem.* **21**, 211–235.

Williams, M. and U'Prichard, D. C. (1984) Drug discovery at the molecular level: A decade of radioligand binding in retrospect. *Ann. Rep. Med. Chem.* **19**, 283–292.

Yamamura, H. I., Enna, S. J., and Kuhar, M. J., eds. (1985) *Neurotransmitter Receptor Binding* 2nd Ed., Raven, New York.

Drug Discovery at the Enzyme Level

Ray W. Fuller and Larry R. Steranka

1. Introduction

This chapter deals with enzymes as targets of drug action. The focus is on inhibition of enzyme activity by drugs, although other actions of drugs on enzymes are possible, such as activation or allosteric modification of enzyme function. This chapter will not deal with enzymes themselves as drugs, e. g., L-asparaginase in the treatment of acute lymphocytic leukemia, papain used for proteolysis in wound debridement, and urokinase used in thrombolytic therapy. Metabolism of drugs by enzymes and inhibition or induction of drug-metabolizing enzymes are other drug-enzyme interactions that can be important considerations in drug development, but our focus will be entirely on drug inhibition of enzymes as a primary mechanism to which drugs owe their efficacy. We will consider how enzyme activity is assayed, how inhibitors are found (or made) and characterized, how enzyme inhibition is demonstrated in vivo, and some specific examples of enzyme inhibitors that are used as drugs.

2. Enzyme Assays In Vitro

2.1. *Detecting Substrate Disappearance or Product Formation*

Enzyme activity is typically measured as the rate of disappearance of one of the substrates or the rate of appearance of one of the products. Often the substrate measured is not the "primary" substrate, but a cofactor or cosubstrate that may be common to multiple enzyme systems, such as nicotinamide adenine dinucleotide (NAD) or nicotinamide adenosine dinucleotide phosphate (NADP), for which there are convenient assay methods. Likewise the product that is measured may be one common to multiple enzymic reactions, such as hydrogen peroxide, carbon dioxide, or ammonia. Specificity in these cases is provided by comparing rates of disappearance or formation with and without the addition of a particular substrate to a reaction mixture sufficiently pure that conversion of a cofactor or formation of a nonspecific product is not supported to more than a negligible extent by endogenous materials.

Numerous methods of detecting and quantifying the substrate or product to be measured are now available. Detailed descriptions of these detection principles are found in reference sources such as *The Methods of Enzymology* series (S. P. Colowick and N. O. Kaplan, editors-in-chief, Academic Press, New York) and *Methods of Enzymatic Analysis* (3rd edition, H.U. Bergmeyer, editor-in-chief, Verlag Chemie, Weinheim). Colorimetric, spectrophotometric, spectrofluorometric, and radiometric methods are frequently used. Chromatographic separation followed by an appropriate detection method is another common assay technique and can provide high specificity. Some colorimetric methods that have been used rely on visual observation and are only semi-quantitative, but automated colorimetric methods can be almost as rapid and much more objective and reliable. Direct spectrophotometric methods lend themselves to continuous monitoring, which has the advantage of checking linearity of reaction rates in every assay. Often these spectroscopic methods are simple, economic, and convenient, but they may not be the most sensitive. When very small amounts of substrate or product must be measured, other detection methods may be required. Fluorometric detection has offered high sensitivity for many molecules that have native fluorescence or that can be converted chemically to fluorophores. Radiometric methods have become especially convenient for reactions in which one or more substrate is available in suitable radioactive form, or radioenzymatic assay or radioimmunoassay techniques can be used. Liquid chromatography with electrochemical detection is one of the newer methods that has been applied to enzyme assays in studying drug effects.

2.2. Reaction Conditions

The in vitro assay conditions should be chosen carefully so that reaction rates are linear over the time period used. This ordinarily requires that all substrates remain at concentrations well above their K_m concentration throughout the assay period. To achieve that, enzyme concentration must be sufficiently low so that conversion of substrate does not significantly reduce substrate concentration during the assay period and so that product does not accumulate to concentrations that are inhibitory to the reaction. These conditions are preferred for accurate measurement of enzyme activity and percentage inhibition by a drug or drug candidate. Nonetheless, these in vitro conditions of assay are artificial in that the concentration of substrate(s) often is low relative to the amount of enzyme present in vivo (Srere, 1967), and other factors such as accessibility of substrate to enzyme are often major determinants of reaction rates in whole cell environments.

Enzyme sources for in vitro screening of potential inhibitors or characterization of inhibitors need not be highly purified as long as other enzymes or constituents present do not interfere in a particular assay. Enzymes of high purity can be used if they are readily available, but partially purified preparations, subcellular particles such as mitochondria or microsomes, whole homogenates of tissues, synaptosomal preparations, or even whole cell preparations may be used. In some cases, two events are studied simultaneously, such as the transport of a substrate into the synaptosome and the subsequent degradation by an enzyme present therein. For example, Ask et al. (1983) studied the oxidative deamination of serotonin, dopamine, and norepinephrine by brain synaptosomes, a process that involved first the active transport of the monoamines selectively into synaptosomes from those respective neurons and then enzymatic attack by mitochondrial monoamine oxidase present within those synaptosomes.

3. Sources of Inhibitors

In setting out to find new inhibitors of a particular enzyme, an early decision that must be made is where to look. Chemical synthesis of new compounds designed to be inhibitors of the enzyme probably is the most common approach. Random or selective screening of existing synthetic compounds can be a source of leads to new structural types of inhibitors. Extraction of naturally occurring inhibitors from animal tissues or plants has been another source. Microbial fermentation products have been sources of many antibiotics, but until relatively recently, less emphasis was placed on trying to find other pharmacologically active agents, such as enzyme inhibitors, from that source.

3.1. Chemical Synthesis

Chemical synthesis of enzyme inhibitors may represent attempts at designing analogs of substrates, products, or transition states of an enzyme, or may represent structural modification of known inhibitors.

3.1.1. Substrate or Product Analogs

Competitive inhibitors are often close analogs of substrates, or in some cases are themselves substrates. An example of the latter case is the use of various catechols to inhibit catecholamine O-methylation. Catechol *O*-methyltransferase is a relatively nonspecific enzyme that O-methylates catechol itself and numerous substituted catechols. Some of these catechols, e. g., 2,3,4-trihydroxybenzylhydrazine (Ro 4-4602) (Baldessarini and Greiner, 1973) and 3,4-dihydroxy-2-methyl-propiophenone (U-0521) (Fahn et al., 1979), have been used as competitive inhibitors of the O-methylation of endogenous catecholamines. α-Methyltyrosine, a close analog of tyrosine, is a competitive inhibitor of tyrosine hydroxylation used therapeutically to inhibit excessive catecholamine formation in pheochromocytoma.

Norepinephrine *N*-methyltransferase inhibitors have been designed from phenylethanolamine substrates by removal of the β-hydroxyl group, which is necessary for substrate activity, but not for attachment to the enzyme. Phenylethylamines are competitive inhibitors of norepinephrine *N*-methyltransferase whose affinity for the enzyme is enhanced by aromatic halogen substituents, and a bicyclic phenylethylamine analog with two chlorine substituents on the phenyl ring, SKF 64139, is a potent, competitive inhibitor of the enzyme with a K_i of 3 n*M* (Bondinell et al., 1983).

Many enzymes are also inhibited by their products, and sometimes product inhibition has physiological significance as a regulatory mechanism. For instance, *S*-adenosylhomocysteine is an inhibitor of several transmethylases that form it via transfer of a methyl group from *S*-adenosylmethionine to a methyl acceptor. The acceptor often is an oxygen or nitrogen atom in small molecules, such as catecholamines, or in larger molecules, such as phospholipids, nucleic acids, or proteins. Analogs of *S*-adenosyl-homocysteine have been of interest as inhibitors of methyl-transferring enzymes (e. g., Borchardt, 1977). The problem of nonspecificity arises since the analogs resemble the common product that inhibits several related enzymes. However, sinefungin and some of its analogs, which are structurally similar to *S*-adenosylhomocysteine, have a considerably greater selectivity as inhibitors of different methyltransferases than does *S*-adenosylhomocysteine itself (Fuller and Nagarajan, 1978), lending credence to the idea that specificity is attainable even among structural analogs of a common enzymatic product.

3.1.2. Transition State Analogs

The design of inhibitors based on the transition state in the enzymatic transformation of substrates is a powerful approach to drug design. Catalysis by enzymes increases reaction rates by many orders of magnitude because the enzyme promotes the development of transition states through which bonds become broken or formed, resulting in the conversion of substrates to products. The substrate molecule in the transition state has very much enhanced affinity for the enzyme. Transition state analogs are stable molecules intended to resemble the substrate portion of the enzymic transition state and thus are expected to bind tightly to the enzyme; that is, to be potent inhibitors (Lienhard, 1980). The design of such analogs requires some understanding of the enzyme mechanism and the chemical nature of the transition state. The design and synthesis of transition state analogs have produced potent inhibitors for enzymes of various classes (Wolfenden, 1976).

3.1.3. Screening

Random screening of chemicals is one means of finding inhibitors of novel structural types that may suggest further synthetic modifications. Since large numbers of compounds have to be studied, rapid enzyme assays are needed, and quantitative precision may be sacrificed for speed. Many screening methods are applicable not only to testing pure chemicals, but also to crude extracts from natural products. Brannon and Fuller (1974) have reviewed some methods for screening for microbial production of pharmacologically active compounds other than antibiotics, including enzyme inhibitors. Illustrative of the simple screening methods ideally suited for use with crude microbial broths or similar preparations was a culture plate assay for inhibitors of tryptophan 5-hydroxylase. The production of a purple pigment by the bacterium *Chromobacterium violaceum* involves 5-hydroxylation of tryptophan, and the appearance of noncolored zones of growth around plate regions to which microbial extracts have been applied can indicate the presence of inhibitors of this enzyme. Umezawa (1982) has listed numerous low-molecular-weight enzyme inhibitors of microbial origin that have been found by screening methods of this general sort. The relative efficiency of screening vs design in the search for enzyme inhibitors can usually be judged best retrospectively.

3.2. Natural Products

There are naturally occurring inhibitors of many enzymes with potential not only as drugs themselves, but as prototypes for inhibitors that would be drug candidates. Naturally occurring inhibitors have been found in animal, plant, and microbial sources. Aprotinin, a proteinase inhibitor prepared commercially from bovine lung, pancreas, and parotid glands, is used therapeutically in the treatment of diseases like hyperfibrinolytic hemorrhage and traumatic-hemorrhagic shock (Fritz and Wunderer, 1983). Physostigmine is an inhibitor of acetylcholinesterase obtained from a plant source, the Calabar bean. Umezawa (1982) described inhibitors of proteases, glycosidases, decarboxylases, hydroxylases, methyltransferases, phosphodiesterases, glyoxalase, and other enzymes that have been obtained from microbial origin. The large numbers of penicillins and cephalosporins, drugs whose development was based on naturally occurring agents with antimicrobial effects, exemplify the use of natural products as drug prototypes (Frere et al., 1980).

4. Characterization of Inhibitors In Vitro

4.1. Reversible or Irreversible

One of the first characteristics to be determined about an enzyme inhibitor is whether it combines reversibly or irreversibly with the enzyme. Probably the most useful and widely used method involves dialysis or ultrafiltration to determine if a low-molecular-weight inhibitor can be separated from the high-molecular-weight enzyme protein. Typically the inhibitor and enzyme are preincubated for a period of time, then the mixture is subjected to dialysis or other means of separation prior to assay of enzyme activity. Whether or not the enzyme activity can be restored reveals the reversible or irreversible nature of the enzyme–inhibitor complex. Controls are (1) enzyme subjected to the same conditions, but without exposure to inhibitor, and (2) inhibitor added at the same concentrations to enzyme and subjected to similar manipulations (such as exposure to cold for the same time as the dialysis period), except for the actual separation technique.

Irreversible inhibitors generally have a time dependency not seen with reversible inhibitors. Percentage inhibition of enzyme activity increases with time of preincubation of inhibitor and enzyme prior to addition of substrate for enzyme assay. Plotting percentage enzyme remaining (on a logarithmic scale) vs time of preincubation gives straight-line plots that can be used to calculate the rate of combination of inhibitor with enzyme.

Ackermann and Potter (1949) described a graphic means of distinguishing reversible and irreversible inhibitors (Fig. 1). Enzyme reaction rate is plotted as a function of amount of enzyme added. In the case of reversible inhibitors, increasing concentrations of the inhibitor produces lines with decreased slope, but passing through the origin on both axes. Irreversible inhibitors, on the other hand,

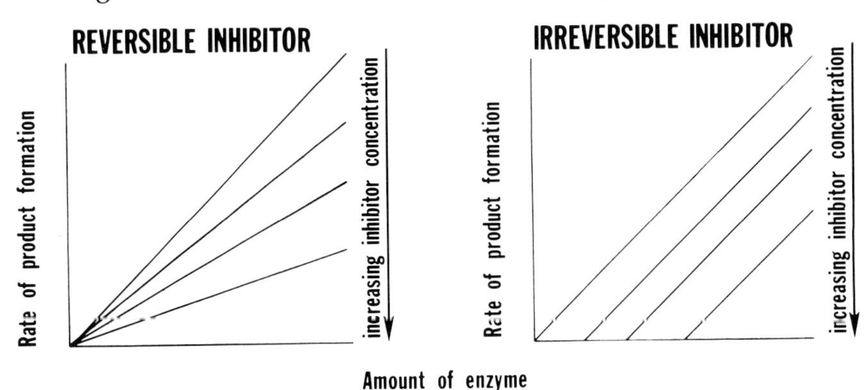

Fig. 1. Ackermann-Potter plots for distinguishing reversible and irreversible inhibitors. The rate of product formation is plotted as a function of the amount of enzyme added. Reversible inhibitors (left) result in lines of decreased slope, but that pass through the origin. Irreversible inhibitors (right) result in lines of unchanged slope with the intercept on the X-axis displaced to the right proportional to the amount of enzyme inactivated.

produce lines that are parallel and intercept the X-axis to the right of the origin by an amount that is proportional to the amount of inhibitor added.

4.1.1. Mechanism-based inhibitors

Rando (1974) has pointed out that irreversible enzyme inhibitors having latent reactive groupings unmasked by the catalytic action of the target enzyme are extraordinarily specific. If the unmasking of the reactive groupings comes about only through the action of one specific enzyme, then only that enzyme is irreversibly inactivated. Transient interactions of the inhibitor with other enzymes or other biological receptors terminate with the disappearance of the drug from tissues, whereas inactivation of the target enzyme remains. How long the inactivation remains depends on the normal turnover rate of the enzyme (the rate of synthesis of new enzyme protein).

Inhibitors of this type often are referred to as "k_{cat}" inhibitors, a term arising from the following reaction:

$$E + S \underset{K_s}{\rightleftharpoons} ES \underset{k_{cat}}{\rightarrow} EP \rightarrow E + P$$

In this case, E represents enzyme, S represents substrate, and P represents product. The enzyme–substrate complex (ES) can dissociate back to free enzyme and substrate or can, through the catalytic action of the enzyme, go to a second intermediate state in the overall enzymatic reaction. This second complex, EP, ordinarily dissociates into free enzyme and product, but in the case of k_{cat} inhibitors, the transformation of ES and EP unmasks a reactive grouping so that the molecule that was once S and potentially dissociates to form P instead reacts covalently with a site on the enzyme molecule to form a stable product that might be called $E–P$, which lacks enzymatic activity.

There are many examples of k_{cat} inhibitors (e. g., *see* Kalman, 1979; Sjoerdsma, 1981; Penning, 1983; Rajashekhar et al., 1984), but not all irreversible inhibitors are k_{cat} inhibitors. If ES simply does not dissociate, or if the dissociation is so slow as to be imperceptible, irreversible inhibition is also the consequence. In that case no catalytic action of the enzyme has been involved. Such inhibitors have been called "pseudoirreversible" inhibitors, since the dissociation is probably not zero, though it may be so low as to indistinguishable from zero experimentally.

The terminology for k_{cat} inhibitors has not been universally agreed upon. Such inhibitors have often been referred to as "suicide substrates," "suicide enzyme inactivators," or "suicide inhibitors." Silverman (1983) has objected to these terms on the grounds that suicide is voluntary and intentional, neither of which is a characteristic of the enzyme inactivation that results from an action of an enzyme on k_{cat} inhibitors. He proposed the acceptance of terms like "mechanism-based inhibitor" or "enzyme-activated inhibitor."

4.2. Competitive, Noncompetitive, or Uncompetitive Inhibitors

Reversible inhibitors can be further characterized kinetically by graphic or statistical treatment of data obtained at selected concentrations of inhibitor and substrate. The most well known method is the Lineweaver-Burk plot, or double reciprocal plot, in which the reciprocal of enzyme velocity is plotted as a function of the reciprocal of substrate concentration (Fig. 2). Competitive inhibitors added at increasing concentrations produce lines of increased slope intersecting on the Y-axis. Noncompetitive inhibitors added at increasing concentrations produce lines of increased slope intersecting on the X-axis. Uncompetitive inhibitors added at increasing concentrations produce parallel lines (unchanged slope). Several other graphic means of analyzing data of these sorts also exist (*see* Webb, 1963). Competitive inhibitors increase the apparent K_m of the substrate with which they compete for a site on the enzyme, whereas noncompetitive inhibitors reduce the maximum velocity (V_{max}) without altering K_m. Uncompetitive inhibitors change both K_m and V_{max}. Usually a combination of graphic and statistical analysis of enzyme kinetic data is preferable. Graphic examination of the data readily reveals any consistent divergence from theoretical expectation, such as departure from linearity. Statistical analysis provides objective measures of K_m, V_{max}, and K_i values with confidence limits. Wilkinson (1961) has described a method of statistical analysis of enzyme kinetic data, and others are available (e. g., Cleland, 1967).

The relative potency of enzyme inhibitors is often expressed by their IC_{50} (or I_{50}) values, the concentration (which should be expressed in molar units) required for 50% inhibition of enzyme activity under standard assay conditions. The type of inhibition (e. g., reversible or irreversible, competitive or noncompetitive) should also be specified. The type of inhibition is often assumed to be

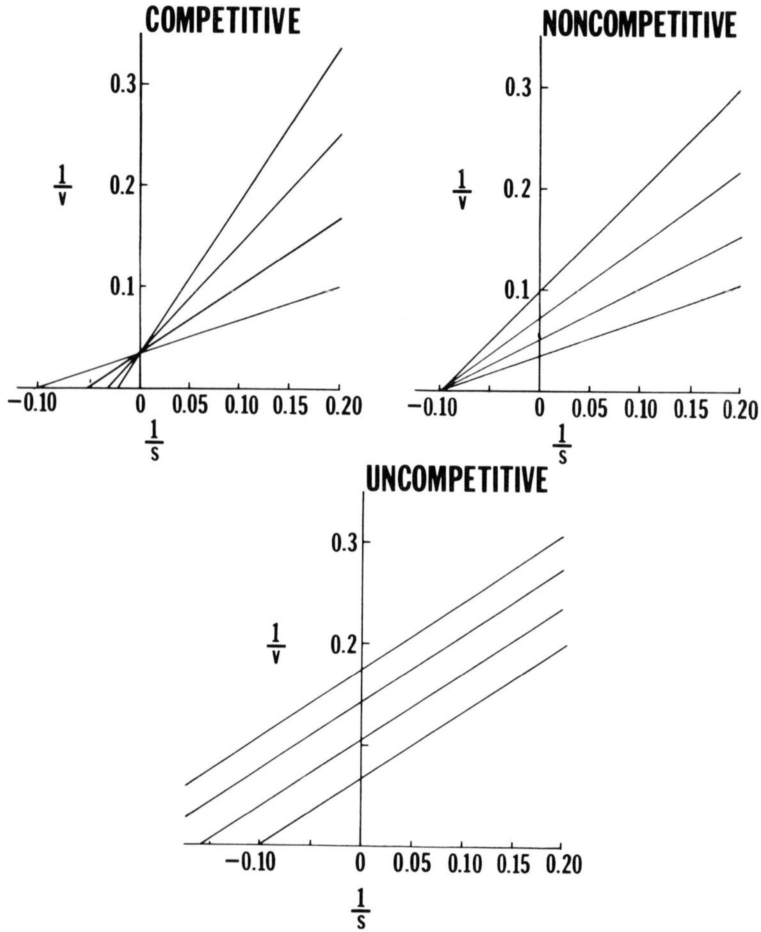

Fig. 2. Lineweaver-Burk plots for distinguishing competitive, noncompetitive, and uncompetitive inhibitors. The reciprocal of enzyme velocity ($1/v$) is plotted against the reciprocal of substrate concentration ($1/s$). The intercept on the X-axis is equal to $-1/K_m$, where K_m is the Michaelis constant (unit is substrate concentration), a measure of the affinity of substrate for the enzyme. The intercept on the Y-axis is equal to $1/V_{max}$, where V_{max} is the theoretical maximum velocity obtainable. Competitive inhibitors increase the apparent K_m value (decrease the apparent affinity of substrate for the enzyme) without changing V_{max}. Noncompetitive inhibitors decrease V_{max} without changing K_m. Uncompetitive inhibitors increase K_m and decrease V_{max}. Mixed kinetic patterns may also be obtained. In each case, the lower-most line represents the control (no inhibitor), and the upper-most line represents the highest concentration of inhibitor.

similar within a chemical class of inhibitors to avoid having to do kinetic studies on each compound. Since IC_{50} values depend on substrate concentration and affinity, K_i values are better for comparing the potencies of reversible enzyme inhibitors. Although it is preferable to determine the K_i for each inhibitor by one of the usual kinetic analyses, a common practice is to calculate the K_i value from the IC_{50} concentration of competitive inhibitors using the following formula (Cheng and Prusoff, 1973):

$$K_i = \frac{IC_{50}}{1 + S/K_m}$$

where S is the substrate concentration. Comparison of K_i values should be independent of the substrate used or its concentration and facilitates comparison of inhibitor potencies from different laboratories or from different chemical series. Cheng and Prusoff (1973) pointed out that K_i is equal to IC_{50} for noncompetitive or uncompetitive inhibitors.

4.3. Specificity

A concern with enzyme inhibitors, as with other kinds of potential drugs, is their specificity. Consider the enzyme catechol *O*-methyltransferase, which O-methylates norepinephrine using *S*-adenosylmethionine as the methyl donor, forming *S*-adenosylhomocysteine as the product. Analogs of *S*-adenosylmethionine or *S*-adenosylhomocysteine can inhibit this enzyme, but they may also inhibit other methyltransferases for which *S*-adenosylmethionine is the methyl donor and *S*-adenosylhomocysteine is the product. Furthermore, norepinephrine (like many other biological substances whose metabolism the drug designer is attempting to influence) also combines with other proteins that are not enzymes, but that are receptors mediating physiologic responses, and norepinephrine can be metabolized by other enzymes, e. g., monoamine oxidase and norepinephrine *N*-methyltransferase. Analogs of norepinephrine may interact with these other proteins (receptors or enzymes), as well as with catechol *O*-methyltransferase. These opportunities for nonspecificity require that proper attention be paid to the specificity of inhibitors that may be found.

The factors that contribute to nonspecificity are general. Design of analogs of substrates or products opens the risk of interactions with other enzymes or receptors for which the substrate or product has affinity. An argument for the transition-state analog approach to inhibitor design is that the drug is synthesized to resemble not the substrate or the product themselves, but the transition state that may be involved in only one type of enzyme action on that particular substrate and perhaps not at all in receptor interactions.

Absolute specificity is rare and may not be attainable, but a satisfactory degree of specificity, so that negligible influences on other enzymes or receptors occur at in vivo doses that are effective in inhibiting the target enzyme, can often be achieved. An example of enzyme inhibitors in which specificity is of particular concern would be inhibitors of peptidases and proteinases. Many different enzymes in this family exist, and inhibition of various of them would have potential therapeutic significance (Barrett, 1980). Angiotensin-converting enzyme inhibitors illustrate the possibility of finding agents that inhibit only one of a family of enzymes whose physiological substrates are sufficiently limited to make the inhibitors useful in clinical therapy (Antonaccio, 1982).

5. Demonstration of Enzyme Inhibition In Vivo

Once enzyme inhibitors have been identified and characterized in vitro, the most important step is to determine if they are capable of reaching their target enzyme and inhibiting its actions in vivo with an efficacy and duration that is acceptable. There are a number of ways for demonstrating efficacy of enzyme inhibitors in vivo, and the choice of methods depends on factors such as the nature of the inhibition, properties of the inhibitor, and knowledge of physiologic roles of the enzyme.

5.1. *Ex Vivo Inhibition*

Probably the most straightforward way of demonstrating enzyme inhibition in vivo is by removing tissues from animals treated with the inhibitor, then assaying enzyme activity in vitro in an appropriate preparation (e. g., homogenate, particulate, or supernatant fractions) from the tissue. In the case of irreversible inhibitors of an enzyme, this method probably provides an accurate assessment of the effectiveness of the inhibitor in vivo. In the case of reversible inhibitors, however, such a procedure amounts to a biological assay of inhibitor present in the tissue preparation. That in itself can be useful, since it reveals the ability of an inhibitor to reach and remain in the tissue containing the enzyme to be inhibited.

However, such a procedure may not give a true indication of the degree of inhibition that occurred in the intact organism. Subcellular compartmentation is destroyed, which might give a falsely low or high estimate of percentage inhibition, and dilution of drug in setting up the enzyme assay will reduce the apparent degree of inhibition from that which occurred in the intact tissue.

5.2. Substrate Conversion In Vivo

The above concerns about destruction of tissue integrity and compartmentation can be avoided by measuring substrate conversion to product in tissues of the living animal instead of in tissue homogenates or subfractions in vitro. The substrate used may be the natural, endogenous substrate for the enzyme, given in pharmacological amounts or in radiolabeled form, or the substrate used may be an artificial (not naturally occurring) compound. For instance, inhibition of monoamine oxidase in vivo has been measured conveniently by measuring the amount of radioactive tryptamine or phenylethylamine remaining in tissues at specified times after the injection of these radioactive amines into rats. Both tryptamine and phenylethylamine occur naturally, but their endogenous concentrations are so low that exogenous administration of the amines (not necessarily in radioactive form) permits measurement of the rate of substrate disappearance after the administration of an enzyme inhibitor.

An example of the use of an artificial substrate is *p*-hydroxyamphetamine, which is metabolized by dopamine β-hydroxylase to *p*-hydroxynorephedrine. Measurement of urinary levels of the latter substance after administration of *p*-hydroxyamphetamine is a means of evaluating dopamine β-hydroxylase inhibition in vivo that is applicable to humans or laboratory animals (Sjoerdsma and von Studnitz, 1963).

An example of the conversion of an endogenous substrate to a product is the rapid metabolism of dopamine by monoamine oxidase once dopamine is released from endogenous storage granules. Ro 4-1284 is a rapid and short-acting monoamine releaser. When Ro 4-1284 is injected into animals, there is an acute release of dopamine and other monoamines from storage granules, exposing the amines to enzymatic degradation by monoamine oxidase. The deaminated metabolites of dopamine, 3,4-dihydroxyphenylacetic acid and homovanillic acid, increase rapidly following the administration of Ro 4-1284. Antagonism of this acute increase in these metabolites has been used as an index of monoamine oxidase inhibition in rat striatum (Fuller and Hemrick-Luecke, 1985).

5.3. Antagonism of Mechanism-Based Inactivation

Since enzyme inactivation by k_{cat} inhibitors requires catalytic activity of the enzyme on the inhibitor molecule, reversible inhibitors can prevent the inactivation by inhibiting enzyme activity during the period that the irreversible inhibitor is available for interaction with the enzyme. Inactivation of the enzyme measured at long time intervals is prevented, and if the reversible inhibitor is short-acting, its inhibition will have disappeared before enzyme activity is assayed. The use of this technique can be demonstrated with monoamine oxidase inhibitors.

Short-acting, reversible inhibitors of MAO were observed to prevent the long-term irreversible inhibition of the enzyme even before it was known that inhibitors of the latter group were k_{cat} inhibitors. Pletscher and Besendorf (1959) reported that harmaline antagonized the inhibition of MAO by irreversible inhibitors of the hydrazide class, and Horita and McGrath (1960) reported that harmine antagonized the inhibition of MAO by hydrazine and non-hydrazine inhibitors. Planz et al. (1973) called attention to the principle that such antagonism is a useful means of demonstrating temporary, reversible inhibition of MAO in vivo. Green and El Hait (1980) used this method for demonstrating the in vivo inhibition of monoamine oxidase by amphetamine and *p*-methoxyamphetamine as well as harmaline. Green and El Hait also emphasized that a highly labile, irreversible inhibitor rapidly cleared from tissues should be used ideally if the protection technique is to give a valid estimate of the inhibitory potency of the reversible inhibitor.

5.4. Substrate Accumulation or Product Depletion

Apart from administering pharmacological amounts of natural substrates, another means of demonstrating enzyme inhibition in vivo is to measure the accumulation of an endogenous substrate for an enzyme or the disappearance of the immediate or later products of an enzyme reaction. For example, Duch et al. (1979) have shown that inhibition of histamine *N*-methyltransferase elevates brain concentrations of histamine, the measure of which provides an index of duration of enzyme inhibition. Monoamine oxidase inhibitors increase the concentration of various monoamines, e. g., serotonin, dopamine, and norepinephrine, in brain and in other tissues, while decreasing the concentration of deaminated metabolites of those monoamines, such as 5-hydroxyindoleacetic acid, 3,4-dihydroxyphenylacetic acid, homovanillic acid, and 3-methoxy-4-hydroxyphenylethyleneglycol. Dopamine β-hydroxylase inhibitors decrease norepinephrine concentrations in all tissues, and in tissues like heart and blood vessels, where dopamine is present mainly or perhaps entirely as a precursor to norepinephrine (unlike brain, where much or most of the dopamine is in dopamine neurons), there is a rapid and several-fold increase in dopamine after dopamine β-hydroxylase inhibition (Fuller et al., 1982).

5.5. Functional Effects

Often there are characteristic functional effects of enzyme inhibitors that can be used instead of or in addition to biochemical measurements to demonstrate in vivo efficacy of the inhibitors. For instance, monoamine oxidase inhibitors given to mice potentiate a behavioral syndrome (hyperactivity, irritability, aggression) induced by L-DOPA. DOPA potentiation in mice has been used as a screening method for monoamine oxidase inhibitors and is probably indicative of inhibition of type A monoamine oxidase (Fuller, 1972), whereas potentiation of phenylethylamine-induced stereotypies in the rat is a behavioral test system for demonstrating inhibition of type B monoamine oxidase in vivo (Ortmann et al., 1984).

Some functional effects are not diagnostic for inhibition of a specific enzyme, but relate directly to an intended therapeutic use of inhibitors. For instance, blood pressure lowering has been one in vivo effect of dopamine β-hydroxylase inhibitors commonly studied, not because blood pressure lowering demonstrates dopamine β-hydroxylase inhibition as the mechanism, but because an antihypertensive effect through reduction of neurogenic tonic is one of the therapeutic targets in the search for dopamine β-hydroxylase inhibitors (Fuller et al., 1977). Aldose reductase inhibitors, of potential use in preventing or treating complications of diabetes, such as cataracts, have been shown to inhibit experimentally induced cataract formation in animals (*see* Lipinski and Hutson, 1984).

6. Specific Examples of Enzyme Inhibitors Used as Drugs

6.1. Monoamine Oxidase Inhibitors

The fortuitous observation of mood elevation in tuberculosis patients given iproniazid and the discovery that iproniazid inhibited monoamine oxidase led to development and use of additional monoamine oxidase-inhibiting drugs in the treatment of depression (Quitkin et al., 1979). Currently, phenelzine and tranylcypromine are marketed in the US for the treatment of depression, and pargyline is marketed for the treatment of hypertension. All three of these drugs are irreversible inhibitors of monoamine oxidase of the k_{cat} type. Their antidepressant activity is believed to result from increased concentrations of neurotransmitter amines, such as serotonin and norepinephrine, in the brain. The mechanism of antihypertensive effect is less well understood, but enhanced brain monoaminergic function may also be involved in that effect.

For more than a decade, it has been recognized that monoamine oxidase is not a single enzyme. At least two forms, types A and B, have been defined based on substrate and inhibitor selectivity. Phenelzine, tranylcypromine and pargyline are all relatively nonselective in blocking both types of the enzyme, whereas some newer inhibitors (such as clorgyline, deprenyl, and LY 51641) are highly selective inhibitors of either type A or B monoamine oxidase.

Despite their potential for interactions with other drugs or with ingested amines that occur in some foods, e. g., cheeses, monoamine oxidase inhibitors continue to be useful in the treatment of psychiatric disorders (Davidson et al., 1984). The greatly improved understanding of monoamine oxidase subtypes and their physiologic functions and enzymatic properties raises the hope that improved inhibitors of monoamine oxidase can be developed as therapeutic agents (*see* Fuller, 1978). A focus in present research on monoamine oxidase inhibitors is the development of reversible, competitive inhibitors whose inhibitory effects might be overcome competitively as endogenous substrate monoamines accumulate (Waldmeier and Baumann, 1983).

6.2. Xanthine Oxidase Inhibitors

Gout occurs as a result of an inflammatory reaction to crystalline deposits of uric acid (the end product of purine catabolism) in joints.

One means of preventing or treating gout is to inhibit the biosynthesis of uric acid, ideally at the terminal step. Allopurinol is a substrate and a competitive inhibitor of xanthine oxidase, the enzyme that forms uric acid from hypoxanthine and xanthine. Allopurinol is a structural analog of hypoxanthine, and its product, oxypurinol, is a structural analog of xanthine. Both allopurinol and oxypurinol are inhibitors of hypoxanthine and xanthine oxidation by xanthine oxidase. Oxypurinol binds very tightly to partially reduced xanthine oxidase in a stoichiometric complex that does not involve a covalent linkage, but is almost irreversible (Elion, 1978). Although allopurinol appears not to inhibit purine biosynthesis directly, total purine protection is reduced, apparently because the oxypurines are reused for nucleic acid biosynthesis and cause feedback inhibition of purine synthesis through an action of purine nucleotides. Some other inhibitors of xanthine oxidase, e. g., thiopurinol, are also effective in reducing uric acid formation. Allopurinol mimics purines at the active site of other enzymes as well, since allopurinol ribonucleoside and allopurinol ribonucleotide are formed as metabolites. This illustrates the point that most endogenous substances interact with more than one enzyme or receptor macromolecule, and structural analogs of these substances may compete at multiple sites. Transition state analogs may have the advantage of more specific interactions, if the transition states differ for various metabolic transformations of an endogenous substrate.

6.3. Carbidopa

Carbidopa, (−)-α-hydrazino-3,4-dihydroxy-α-methylhydrocinnamic acid, is the hydrazino analog of α-methyldopa. It is a potent, competitive inhibitor of the aromatic L-amino acid decarboxylase with very little ability to cross the blood–brain barrier (Porter et al., 1962). This decarboxylase is responsible for the conversion of the catecholamine precursor, L-DOPA, to dopamine and for the conversion of L-5-hydroxytryptophan to serotonin. Catecholamines and serotonin do not cross the blood–brain barrier, so the precursor amino acids are often administered to increase brain levels of the amines. Normally the amino acids are partially metabolized by decarboxylation in peripheral tissues, resulting in diminished availability to the brain and pharmacologic effects secondary to the formation of catecholamines and serotonin peripherally. Pretreatment with carbidopa, by preventing decarboxylation in peripheral tissues, where it inhibits decarboxylase, but not in the brain (which it does not reach in sufficient amounts to be effective), increases the amount of precursor amino acid that reaches the brain and minimizes side effects caused by peripheral formation of the bioactive amines. Carbidopa is used in combination with L-DOPA for the treatment of Parkinson's disease, where L-DOPA has therapeutic effects because of restoration of dopamine levels in the nigrostriatal neurons that degenerate to cause the disease. The combination of carbidopa and L-DOPA is more effective and better tolerated than L-DOPA alone.

In addition to carbidopa, benserazide (Ro 4-4602) is another decarboxylase inhibitor that penetrates poorly into the brain. Although not used clinically, benserazide is effective in laboratory animals in much the same way as carbidopa to inhibit amino acid decarboxylation in peripheral tissues resulting in greater availability of the amino acid to the brain (Bartholini and Pletscher, 1969).

7. Summary

The design or discovery of enzyme inhibitors as an approach to finding drugs is a part of modern pharmaceutical sciences. Not all drugs that act as enzyme inhibitors were developed in that way. Aspirin stands as an example of a drug used very widely for many years before inhibition of prostaglandin synthetase was reported to be its mechanism of action (*see* Moncada and Vane, 1980). Development of drugs based on functional effects continues to predate the discovery that some of them act through enzyme inhibition. But the screening and design of drug candidates as enzyme inhibitors offers one practical approach to drug developers that is now a proven one and that can be done efficiently with modern technology for accurate and rapid, often automated, measurement of enzyme activity. The continued development of methodologies for demonstrating inhibition of enzyme activity in vivo facilitates the testing of enzyme inhibitors as drugs in laboratory animals and, especially with the advent of noninvasive techniques, permits parallel demonstration of drug efficacy in humans. We expect that the number of drugs that act through inhibition of specific enzymes that have key roles in human physiology will grow.

References

Ackermann, W. W. and Potter, V. R. (1949) Enzyme inhibition in relation to chemotherapy. *Proc. Soc. Exp. Biol. Med.* **72**, 1–9.

Antonaccio, M. J. (1982) Angiotensin converting enzyme (ACE) inhibitors. *Ann. Rev. Pharmacol. Toxicol.* **22**, 57–87.

Ask, A.-L., Fagervall, I., and Ross, S. B. (1983) Selective inhibition of monoamine oxidase in monoaminergic neurons in the rat brain. *Naunyn-Schmiedeberg's Arch. Pharmacol.* **324**, 79–87.

Baldessarini, R. J. and Greiner, E. (1973) Inhibition of catechol-*O*-methyl transferase by catechols and polyphenols. *Biochem. Pharmacol.* **22**, 247–256.

Barrett, A. J. (1980) Proteinase Inhibitors: Potential Drugs?, in *Enzyme Inhibitors as Drugs* (Sandler, M., ed.) University Park Press, Baltimore, Maryland.

Bartholini, G. and Pletscher, A. (1969) Effect of various decarboxylase inhibitors on the cerebral metabolism of dihydroxyphenylalanine. *J. Pharm. Pharmacol.* **21**, 323–324.

Bondinell, W. E., Chapin, F. W., Frazee, J. S., Girard, G. R., Holden, K. G., Kaiser, C., Maryanoff, C., and Perchonock, C. D. (1983) Inhibitors of phenylethanolamine *N*-methyltransferase and epinephrine biosynthesis: A potential source of new drugs. *Drug Metab. Rev.* **14**, 709–721.

Borchardt, R. T. (1977) Synthesis and Biological Activity of Analogues of Adenosylhomocysteine as Inhibitors of Methyltransferases, in *The Biochemistry of S-Adenosylmethionine* (Salvatore, F., Borek, E., Zappia, V., Williams-Ashman, H. G., and Schlenk, F., eds.) Columbia University Press, New York.

Brannon, D. R. and Fuller, R. W. (1974) Microbial production of pharmacologically active compounds other than antibiotics. *Lloydia* **37**, 134–146.

Cheng, Y.-C. and Prusoff, W. H. (1973) Relationship between the inhibition constant (K_i) and the concentration of inhibitor which causes 50 per cent inhibition (I50) of an enzymatic reaction. *Biochem. Pharmacol.* **22**, 3099–3108.

Cleland, W. W. (1967) The statistical analysis of enzyme kinetic data. *Adv. Enzymol.* **29**, 1–32.

Davidson, J., Zung, W. W. K., and Walker, J. I. (1984) Practical aspects of MAO inhibitor therapy. *J. Clin. Psychiatr.* **45**, 81–84.

Duch, D. S., Bowers, S., Edelstein, M., and Nichol, C. A. (1979) Histamine: Elevation of Brain Levels by Inhibition of Histamine N-methyl Transferase, in *Transmethylation* (Usdin, E., Borchardt, R. T., and Creveling, C. R., eds.) Elsevier/North-Holland, New York, Amsterdam, London.

Elion, G. B. (1978) Allopurinol and Other Inhibitors of Urate Synthesis, in *Handbook of Experimental Pharmacology* vol. 51 *Uric Acid* (Kelley, W. N. and Weiner, I. M., eds.) Springer-Verlag, Berlin.

Fahn, S., Comi, R., Snider, S. R., and Prasad, A. L. N. (1979) Effect of a catechol-*O*-methyl transferase inhibitor, U-0521, with levodopa administration. *Biochem. Pharmacol.* **28**, 1221–1225.

Frere, J. M., Duez, C., Dusart, J., Coyette, J., Leyh-Bouille, M., Ghuysen, J. M., Dideberg, O., and Knox, J. (1980) Mode of Action of β-Lactam Antibiotics at the Molecular Level, in *Enzyme Inhibitors as Drugs* (Sandler, M., ed.) University Park Press, Baltimore, Maryland.

Fritz, H. and Wunderer, G. (1983) Biochemistry and applications of aprotinin, the kallikrein inhibitor from bovine organs. *Arzneimittel-Forsch.* **33(I)**, 479–494.

Fuller, R. W. (1972) Selective inhibition of monoamine oxidase. *Adv. Biochem. psychopharmacol.* **5**, 339–354.

Fuller, R. W. (1978) Selectivity among monoamine oxidase inhibitors and its possible importance for development of antidepressant drugs. *Progr. Neuro-Psychopharmacol.* **2**, 303–311.

Fuller, R. W. and Hemrick-Luecke, S. K. (1985) Inhibition of types A and B monoamine oxidase by 1-methyl-4-phenyl-1,2,3,6-tetrahydropyridine. *J. Pharmacol. Exp. Ther.* **232**, 696–701.

Fuller, R. W. and Nagarajan, R. (1978) Inhibition of methyltransferases by some new analogs of *S*-adenosylhomocysteine. *Biochem. Pharmacol.* **27**, 1981–1983.

Fuller, R. W., Ho, P. P. K., Matsumoto, C., and Clemens, J. A. (1977) New inhibitors of dopamine β-hydroxylase. *Adv. Enz. Regul.* **15**, 267–281.

Fuller, R. W., Snoddy, H. D., and Perry, K. W. (1982) Dopamine accumulation after dopamine β-hydroxylase inhibition in rat heart as an index of norepinephrine turnover. *Life Sci.* **31**, 563–570.

Green, A. L. and El Hait, M. A. S. (1980) A new approach to the assessment of the potency of reversible monoamine oxidase inhibitors *in vivo*, and its application to (+)-amphetamine, *p*-methoxyamphetamine and harmaline. *Biochem. Pharmacol.* **29**, 2781–2789.

Horita, A. and McGrath, W. R. (1960) The interaction between reversible and irreversible monoamine oxidase inhibitors. *Biochem. Pharmacol.* **3**, 206–211.

Kalman, T. I. (1979) *Drug Action and Design: Mechanism-Based Enzyme Inhibitors.* Elsevier/North-Holland, New York, Amsterdam, Oxford.

Lienhard, G. E. (1980) Transition-State Analogues, in *Enzyme Inhibitors as Drugs* (Sandler, M., ed.) University Park Press, Baltimore, Maryland.

Lipinski, C. A. and Hutson, N. J. (1984) Aldose reductase inhibitors as a new approach to the treatment of diabetic complications. *Ann. Repts. Med. Chem.* **19**, 169–177.

Moncada, S. and Vane, J. R. (1980) Inhibitors of Arachidonic Acid Metabolism, in *Enzyme Inhibitors as Drugs* (Sandler, M., ed.) University Park Press, Baltimore, Maryland.

Ortmann, R., Schaub, M., Felner, A., Lauber, J., Christen, P., and Waldmeier, P. C. (1984) Phenylethylamine-induced stereotypies in the rat: A behavioral test system for assessment of MAO-B inhibitors. *Psychopharmacology* **84**, 22–27.

Penning, T. M. (1983) Design of suicide substrates: An approach to the development of highly selective enzyme inhibitors as drugs. *Trends Pharmacol. Sci.* **4**, 212–217.

Planz, G., Palm, D., and Quiring, K. (1973) On the evaluation of weak and reversible inhibitors of monoamine oxidase in vivo and in vitro. *Arzneimittel-Forsch.* **23**, 281–285.

Pletscher, A. and Besendorf, H. (1959) Antagonism between harmaline and long-acting monoamine oxidase inhibitors concerning the effect on 5-hydroxytryptamine and norepinephrine metabolism in the brain. *Experientia* **15**, 25–26.

Porter, C. C., Watson, L. S., Titus, D. C., Totaro, J. A., and Byer, S. S. (1962) Inhibition of the dopa decarboxylase by the hydrazino analog of α-methyldopa. *Biochem. Pharmacol.* **11**, 1067–1077.

Quitkin, F., Rifkin, A., and Klein, D. F. (1979) Monoamine oxidase inhibitors. *Arch. Gen. Psychiat.* **36**, 749–760.

Rajashekhar, B., Fitzpatric, P. F., Colombo, G., and Villafranca, J. J. (1984) Synthesis of several 2 substituted 3-(p-hydroxyphenyl)-1-propenes and their characterization as mechanism based inhibitors of dopamine β-hydroxylase. *J. Biol. Chem.* **259**, 6925–6930.

Rando, R. R. (1974) Chemistry and enzymology of k_{cat} inhibitors. *Science* **185**, 320–324.

Silverman, R. B. (1983) Objection to terminology used in special reports. *Chem. Eng. News* **61** (43), 2.

Sjoerdsma, A. (1981) Suicide enzyme inhibitors as potential drugs. *Clin. Pharmacol. Ther.* **30**, 3–22.

Sjoerdsma, A. and von Studnitz, W. (1963) Dopamine-β-oxidase activity in man, using hydroxyamphetamine as substrate. *Br. J. Pharmacol. Chemother.* **20**, 278–284.

Srere, P. A. (1967) Enzyme concentrations in tissues. *Science* **158**, 936–937.

Umezawa, H. (1982) Low-molecular-weight enzyme inhibitors of microbial origin. *Ann. Rev. Microbiol.* **36**, 75–99.

Waldmeier, P. C. and Baumann, P. A. (1983) Effects of CGP 11305A, a new reversible and selective inhibitor of MAO A, on biogenic amine levels and metabolism in the rat brain. *Naunyn Schmiedebergs Arch. Pharmacol.* **324**, 20–26.

Webb, J. L. (1963) *Enzyme and Metabolic Inhibitors.* Academic, New York.

Wilkinson, G. N. (1961) Statistical estimations in enzyme kinetics. *Biochem. J.* **80**, 324–332.

Wolfenden, R. (1976) Transition state analog inhibitors and enzyme catalysis. *Ann. Rev. Biophys. Bioeng.* **5**, 271–306.

EEG, EEG Power Spectra, and Behavioral Correlates of Opioids and Other Psychoactive Agents

GERALD A. YOUNG, OKSOON HONG, AND NAIM KHAZAN

1. Electroencephalography (EEG)—An Overview

The discovery of electrical potentials of the brain is thought to have been made by Caton (1875), who presented the results of his research with rabbits and monkeys to the British Medical Association in Edinburgh. Half a century later in Jena, Austria, Hans Berger (1929) discovered human brain waves and, hence, Berger is recognized as the father of electroencephalography, the recording of oscillations in the potential differences between two points in the brain (Berger, 1937).

Although the generation of time-dependent EEG waveforms is a complex phenomenon, the neurophysiological events that contribute to EEG recordings have been thoroughly studied (Speckmann and Elger, 1982). Axonal action potentials are associated with intracellular and extracellular ionic flows that generate extracellular field potentials. If appropriate electronic instrumentation is used, continual analog measurements of these extracellular field potentials result in an EEG recording. EEG activities vary in frequency and amplitude. The clinically relevant EEG frequency range is approximately from 0.1 to 100 Hz (Niedermeyer, 1982). This frequency range is usually divided into frequency bands of 0–4 Hz (delta), 4–8 Hz (theta), 8–13 Hz (alpha), and 13 Hz and higher (beta).

2. Use of EEG and EEG Power Spectra in Drug Classification

2.1. Clinical Use of EEG Quantification in the Study of Psychoactive Drugs

The raw EEG has been used extensively to study and assess effects of drugs upon the brain (Brazier, 1964). Based upon visual evaluations of clinical EEG recordings, it was established that psychoactive drugs produce discrete changes in EEG activity (Itil, 1961; Fink, 1964). Although this approach provided pertinent information, recent computer analysis techniques have been developed that provide quantitative characterization of EEG activities, which have allowed experimenters to more easily assess both subtle and complex EEG changes produced by drugs and chemicals. One of these techniques provides a periodic analysis of an EEG sample by the measurement of the time intervals between successive zero baseline crossings of the EEG waveforms (Stein et al., 1949). Furthermore, EEG frequency characteristics can be separated from EEG amplitude characteristics. A second technique generates EEG power spectral density arrays based upon the estimation of a complex Fourier series from the sample (Grass and Gibbs, 1938). Later, a fast Fourier transformation (FFT) expressed by a single algorithm provided a means to derive EEG power spectra very efficiently with high-speed digital computers (Walter, 1963; Cooley and Tukey, 1965).

The first utilization of the spectral analysis of EEG to assess drug effects was a clinical study carried out by Gibbs and Maltby (1943). It was observed that barbiturates and morphine produced shifts in cortical EEG to slower frequencies, whereas caffeine, amphetamine, and epinephrine produced shifts to faster frequencies. Consequently, Fink, Itil, and their colleagues proposed a drug classification scheme based upon the periodic analysis of cortical EEG and characteristic changes that occur following drug administration to humans in a clinical setting. They originally categorized psychoactive drugs into six groups (Fink et al., 1958; Itil, 1968) based upon differences in drug-induced EEG effects. Thereafter, the Fink and Itil groups continued their efforts in characterizing newly available psychoactive drugs by delineating their EEG effects with the use of periodic analyses (Itil, 1969, 1971; Itil et al., 1970, 1977,

1978; Fink et al., 1977; Fink and Irwin, 1980a,b). For example, mianserin, a serotonin antagonist, was first classified as an antidepressant using their quantitative EEG model; this finding has been confirmed in the clinic and it has been marketed as an antidepressant in Europe for many years (Itil et al., 1972). More recently, the previous drug classification data have been reanalyzed using both periodic analysis and EEG power density analysis and the degree of replicability, reliability, and validity of this data has been established (Itil et al., 1981, 1982).

In other human studies of psychoactive agents employing spectral analyses of cortical EEG, methadone was found to increase spectral power in the 0–10 Hz range (Kay, 1975), alcohol increased 0–10 Hz activity and abolished alpha EEG activity, and the neuroleptic perphenazine increased slower frequencies and reduced the spectral peak in the alpha band (8–13 Hz) (Rosadini et al., 1977). In patients in which chlorpromazine produced marked side effects, it also produced decreases in alpha band spectral power and increases in theta band (4–8 Hz) spectral power (Laurian et al., 1981). The anxiolytic diazepam increased delta (0–4 Hz) and beta (13 Hz and higher) activities, slowed the alpha peak frequency, and decreased the amplitude of the alpha band spectral peak (Matejcek, 1978). Estimates of equipotent doses of the anxiolytics temazepam and nitrazepam were based upon peak EEG power spectral effects (Matejcek et al., 1983). In studies of antidepressants, typical and atypical antidepressants were reported to induce different EEG profiles (Saletu, 1982). Thus, different classes of psychoactive drugs produce very characteristic EEG power spectral profiles, and new psychotropic drugs can be categorized based upon similarities between their EEG profiles and those in a reference library of previous data obtained with a range of reference agents.

2.2. Pharmaco-EEG Correlates of Psychoactive Agents in Experimental Animals

Several antianxiety agents have been studied using EEG spectral analysis. The two benzodiazepines, diazepam and flurazepam, increased cortical EEG spectral power above 20 Hz when administered orally in the cat. Moreover, diazepam also decreased slower frequency power, whereas flurazepam increased spectral power in the middle frequencies (Schallek and Johnson, 1976). Gehrmann and Killam (1978) found that chlordiazepoxide and clonazepam produced moderate increases in cortical EEG spectral power in the 8–32 Hz sub-band in Macaca mulatta monkeys, whereas meprobamate produced increases in cortical EEG spectral power in the 16–64 Hz sub-band. Diazepam, nitrazapem, flunitrazepam, and zopiclone produced increases in cortical EEG spectral power in the beta frequencies (13 Hz and higher) along with the appearance of sensorimotor sleep spindles in the rat (Depoortere et al., 1983). The above data suggest that, in general, antianxiety agents produce increases in relatively higher-frequency EEG spectral power.

The antipsychotic chlorpromazine was found to increase cortical EEG spectral power in the slower frequency range and to decrease spectral power at the faster frequencies in the Macaca mulatta monkey (Gehrmann and Killam, 1975, 1976). Orally administered fluphenazine in dogs increased occipital EEG spectral power in the 0–20 Hz range with a peak at 16 Hz 4 h posttreatment (Robert et al., 1978).

The tricyclic antidepressant amitriptyline was observed to increase cortical EEG spectral power in the lower frequency range and to decrease spectral power at the faster frequencies in the Macaca mulatta monkey (Gehrmann and Killam, 1975, 1976). The antidepressants imipramine, amitriptyline, desipramine, and mianserin increased cortical EEG spectral power in the 4.7–7.5 Hz range in dogs (Frankenheim, 1982). Thus, antipsychotics and antidepressants appear, in general, to produce increases in relatively lower-frequency EEG spectral power. Any differences in power spectral characteristics produced by antipsychotics and antidepressants would need to be delineated in studies in which the species, electrode locations, and power spectral parameters were the same.

Morphine produced naloxone-reversible increases in 8–16 Hz cortical EEG activity in the dog (Pickworth et al., 1982). Spectral power in the 5–7 Hz band was found to be a good indicator of the duration of morphine effect in the rat (Bronzino et al., 1982). The mixed agonist–antagonist opioid analgesics buprenophine in the rat (Kareti et al., 1980) and bremazocine in the dog (Freye et al., 1983) produced increases in EEG spectral power in the delta (0–4 Hz) and theta (4–8 Hz) frequency bands. The synthetic opioid peptide FK 33-824 also produced increases in EEG spectral power in the delta and theta frequency bands (Freye et al., 1982). Thus, different classes of psychoactive drugs have also been shown to produce very characteristic power spectral EEG profiles in preclinical studies.

3. Delineation of Opioid Pharmacodynamic Properties with EEG and Behavior

3.1. Introduction

As depicted in the previous section, a prevalent utilization of EEG and EEG power spectral techniques in pharmacology has been to categorize different classes of CNS-active agents by their differential qualitative and quantitative effects on EEG and associated power spectral parameters. Furthermore, effects of new CNS-active agents on EEG and associated power spectra have been compared to those already established with previous data. In our laboratory, however, drug-induced changes in EEG and EEG spectral power have been used to delineate pharmacodynamic properties of CNS-active agents (Khazan, 1975; Khazan and Young, 1980). For example, pharmacological phenomena such as tolerance, cross-tolerance, physical dependence, receptor selectivity, and stereospecificity have been investigated using EEG and EEG power spectra. Therefore, examples of this unique utilization of EEG and EEG power spectra are illustrated in this section for opioids and in subsequent sections for ethanol, Δ^9-THC, and certain neurotoxins.

3.2. Differential Effects of Mu, Kappa, and Sigma Agonists on EEG, EEG Power Spectra, and Behavior

Martin et al. (1976) and Gilbert and Martin (1976) originally proposed, based upon in vivo findings of acute and chronic opioid effects on neurophysiological and behavioral parameters in the chronic spinal dog preparation, that different opioid agonists appear to activate different populations of receptors. Thus, Martin et al. (1976) described morphine as a prototypic opioid acting on mu receptors producing miosis, bradycardia, hypothermia, depressed nociception, and indifference to environmental stimuli; ketocyclazocine acting on kappa receptors producing miosis, flexor reflex depression, sedation, and no change in pupil size; and SKF 10,047 (*N*-allylnormetazocine) acting on sigma receptors producing mydriasis, tachycardia, tachypnea, and mania, and producing abstinence in morphine-dependent dogs. Since then, many additional differences in the in vivo and in vitro effects of these prototypic opioid agonists on several neurophysiological, behavioral, and biochemical parameters have been delineated.

For example, in studies of the effects of various opioids on flurothyl-induced seizures in the rat, morphine and other mu opioid agonists demonstrated dose-related anticonvulsant effects that were antagonized by naloxone (Cowan et al., 1979). Tolerance developed to these anticonvulsant effects upon chronic treatment. In contrast, neither ketocyclazocine nor ethylketocyclazocine (EKC), kappa agonists, or related opioids had significant effects on flurothyl-induced seizure thresholds. Furthermore, SKF 10,047 demonstrated a dose-related anticonvulsant effect like that of morphine, but, in contrast to morphine, no tolerance developed to this effect and naloxone was ineffective as an antagonist, even at a 10 mg/kg dose.

In studies in which both cortical EEG activity and behavior were assessed, morphine, ketocyclazocine, and EKC were reported to produce similar "EEG-behavior dissociation" in the dog (Pickworth and Sharpe, 1979). However, morphine reportedly increased the amount of sleep and decreased body temperature, heart rate, and respiratory rate; whereas ketocyclazocine and EKC had no effect on sleep and increased body temperature, heart rate, and respiratory rate. The morphine-induced effects were reported to be more sensitive to naloxone antagonism than those of either ketocyclazocine or EKC. Furthermore, Tortella et al. (1980) reported that in the rat both morphine and EKC produced similar biphasic EEG and behavioral profiles.

Intravenously administered morphine to freely moving rats produced high-voltage cortical EEG bursts (Khazan et al., 1967) associated with increases in EEG spectral power in the 0–10 Hz range (Fig. 1) (Young et al., 1981). Ketocyclazocine produced high-voltage cortical EEG bursts associated with a predominant spectral peak in the 5–8 Hz band. SKF 10,047 produced desynchronized cortical EEG along with frequent theta wave activity; associated EEG power spectra consisted of the lowest power, peaking at about 7.5 Hz.

Furthermore, when EEG and behavioral effects of opioid enantiomers were assessed, (−)-methadone (mu agonist) produced increases in spectral power over the 0–10 Hz range, whereas (−)-ketocyclazocine produced increases in the 5–8 Hz band as a predominant peak (Fig. 2) (Young and Khazan, 1984a). The (+-enantiomers of methadone and ketocyclazocine were inactive. (+)-SKF 10,047 produced a predominant spectral peak in the 7–9 Hz band that was associated with aroused, bizarre behavior that suggested psychotomimetic effects. The effects of morphine on EEG and EEG power spectra were more sensitive to antagonism by naloxone than those produced by ketocyclazocine. The effects of

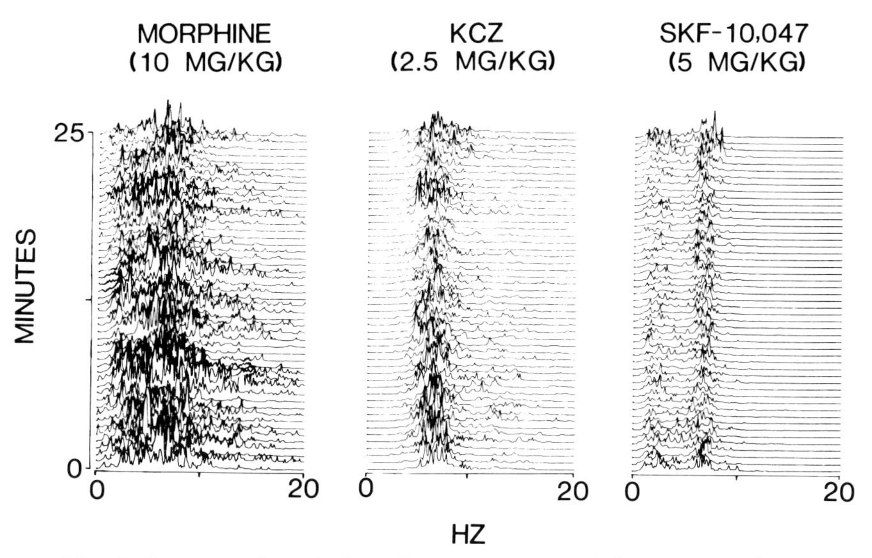

Fig. 1. Sequential cortical EEG power spectra following iv administration of morphine (10 mg/kg), ketocyclazocine (2.5 mg/kg), and SKF 10,047 (5 mg/kg) (reprinted from Young et al., 1981, with permission).

(±)-SKF 10,047 and (+)-SKF 10,047 were not antagonized by naloxone (10 mg/kg), whereas the effects of (−)-SKF 10,047 were only partially antagonized. (−)-SKF 10,047 also blocked EEG and behavioral effects of morphine in naive rats and precipitated withdrawal in morphine-dependent rats, whereas (+)-SKF 10,047 did not (Khazan et al., 1984). The above findings lend further support to the theory of multiple opioid receptors (Martin et al., 1976; Lord et al., 1977) and to their stereospecificity.

3.3. Tolerance and Cross-Tolerance Among Opioids

In comparing the development of tolerance to the effects of morphine and ethylketocyclazocine (EKC) on EEG, EEG power spectra, and behavior, the possibility of cross-tolerance was assessed (Young and Khazan, 1984b). In nontolerant rats, 10 mg/kg (iv) injections of morphine and EKC produced biphasic EEG and behavioral profiles lasting for 3 and 2 h, respectively. In both cases, a stuporous phase, associated with high-voltage cortical EEG bursts, was followed by a hyperactive phase, associated with low-voltage desynchronized EEG. However, power spectra derived from epochs of EEG bursting produced by morphine and EKC were qualitatively

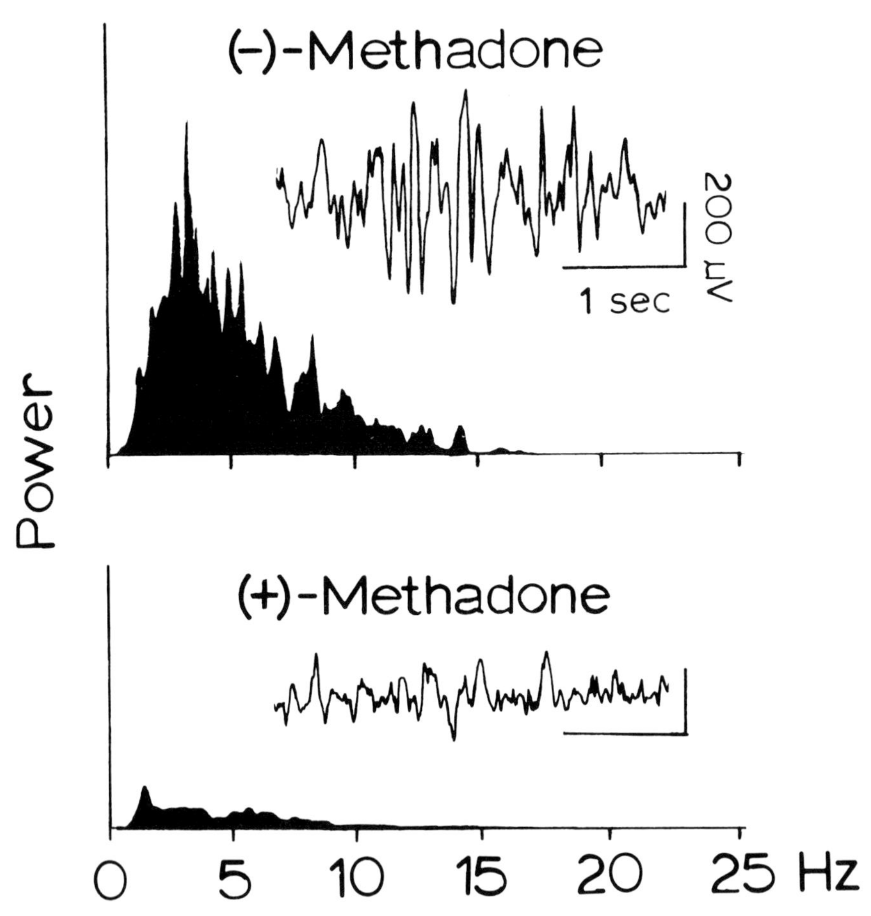

Fig. 2. Cortical EEG recordings and related power spectra after acute iv administration of (−)- and (+)-methadone (1.0 mg/kg) in an individual rat. (reprinted from Young and Khazan, 1984a, with permission).

different (Fig. 3). One group of rats was given a series of automatic, iv injections of morphine, and a second group received EKC. Following chronic treatment, the duration of the biphasic EEG and behavioral profiles produced by morphine and EKC were both significantly reduced. In both cases, the intensity of EEG bursting was also reduced, as reflected by significant quantitative reductions in EEG power spectral densities (Fig. 3).

In another study, effects of morphine (5–80 mg/kg) and methadone (0.5–4.0 mg/kg) were assessed in both naive and *1*-alpha-acetylmethadol (LAAM)-maintained rats (Lukas et al., 1982).

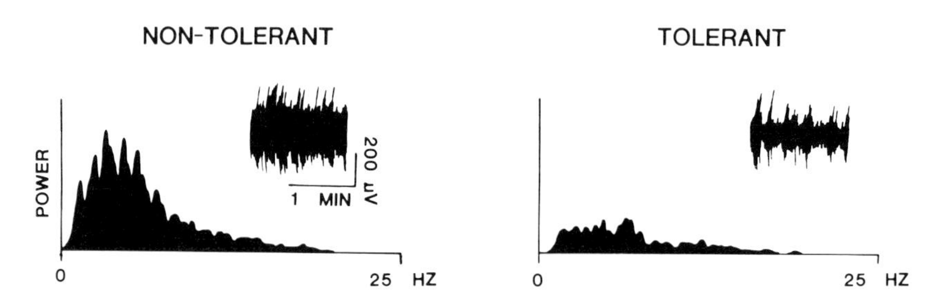

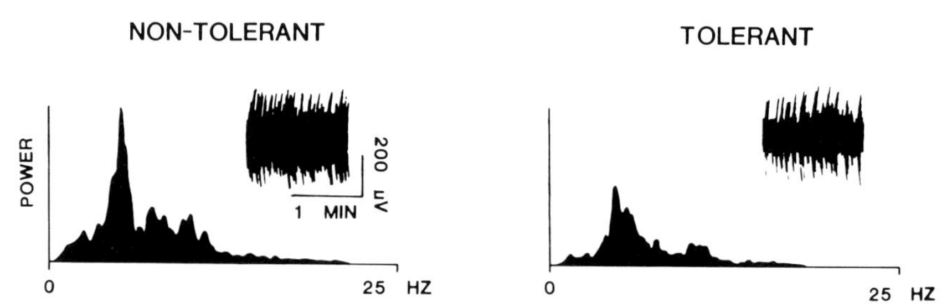

Fig. 3. Cortical EEG samples and associated EEG power spectra are shown during opioid-induced behavioral stupor. Effects of morphine are shown in the top of the figure during the nontolerant and tolerant states. Similar data are shown after the administration of ethylketocyclazocine in the bottom of the figure (reprinted from Young and Khazan, 1984b, with permission).

LAAM maintenance consisted of programmed iv injections every 3 h for 2 wk at a final dose of either 2 or 8 mg/kg/day. At the end of the 2 wk, each rat received a single iv injection of either morphine or methadone, and the resultant EEG was quantified by using power spectral analysis, the EEG being correlated with overt behavioral changes. The morphine-induced increases in EEG amplitude and EEG spectral power were markedly attenuated in rats that were maintained on the low dose of LAAM (Fig. 4). The duration of the stuporous phase after morphine challenge was shortened in the low-dose LAAM-maintained rats. In addition, maintenance on the higher dose of LAAM produced further significant decreases

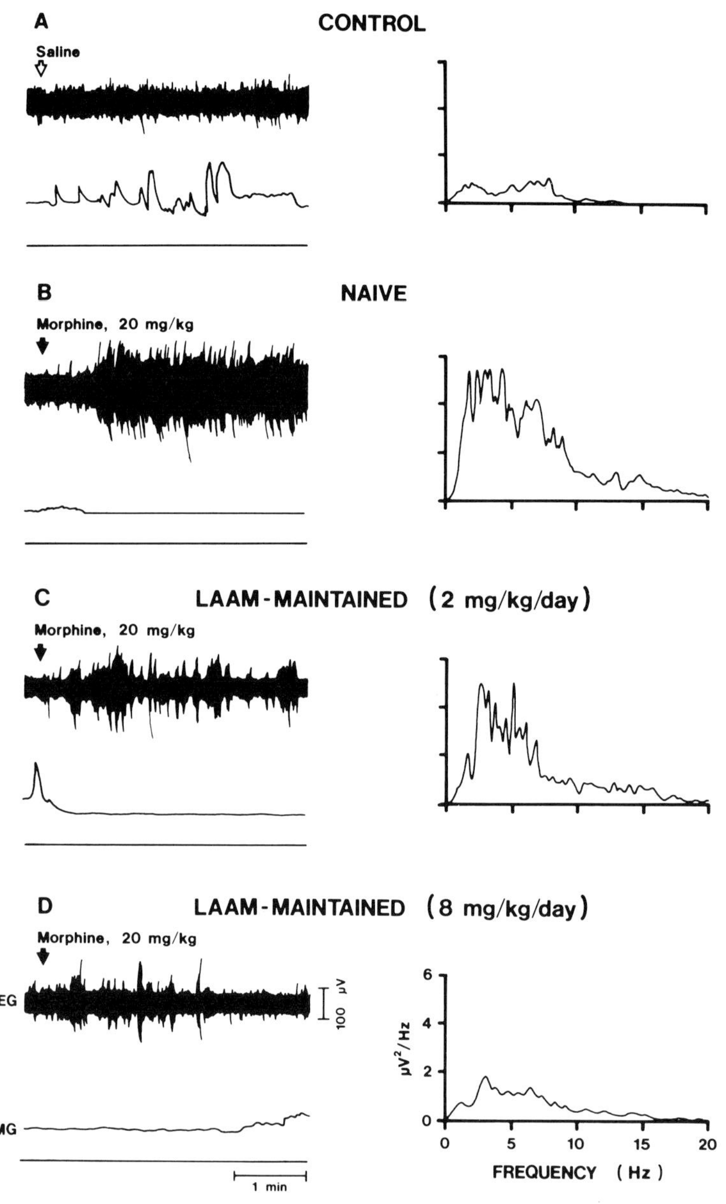

Fig. 4. Effects of morphine on the direct EEG and integrated EMG and the corresponding EEG power spectrum in naive and LAAM-maintained rats. Morphine was given iv over a 15-s interval. The averaged power spectrum was taken for the first 1 min of high-amplitude EEG activity (reprinted from Lukas et al., 1982, with permission).

in these effects of morphine. On the other hand, whereas low-dose LAAM-maintained rats also exhibited some degree of cross-tolerance to methadone, maintenance on a higher dose of LAAM failed to further attenuate the EEG and behavioral effects of methadone. These results suggest that EEG and behavioral cross-tolerance in LAAM-maintained rats is more complete to morphine than to methadone.

3.4. Differential Abstinence Syndromes Between Rats Chronically Treated with Mu and Kappa Opioid Agonists

In a comparative study, one group of rats was chronically administered iv morphine, and a second group received chronic injections of iv EKC (Young and Khazan, 1985). Morphine abstinence was associated with REM sleep suppression, wet-dog shakes, diarrhea, ptosis, piloerection, irritability, and a decline in EEG spectral power during SWS episodes (Fig. 5). In contrast, the EKC abstinence syndrome was associated only with increases in wet-dog shakes.

Protracted symptoms of abstinence were assessed in other rats that were chronically treated with morphine or EKC as above (Young and Khazan, 1986). Fifteen days and 1 mo after withdrawal of these opioids, these two groups of posttolerant rats were challenged with morphine and EKC, respectively. In morphine posttolerant rats, morphine challenges (10 mg/kg, iv) produced primarily EEG and behavioral arousal for 2–3 h. This is in contrast to control rats in which morphine challenges produced a biphasic EEG and behavioral response consisting of 60–90 min of depression followed by 60–90 min of arousal. In EKC posttolerant rats, EKC challenges (10 mg/kg, iv) produced a "naive-like" biphasic EEG and behavioral response consisting of about 60 min of depression followed by about 60 min of arousal; EKC challenges in control rats produce similar effects. Thus, protracted effects on EEG and behavior were evident in morphine posttolerant rats, but not in EKC posttolerant rats. These results further extended the range of pharmacodynamic differences observed between these two prototype agonists for the mu and kappa receptors.

3.5. Characteristic Changes in REM Sleep EEG Frequencies During Opioid Self-Administration

Power spectral analyses were used to study changes in cortical EEG during morphine self-administration (Young et al., 1978a). As

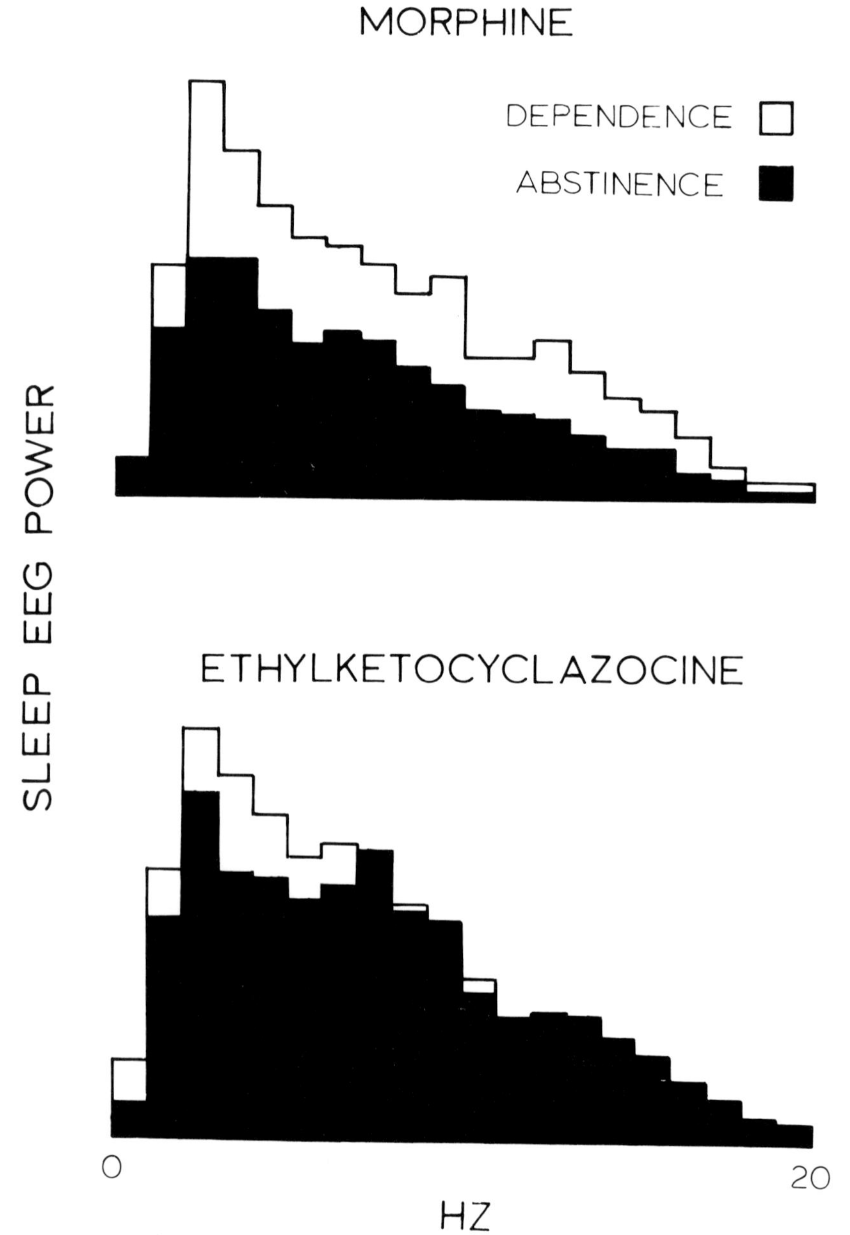

Fig. 5. Mean sleep EEG power spectra are shown during dependence on morphine and ethylketocyclazocine and during the 8th h of spontaneous abstinence. Sleep EEG power is presented as a function of 1-Hz intervals (reprinted from Young and Khazan, 1985, with permission).

time progressed from a morphine self-injection toward another injection, a significant spectral shift of the EEG to lower frequencies occurred during successive REM sleep episodes (Fig. 6). Each morphine self-injection reinstated the predominance of higher frequencies in the EEG spectra. The EEG changes that preceded lever pressing may reflect changes in morphine plasma levels and in the state of the CNS that precedes drug-seeking behavior.

In a comparative study, changes in EEG power spectra derived from successive REM sleep EEG episodes during self-administration of five opioids were delineated; those opioids being morphine, methadone, LAAM, N-LAAM, and DN-LAAM (Steinfels et al., 1980). During self-administration of these opioids, the first REM sleep episode following an injection had the faster peak EEG frequency. The peak EEG frequencies declined in a linear fashion toward the next injection. Differences in the slopes of the linear peak EEG frequency declines of the different opioids appeared to correlate with differences in the pharmacodynamic profiles. For example, the slopes were significantly steeper for morphine and methadone when compared to LAAM, N-LAAM, and DN-LAAM; the durations of action of morphine and methadone are significantly shorter than those of LAAM, N-LAAM, and DN-LAAM.

4. Pharmacodynamics of Other Psychotropic Drugs

4.1. EEG and Behavioral Effects of Ethanol

The concurrent effects of acute ethanol administration upon the EEG and behavior in freely moving animals have been primarily studied in the cat. In unrestrained cats small doses of ethanol produced cortical EEG and behavioral activation, intermediate doses produced a dose-related lowering of cortical EEG frequencies that were associated with behavioral intoxication, whereas larger doses produced eventual decreases in cortical EEG amplitude along with behavioral coma (Horsey and Akert, 1953; Perrin et al., 1974). In one study with freely moving rats, ip-administered ethanol produced a dose-related lowering of cortical EEG frequencies associated with "large irregular amplitude" (Prado de Carvalho and Izquierdo, 1977). Previous exposure to ethanol in rats was also reported to produce REM sleep disruption (Gitlow et al., 1973).

In studies utilizing computer-assisted analyses of EEG in the cat, smaller doses of ethanol produced increases in higher-frequency spectral power, whereas larger doses increased lower-frequency

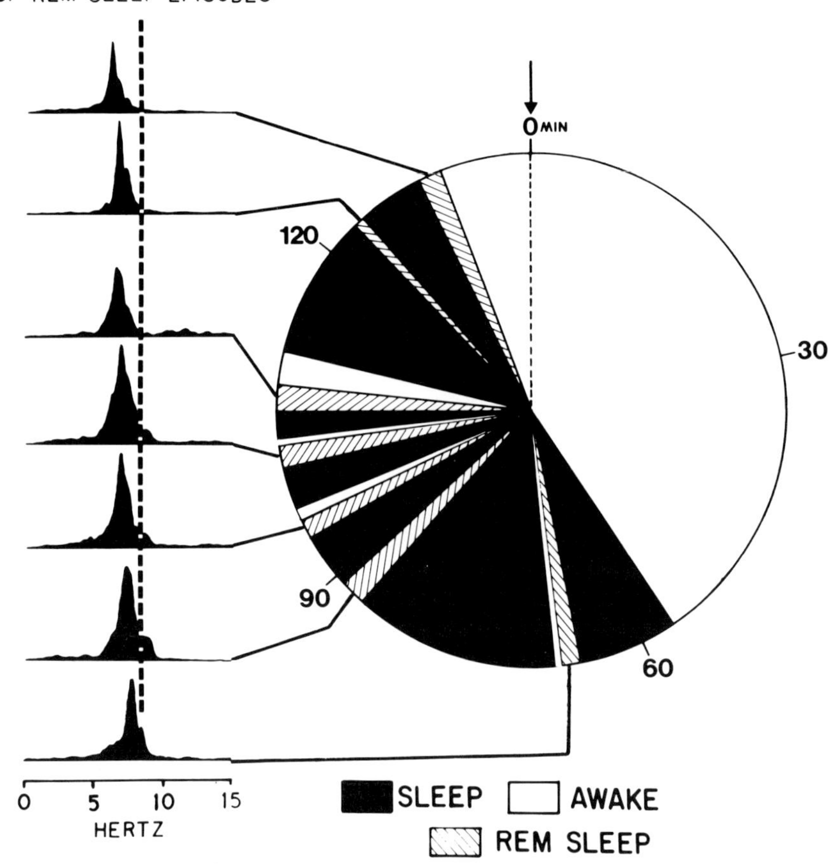

Fig. 6. The pattern of distribution of sleep–awake behavior in an individual morphine-dependent rat during a morphine interinjection interval is shown in the right portion of the figure. Successive power spectra derived from cortical EEG samples during each occurrence of REM sleep are shown in the left portion of the figure. The successive occurrence of REM sleep proceed from the bottom to the top of the figure. Spectral power is presented as a function of frequency (reprinted from Young et al., 1978b, with permission).

spectral power and decreased higher-frequency spectral power (Sauerland and Harper, 1970; Dolce and Decker, 1972). When cortical EEG during slow-wave sleep in the cat (Bronzino et al., 1973), mouse (Moreley and Bradley, 1977), or rat (Young et al., 1978b) has been subjected to spectral analysis, the resulting power spectra have consisted of relatively more lower-frequency than higher-frequency spectral power. These power spectral characteristics of the EEG in slow-wave sleep appear to be qualitatively similar to those associated with ethanol administration in the cat (Sauerland and Harper, 1970; Dolce and Decker, 1972).

Cortical EEG and EEG power spectra during slow-wave sleep were compared with those induced by acute ethanol administration in the rat (Wolf et al., 1981). Ethanol produced dose-related (1, 2, and 4 mg/kg, iv) increases in the mean duration of synchronous EEG episodes. The EEG effects of ethanol were compared to those of slow-wave sleep. Differences in EEG waveforms between the EEG in slow-wave sleep and ethanol-induced EEG synchrony were detected. Dose-related increases by ethanol in 0–4 Hz spectral power and decreases in 8–13 Hz spectral power were found. It was, therefore, concluded that there were discrete differences between the normal EEG in slow-wave sleep and ethanol-induced EEG synchrony in the rat.

Furthermore, after acute ethanol administration, dose-dependent linear declines in blood ethanol concentration were found (Young et al., 1982). Ethanol-induced increases in EEG spectral power in the 0–4 Hz band persisted long after blood ethanol levels had declined to 0; therefore, no correlation was found (Fig. 7). Acute ethanol administration also produced an initial drop in 8–13 Hz spectral power. Then, as blood ethanol levels declined, 8–13 Hz spectral power increased toward normal; a significant negative linear correlation was found.

4.2. EEG and Behavioral Effects of Cannabinoids

Several tetrahydrocannabinol (THC) analogs, including Δ^9-THC and Δ^8-THC, were found to induce a flattening of cortical and hippocampal EEGs and to produce trains of high-voltage EEG bursts in the cat, rabbit, and rat (Lipparini et al., 1969). Other investigators have reported similar EEG effects produced by Δ^9-THC in the cat (Hockman et al., 1971) and rabbit (Fujimori and Himwich, 1973; Fujimori et al., 1973), by marijuana extract, Δ^9-THC, or Δ^8-THC in the rat (Masur and Khazan, 1970; Colasanti and Khazan, 1971; Moreton and Davis, 1973), and by Δ^9-THC in the rhesus monkey (Martinez et al., 1972; Heath et al., 1980). Marijuana extract, Δ^9-THC, and Δ^8-THC have also been reported to produce high-

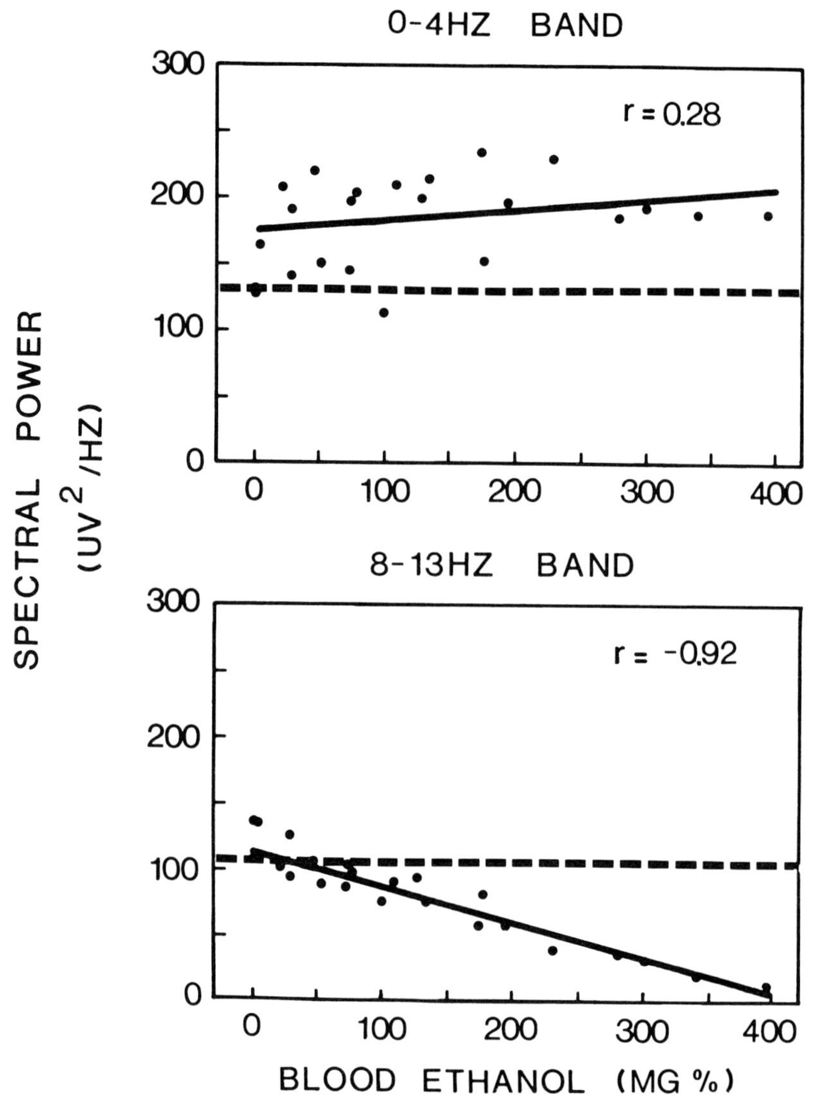

Fig. 7. Correlations of blood ethanol levels with EEG spectral power alterations after acute ethanol administration. Values are the means for each rat for consecutive 30-min blocks at the 1 and 2 g/kg doses and for 1-h blocks at the 4 k/kg dose. For both the 0–4 Hz and 8–13 Hz bands, time-correlated points were combined for the three ethanol doses. Linear regression lines and correlation coefficients are shown. The dashed lines represent the mean power spectral values during normal slow-wave sleep (reprinted from Young et al., 1982, with permission).

voltage EEG bursts during occurrences of REM sleep in the rat (Masur and Khazan, 1970; Moreton and Davis, 1973).

In subsequent studies using power spectral analysis, it was discovered that Δ^9-THC administration in the rat produced characteristic EEG changes (Buonamici et al., 1982). Intraperitoneally administered Δ^9-THC (5 and 10 mg/kg) produced a reduction in peak-to-peak voltage of the desynchronized cortical EEG during wakefulness (Fig. 8). Associated spectral power was reduced to about 50% of control during the first hour after injection of Δ^9-THC and gradually returned toward the control value over an 8-h period. Occurrences of Δ^9-THC-induced high-voltage EEG bursts, overriding the reduced EEG tracing, were associated with an EEG spectral peak at 6 Hz. The first few SWS episodes appearing after Δ^9-THC administration were associated with more slow-frequency waveforms and more slow-frequency spectral power than with control SWS episodes. During control REM sleep episodes, an EEG theta wave pattern, with an associated spectral peak at about 8 Hz, was characteristic. Conversely, the first few REM sleep episodes emerging after Δ^9-THC administration contained high-voltage bursts, the related power spectra of which had two peaks at about 7 and 11 Hz.

5. Use of EEG Power Spectra in Neurotoxicology Studies

It has been suggested that the same approach utilized in the study of effects of CNS drugs on EEG and behavior can be utilized for the study of the effects of neurotoxic substances (Khazan and Young, 1980). A similar suggestion was recently advanced by Benignus (1984). Although our major research effort has dealt mainly with the opioids and other CNS active agents, collaborative studies carried out with C. Eccles of the neurotoxicology laboratory (University of Maryland School of Pharmacy) have demonstrated that EEG changes can be detected after exposure to lead (Pb), trimethyltin (TMT), and kainic acid.

EEG activities and seizure responsiveness were examined in rats exposed to lead via the dam's milk from birth to the time of weaning (McCarren and Eccles, 1983). In the prekindled state, electrically induced hippocampal afterdischarges (ADs) of Pb-treated rats were more sensitive to the effects of a phenytoin challenge. A phenytoin-induced increase in AD duration was exacerbated in

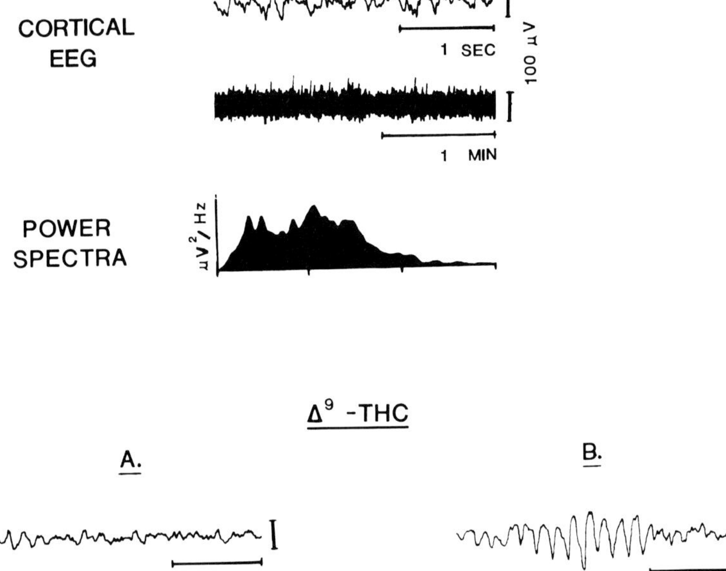

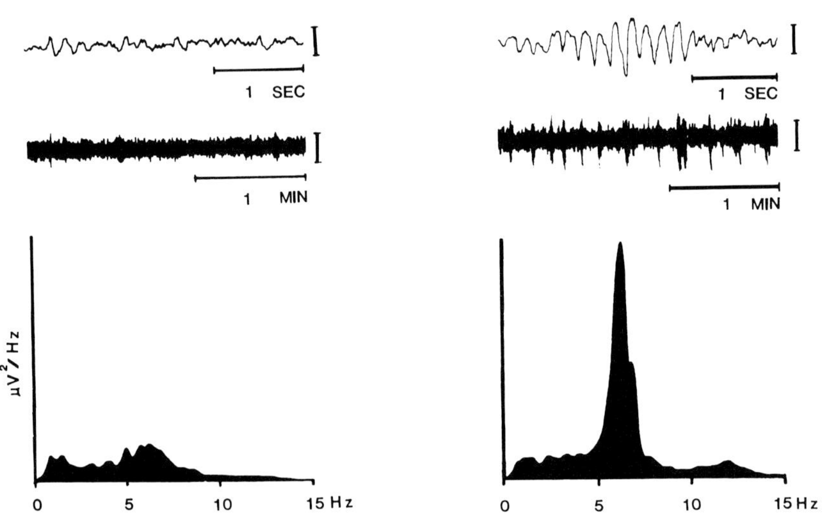

Fig. 8. Direct cortical EEG recordings and related EEG power spectra during wakefulness before (control) and after Δ^9-THC administration (10 mg/kg ip) in an individual rat. Data in panel A (bottom of figure) illustrates Δ^9-THC-induced reduction of both desynchronized cortical EEG and related EEG power spectra. Data in panel B illustrates Δ^9-THC-induced high-voltage EEG bursts and related power spectra (reprinted from Buonamici et al., 1982, with permission).

Pb-treated groups. Pb-treated rats subjected to hippocampal kindling as adults were not found to have altered kindling rates, but differed from control in several other respects (McCarren et al., 1984). In contrast to the low-level lead groups and control, the high-level lead group did not display an increase in AD duration with kindling. A clear dose-related effect of lead on the spectra of the kindled AD was characterized by greater power in the low-frequency bands (Fig. 9). These data indicated that long-lasting effects can occur following a neonatal lead exposure and are detectable with the use of EEG parameters.

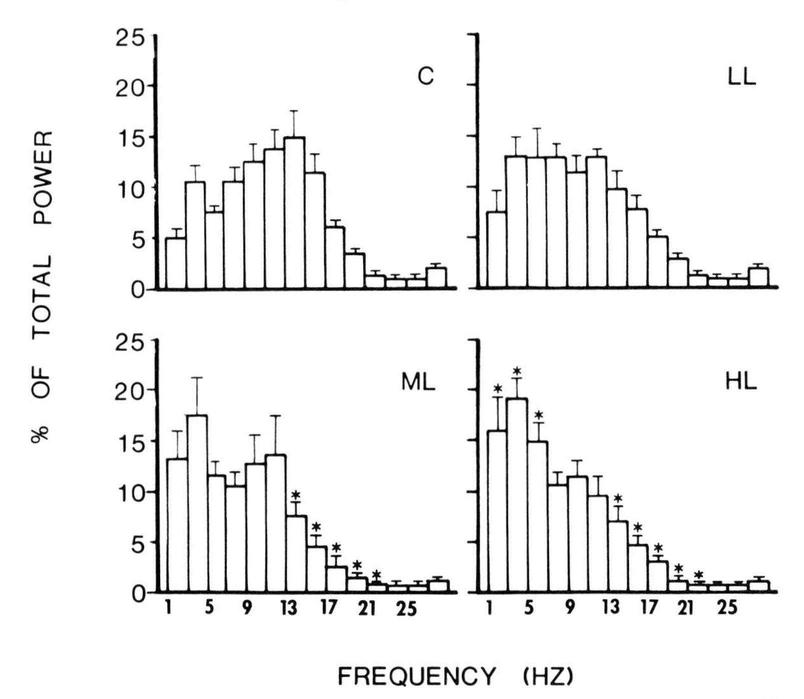

Fig. 9. Power spectra of the kindled primary afterdischarge: effect of lead treatment. Values are mean ± SEM values of the percentage of total power in each 2-Hz band (final band includes frequencies 27–35 Hz). *A significant difference from control; c, control; LL, low lead; ML, medium lead; HL, high lead (reprinted from McCarren et al., 1984, with permission).

More recently, using this EEG-EMG rat model, EEG and associated behavioral effects have been investigated in rats sustaining limbic system damage. Trimethyltin (TMT), an agent with rather specific toxicity for limbic system neurons, produces alterations in EEG spectra as early as 1 d after systemic administration (Stratton et al., 1983). Peak frequencies during REM sleep were significantly lower than control values for at least 28 d. Alterations in SWS spectra appeared after the changes in REM sleep spectra were already manifest. SWS spectra of TMT-treated animals contained more relative power in the 0–4 Hz band on d 14 and in the 4–8 Hz band on d 7 than control SWS EEG spectra. Furthermore, TMT-treated animals displayed severely disrupted sleep–awake cycles in the first 2 wk after TMT administration.

In order to determine if these effects were caused by the loss of CA3 pyramidal cells, lesions in the CA3 field of the hippocampus were produced by local injection of 0.1 μg of kainic acid in another group of rats (Stratton et al., 1985). In kainic-acid-treated rats, the mean peak theta frequency during REM sleep episodes was significantly lower, an effect similar to that of TMT. Kainic acid treatment, however, produced no alterations in SWS EEG power spectra. The differences in the EEG effects of TMT and kainic acid treatments on SWS spectra may be caused by a relatively more extensive neurotoxic effect produced by TMT than that produced by kainic acid.

6. Summary and Conclusions

In both clinical and preclinical studies, different classes of psychoactive drugs have been shown to produce very characteristic EEG power spectral profiles. New psychoactive drugs can be categorized based upon similarities between their EEG profiles and those in a reference library of previous data obtained with a range of reference agents. For example, mianserin, a serotonin antagonist, was first classified as an antidepressant using EEG periodic analysis. This finding has been confirmed in the clinic and it has been marketed as an antidepressant in Europe for many years (Itil et al., 1972).

Drug-induced changes in EEG and EEG spectral power have also been used to delineate pharmacodynamic properties of CNS-active agents. For example, in comparative studies of the opioids morphine (mu agonist), ketocyclazocine (kappa agonist), and SKF 10,047 (sigma agonist), differential and stereospecific effects on EEG, EEG power spectra, and behavior were shown. Chronic treatment with morphine and EKC produced tolerance to their corresponding EEG, EEG power spectral, and behavioral effects. LAAM-tolerant rats were cross-tolerant to the EEG power spectral and behavioral effects of morphine, and were less cross-tolerant to effects of methadone. Morphine abstinence in morphine-tolerant rats were associated with behavioral withdrawal signs and a decline in EEG spectral power during sleep episodes. In contrast, the EKC abstinence syndrome was associated with minor abstinence signs. Characteristic changes in EEG frequencies of successive REM sleep episodes during the interinjection intervals of opioid self-administration occurred.

Moreover, characteristic pharmacodynamic changes were produced by ethanol and Δ^9-THC and the neurotoxic substances lead (Pb), trimethyltin (TMT), and kainic acid on EEG, EEG power spectra, and behavior.

Thus, EEG power spectral analysis has proved to be a powerful tool that has aided significantly in the evaluation of a wide variety of psychoactive drugs. Given the general problems associated with designing preclinical models predictive of human activity in the psychotropic area (*see* Williams and Malick; Taylor et al., this volume), it would appear that EEG should continue to be an important tool in the characterization and development of CNS-active drugs.

Acknowledgments

We wish to especially acknowledge the following colleagues for their vital contribution to our research efforts: Drs. Matilde Buonamici, Sukehiro Chiba, Brenda Colasanti, Christine Eccles, Esam El-Fakahany, Scott Lukas, Leonard Meltzer, J. Edward Moreton, George Steinfels, and Frank Tortella. Special thanks are due to Ms. Pat Tretter for preparing and typing the manuscript. (Supported by National Institute on Drug Abuse Grant DA-01050.)

References

Benignus, V. A. (1984) EEG as a cross species indicator of neurotoxicity. *Neurobehav. Toxicol. Teratol.* **6**, 473–783.

Berger, H. (1929) Uber da elektrenkephalogramm des meschen. *Arch. Psychiat. Nervenkrankh.* **87**, 527–570.

Berger, H. (1937) On the electroencephalogram of man. Twelfth report. *Arch. Psychiat. Nervenkrankh.* **106**, 165–187.

Brazier, M. A. B. (1964) The effect of drugs on the electroencephalogram of man. *Clin. Pharmacol. Ther.* **5**, 102–116.

Bronzino, J. D., Brusseau, J. N., Stern, W. C., and Morgane, P. J. (1973) Power density spectra of cortical EEG of the cat in sleep and waking. *Electroenceph. Clin. Neurophysiol.* **35**, 187–191.

Bronzino, J. D., Kelly, M. L., Cordova, C., Gudz, M., Oley, N., Stern, W. C., and Morgane, P. J. (1982) Amplitude and spectral quantification of the effects of morphine on the cortical EEG of the rat. *Electroenceph. Clin. Neurophysiol.* **53**, 14–26.

Buonamici, M., Young, G. A., and Khazan, N. (1982) Effects of acute Δ^9-THC administration on EEG and EEG power spectra in the rat. *Neuropharmacology* **21**, 825–829.

Caton, R. (1875) The electric currents of the brain. *Br. Med. J.* **2**, 278.

Colasanti, B. and Khazan, N. (1971) Changes in EEG voltage output of the sleep-awake cycle in response to tetrahydrocannabinols in the rat. *Pharmacologist* **13**, 246.

Cooley, J. W. and Tukey, J. W. (1965) An algorithm for machine calculation of the complex Fourier series. *Math. Comput.* **19**, 297–301.

Cowan, A., Geller, E. B., and Adler, M. W. (1979) Classification of opioids on the basis of change in seizure threshold in rats. *Science* **206**, 465–467.

Depoortere, H., Decobert, M., and Honorie, L. (1983) Drug effects on the EEG of various species of laboratory animals. *Neuropsychobiology* **9**, 244–249.

Dolce, G. and Decker, H. (1972) The effects of ethanol on cortical and subcortical electrical activity in cats. *Res. Commun. Chem. Path. Pharmacol.* **3**, 523–534.

Fink, M. (1964) A selected bibliography of electroencephalography in human psychopharmacology, 1951–1962. *Electroencephal. Clin. Neurophysiol.* (suppl. 23).

Fink, M. and Irwin, P. (1980a) EEG and behavioral profile of flutroline (CP-36,584), a novel antipsychotic drug. *Psychopharmacology* **72**, 67–71.

Fink, M. and Irwin, P. (1980b) EEG and behavioral effects of pirenzepine in normal volunteers. *Scan. J. Gasroenterol.* (suppl.) **15**, 39–46.

Fink, M., Irwin, P., Gastpar, M., and DeRitter, J. J. (1977) EEG, blood level, and behavioral effects of the antidepressant mianserin (ORG GB-94). *Psychopharmacology* **54**, 249–254.

Fink, M., Shapiro, D. M., Hickman, C., and Itil, T. (1958) Digital Computer EEG Analyses in Psychopharmacology, in *Computers and Electronic Devices in Psychiatry* (Kline, N. and Laska, E., eds.) Grune and Stratton, New York.

Frankenheim, J. (1982) Effects of antidepressants and related drugs on the quantitatively analyzed EEGs of beagle dogs. *Drug Dev. Res.* **2**, 197–213.

Freye, E., Hartung, E., and Schenk, G. K. (1983) Bremazocine: An opiate that induces sedation and analgesia without respiratory depression. *Anesth. Analg.* **62**, 483–488.

Freye, E., Schenk, G. K., and Hartung, E. (1982) Naloxone-resistant EEG slowing induced by the synthetic opioid peptide FK 33-824 in the 4th cerebral ventricle of the dog. *EEG EMG* **13**, 129–132.

Fujimori, M. and Himwich, H. E. (1973) Δ^9-Tetrahydrocannabinol and the sleep–wakefulness cycle in rabbits. *Physiol. Behav.* **11**, 291–295.

Fujimori, M., Trusty, D. M., and Himwich, H. E. (1973) Δ^9-Tetrahydrocannabinol: Electroencephalographic changes and autonomic responses in the rabbit. *Life Sci.* **12**, 553–563.

Gehrmann, J. E. and Killam, K. F., Jr. (1975) EEG Changes Following the Administration of Sedative-Hypnotic Drugs, in *Hypnotics: Methods of Development and Evaluation* (Kagan, F., Harwood, T., Rickels, K., Rudzik, A. D., and Sorer, H., eds.) Spectrum, New York.

Gehrmann, J. E. and Killam, K. F., Jr. (1976) Assessment of CNS drug activity in rhesus monkeys by analysis of the EEG. *Fed. Proc. Fed. Amer. Soc. Exp. Biol.* **35**, 2258–2263.

Gehrmann, J. E. and Killam, K. F., Jr. (1978) Studies of central functional equivalence. I. Time varying distribution of power in discrete frequency bands of the EEG as a function of drug exposure. *Neuropharmacology* **17**, 747–759.

Gibbs, F. A. and Maltby, G. L. (1943) Effects on electrical activity of cortex of certain depressant and stimulant drugs—barbiturates, morphine, caffeine, benzedrine and adrenalin. *J. Pharmacol. Exp. Ther.* **78**, 1–10.

Gilbert, P. E. and Martin, W. R. (1976) The effects of morphine- and nalorphine-like drugs in the nondependent, morphine-dependent and cyclazocine-dependent chronic spinal dog. *J. Pharmacol. Exp. Ther.* **198**, 66–82.

Gitlow, S. E., Bentkover, S. H., Dziedzic, S. W., and Khazan, N. (1973) Persistence of abnormal REM sleep response to ethanol as a result of previous ethanol ingestion. *Psychopharmacologia* **33**, 135–140.

Grass, A. M. and Gibbs, F. A. (1938) Fourier transform of the electroencephalogram. *J. Neurophysiol.* **29**, 306–310.

Heath, R. E., Fitzjarrell, A. T., Frontana, C. J., and Garey, R. E. (1980) *Cannabis sativa*: Effects on brain function and ultrastructure in rhesus monkeys. *Biol. Psychiat.* **15**, 657–690.

Hockman, C. H., Perrin, R. G., and Kalant, H. (1971) Electroencephalographic and behavioral alterations produced by Δ^1-tetrahydrocannabinol. *Science* **172**, 968–970.

Horsey, W. J. and Akert, K. (1953) The influence of ethylalcohol on the spontaneous electrical activity of the cerebral cortex and subcortical structures of the cat. *Q. J. Stud. Alcohol* **14**, 363–377.

Itil, T. M. (1961) Die Veranderungen der Pentothal-Reaktion im Elecktroencephalogramm Bie Psychosen Unter der Behandlung mit Psychotropen Drogen, in *Third World Congress of Psychiatry* University of Toronto Press, Toronto, Canada.

Itil, T. M. (1968) Electroencephalography and Pharmacopsychiatry, in *Clinical Psychopharmacology, Modern Problems in Pharmacopsychiatry* (Freyhan, F. A., Petrolowitsch, N., and Pichot, P. E., eds.) Karger, Basel, New York.

Itil, T. M. (1969) Anticholinergic drug induced sleep-like EEG pattern in man. *Psychopharmacologia* **14**, 383–393.

Itil, T. M. (1971) Quantitative Pharmaco-Electroencephalography in Assessing New Anti-Anxiety Agents, in *Advances in Neuro-Psychopharmacology* (Vinar, O., Votava, Z., and Bradley, P. B., eds.) North-Holland, Amsterdam.

Itil, T. M., Menoh, G. N., and Itil, K. Z. (1982) Computer EEG drug data base in psychoharmacology and in drug development. *Psychopharmacol. Bull.* **18**, 165–172.

Itil, T. M., Polvan, N., and Hsu, W. (1972) Clinical and EEG effects of GB-94, a tetracyclic antidepressant (EEG model in discovery of a new psychotropic drug). *Curr. Ther. Res.* **14**, 395–413.

Itil, T. M., Gannon, P., Hsu, W., and Klingenberg, H. (1970) Digital computer analyzed sleep and resting EEG during haloperidol treatment. *Am. J. Psychiat.* **127**, 462–471.

Itil, T. M., Shapiro, D., Schneider, J. J., and Francis, I. B. (1981) Computerized EEG as a predictor of drug response in treatment resistant schizophrenics. *J. Nerv. Men. Dis.* **169**, 629–637.

Itil, T. M., Reisberg, B., Patterson, C., Amin, A., Wadud, A., and Herrman, W. M. (1978) Pipotiazine palmitate, a long-acting neuroleptic: Clinical and computerized EEG effects. *Curr. Ther. Res.* **24**, 689–707.

Itil, T. M., Seaman, P. S., Huque, M., Mukhopadhyay, S., Blasucci, D., Tat Ng, K., and Ciccone, P. E. (1977) The clinical and quantitative EEG effects and plasma levels of fenobam (McN-3377) in subjects with anxiety: An open rising dose tolerance and efficacy study. *Curr. Ther. Res.* **24**, 708–724.

Kareti, S., Moreton, J. E., and Khazan, N. (1980) Effects of buprenorphine, a new narcotic agonist-antagonist analgesic on the EEG, power spectrum and behavior in the rat. *Neuropharmacology* **19**, 195–201.

Kay, D. C. (1975) Human sleep and EEG through a cycle of methadone dependence. *Electroenceph. Clin. Neurophysiol.* **38**, 35–43.

Khazan, N. (1975) The Implication and Significance of EEG and Sleep-Awake Activity in the Study of Experimental Drug Dependence on Morphine, in *Methods in Narcotic Research (Modern Pharmacology-Toxicology)* (Ehrenpreis, S. and Neidle, A., eds.) Marcel Dekker, New York.

Khazan, N. and Young, G. A. (1980) Use of Neurophysiology in the Study of Drugs and Chemicals, in *The Effects of Foods and Drugs on the Development and Function of the Nervous System: Methods for Predicting Toxicity* (Gryder, R. M. and Frankos, V. H., eds.) HHS publication No. (FDA) 80-1076. Superintendent of Documents, US Government Printing Office, Washington, DC.

Khazan, N., Weeks, J. R., and Schroeder, L. A. (1967) Electroencephalographic, electromyographic and behavioral correlates during a cycle of self-maintained morphine addiction in the rat. *J. Pharmacol. Exp. Ther.* **155**, 521–531.

Khazan, N., Young, G. A., El-Fakahany, E. E., Hong, O., and Calligaro, D. (1984) Sigma receptors mediate the psychotomimetic effects of *N*-allylnormetazocine (SKF-10,047), but not its opioid agonistic-antagonistic properties. *Neuropharmacology* **23**, 983–987.

Laurian, S., Lee, P. K., Baumann, P., Perey, M., and Gaillard, J. M. (1981) Relationship between plasma-levels of chlorpromazine and effects on EEG and evoked potentials in healthy volunteers. *Pharmacopsychiatria* **14**, 199–204.

Lipparini, F., Scotti de Carolis, A., and Longo, V. G. (1969) A neuropharmacologic investigation of some transtetrahydrocannabinol derivatives. *Physiol. Behav.* **4**, 527–532.

Lord, J. A. H., Waterfield, A. A., Hughes, J., and Kosterlitz, H. W. (1977) Endogenous opioid peptides: Multiple agonists and receptors. *Nature* (Lond.) **267**, 495–499.

Lukas, S. E., Moreton, J. E., and Khazan, N. (1982) Differential electroencephalographic and behavioral cross-tolerance to morphine and methadone in the 1-α-acetylmethadol (LAAM)-maintained rat. *J. Pharmacol. Exp. Ther.* **220**, 561–567.

Martin, W. R., Eades, C. G., Thompson, J. A., Huppler, R. E., and Gilbert, P. E. (1976) The effects of morphine- and nalorphine-like drugs in the nondependent and morphine-dependent chronic spinal dog. *J. Pharmacol. Exp. Ther.* **197**, 517–532.

Martinez, J. E., Stadnicki, S. W., and Schaeppi, U. H. (1972) Δ^9-Tetrahydrocannabinol: Effects on EEG and behavior in rhesus monkeys. *Life Sci.* **11**, 643–651.

Masur, J. and Khazan, N. (1970) Induction by *Cannabis sativa* (marijuana) of rhythmic spike discharges overriding REM sleep electrocorticogram in the rat. *Life Sci.* **9**, 1275–1280.

Matejcek, M. (1978) Methodological Consideration in Pharmaco-Encephalography, in *Neuropsychopharmacology* (Keniker, P., Radauco-Thomas, C., and Villeneuve, A., eds.) Pergamon, New York.

Matejcek, M., Neff, G., Abt, K., and Wehrli, W. (1983) Pharmaco-EEG and psychometric study of the effect of single doses of temazepam and nitrazepam. *Neuropsychobiology* **9**, 52–65.

McCarren, M. and Eccles, C. U. (1983) Neonatal lead exposure in rats. II. Effects on the hippocampal afterdischarge. *Neurobehav. Toxicol. Teratol.* **5**, 533–540.

McCarren, M., Young, G. A., and Eccles, C. U. (1984) Spectral analysis of kindled hippocampal afterdischarges in lead-treated rats. *Epilepsia* **25**, 53–60.

Moreley, B. J. and Bradley, R. J. (1977) Spectral analysis of mouse EEG after the administration of *N,N*-dimethyltryptamine. *Biol. Psychiat.* **12**, 757–769.

Moreton, J. E. and Davis, W. M. (1973) Electroencephalographic study of the effects of tetrahydrocannabinols on sleep in the rat. *Neuropharmacology* **12**, 897–907.

Niedermeyer, E. (1982) The EEG Signal: Polarity and Field Determination, in *Electroencephalography: Basic Principles, Clinical Applications and Related Fields* (Niedermeyer, E. and Lopes da Silva, F., eds.) Urban and Schwarzenberg, Baltimore.

Perrin, R. G., Hockman, C. H., Kalant, H., and Livingston, K. E. (1974) Acute effects of ethanol on spontaneous and auditory evoked electrical activity in cat brain. *Electroenceph. Clin. Neurophysiol.* **36**, 19–31.

Pickworth, W. B. and Sharpe, L. G. (1979) EEG-behavioral dissociation after morphine- and cyclazocine-like drugs in the dog: Further evidence for two opiate receptors. *Neuropharmacology* **18**, 617–622.

Pickworth, W. B., Sharpe, L. G., and Gupta, V. N. (1982) Morphine-like effects of clonidine on the EEG, slow wave sleep and behavior in the dog. *Eur. J. Pharmacol.* **81**, 551–557.

Prado de Carvalho, L. and Izquierdo, I. (1977) Changes in the frequency of electroencephalographic rhythms of the rat caused by single, intraperitoneal injections of ethanol. *Arch. Int. Pharmacodyn. Ther.* **229**, 157–162.

Robert, T. A., Daigneault, E. A., and Hagardorn, A. N. (1978) Relationship between fluphenazine plasma concentration and electroencephalographic alterations. *Commun. Psychopharmacol.* **2**, 467–474.

Rosadini, G., Cavazza, B., Rodriquez, G., Sannita, W. G., and Siccardi, A. (1977) Computerized EEG analysis for studying the effect of drugs on the central nervous system. *Int. J. Clin. Pharmacol.* **15**, 519–525.

Saletu, B. (1982) Pharmaco-EEG profiles of typical antidepressants. *Adv. Biochem. Psychopharmacol.* **32**, 257–268.

Sauerland, E. K. and Harper, R. M. (1970) Effects of ethanol on EEG spectra of the intact brain and isolated forebrain. *Exp. Neurol.* **27**, 490–496.

Schallek, W. and Johnson, T. C. (1976) Spectral density analysis of the effects of barbiturates and benzodiazepines on the electrocorticogram of the squirrel monkey. *Arch. Int. Pharmacodyn.* **223**, 301–310.

Speckmann, E.-J. and Elger, C. E. (1982) Neurophysiological Basis of the EEG and of D.C. Potentials, in *Electroenceophalography: Basic Principles, Clinical Applications and Related Fields* (Niedermeyer, E.and Lopes da Silva, F., eds.) Urban and Schwarzenberg, Baltimore.

Stein, S. N., Goodwin, C. W., and Garvin, J. S. (1949) A brain wave correlator and preliminary studies. *Trans. Am. Neurol. Assoc.* **74**, 197–198.

Steinfels, G. F., Young, G. A., and Khazan, N. (1980) Opioid self-administration and REM sleep EEG power spectra. *Neuropharmacology* **19**, 69–74.

Stratton, K. R., Young, G. A., and Eccles, C. U. (1983) Trimethyltin administration alters cortical and hippocampal EEG power spectra during slow-wave and rapid eye movement sleep. *Soc. Neurosci. Absts.* **9**, 1247.

Stratton, K., Young, G., and Eccles, C. (1985) Kainic acid lesions of the hippocampus mimic trimethyltin effects on REM sleep but not slow-wave sleep. *Fed. Proc.* **44**, 743.

Tortella, F. C., Cowan, A., and Adler, M. W. (1980) EEG and behavioral effects of ethylketocyclazocine, morphine and cyclazocine in rats: Differential sensitivities towards naloxone. *Neuropharmacology* **19**, 845–850.

Walter, D. L. (1963) Spectral analysis for electroencephalogram: Mathematical determination of neurophysiological relationships from records of limited duration. *Exp. Neurol.* **8**, 155–181.

Wolf, D. L., Young, G. A., and Khazan, N. (1981) Comparison between ethanol-induced and slow-wave sleep synchronous EEG activities utilizing spectral analyses. *Neuropharmacology* **20**, 687–692.

Young, G. A. and Khazan, N. (1984a) Differential neuropharmacological effects of mu, kappa and sigma opioid agonists on cortical EEG power spectra in the rat. Stereospecificity and naloxone antagonism. *Neuropharmacology* **23**, 1161–1165.

Young, G. A. and Khazan, N. (1984b) Differential tolerance and cross-tolerance to repeated daily injections of mu and kappa opioid agonists in the rat. *Neuropharmacology* **23**, 505–509.

Young, G. A. and Khazan, N. (1985) Comparison of abstinence syndromes following chronic administration of mu and kappa opioid agonists in the rat. *Pharmacol. Biochem. Behav.* **23**, 457–460.

Young, G. A. and Khazan, N. (1986) Differential protracted effects of opioid challenges in mu and kappa post-tolerant rats. *Eur. J. Pharmacol.* **125**, 265–271.

Young, G. A., Neistadt, L., and Khazan, N. (1981) Differential neuropharmacological effects of mu, kappa and sigma opioid agonists on cortical EEG power spectra in the rat. *Res. Commun. Psychol. Psychiat. Behav.* **6**, 365–377.

Young, G. A., Wolf, D. L., and Khazan, N. (1982) Relationships between blood ethanol levels and ethanol-induced changes in cortical EEG power spectra in the rat. *Neuropharmacology* **21**, 721–723.

Young, G. A., Steinfels, G. F., Khazan, N., and Glaser, E. M. (1978a) Morphine self-administration and EEG power spectra in the rat. *Pharmacol. Biochem. Behav.* **9**, 525–527.

Young, G. A., Steinfels, G. F., Khazan, N., and Glaser, E. M. (1978b) Cortical EEG power spectra associated with sleep-awake behavior in the rat. *Pharmacol. Biochem. Behav.* **8**, 89–91.

Immunopharmacological Approaches to Drug Development

STEVEN C. GILMAN AND ALAN J. LEWIS

1. Introduction

Knowledge of the immunologic system has increased tremendously over the last few decades. This has prompted the development of immunopharmacology, a new discipline that began to flourish in the 1970s, and that broadly addresses the effects of synthetic and natural products on the immune response (Mullen, 1979). The field of immunopharmacology encompasses the development of immunologically based assays and their research and clinical applications, the preparation and utilization of monoclonal antibodies as pharmacological tools or therapeutic agents, the pharmacological regulation of immune responsiveness, and immunotoxicology. The immunopharmacologist must conjoin fundamental pharmacology with modern immunology in an attempt to determine the biological and biochemical effects of synthetic and natural products on immune responsiveness, elucidate the mechanism(s) through which such effects are mediated, and evaluate the clinical potential of such agents.

It is impossible to cover this entire field in the present chapter; indeed entire books devoted exclusively to immunopharmacology have been published (Rosenthale and Mansmann, 1975; Hadden et al., 1977; Gibson et al., 1983b). In the present chapter, the focus is on disease-related immunological deficiencies and a discussion on the types of immunopharmacological agents that would be appropriate for correcting these deficits, as well as the experimental procedures used to discover and develop such therapeutic agents. Rather than exhaustively reviewing the specific experimental and clinical data, it is our intent to take a conceptual approach using selected experimental data to highlight salient points.

2. The Immune System

The immune system consists of a heterogenous group of lymphoid cells whose collective function is to provide adequate protective mechanisms against potentially harmful foreign substances (i. e., bacteria, viruses, and so on) and prevent or limit the abnormal growth of host tissues (i. e., malignant cells). Normal and appropriate immune function is necessary for homeostasis; either an insufficient immune response (immune deficiency) or a misdirected immune response (autoimmunity, allergy) can result in severe and life-threatening disease.

2.1. Anatomy

Scattered throughout the body are collections of lymphoid cells organized into various lymphoid tissues whose structural integrity varies dramatically in complexity (Kay et al., 1979). The three central components of the lymphoid system are the stem cells, the central lymphoid organs, and the peripheral lymphoid systems. Stem cells originate in the bone marrow and are the progenitors of all classes of lymphoid cells. The central lymphoid organs—the thymus and bursa of Fabricius (or ''bursal equivalent'' in mammals)—are responsible for promoting the development of mature T and B cells, respectively. The peripheral lymphoid system represents those lymphocyte populations that have been processed by the central lymphoid organs that are directly responsible for the immune response observed upon antigen exposure, including the spleen, lymph nodes, gut-associated lymphoid tissues (GALT), and bronchial-associated lymphoid tissues (BALT). The organs of this system have specialized functions and structure, and the cells therein can interact with each other via blood and lymphatic circulatory systems (Ford, 1975).

2.2. Cellular Components and Their Functions

Cells of immunological importance include polymorphonuclear leukocytes (PMN), granulocytes (basophils, eosinophils, mast cells), monocytes/macrophages, and lymphocytes. Substantial heterogeneity exists within each cell type. Different subsets of T and B lymphocytes can be identified based on the expression of certain surface antigens that are defined by monoclonal or polyclonal antibodies (Reinherz and Schlossman, 1980). For example, although all human peripheral T cells express the membrane antigen identified by OKT3 monoclonal antibody, subpopulations of these cells express either the OKT4 or OKT8 antigens. These antigenic markers correlate with cell function, OKT4 being associated with ''helper'' T cell function and OKT8 with suppressor/cytoxic cell function (see below). However, the relationship between surface antigen expression and T cell function is not absolute, since both $OKT4^+$ and $OKT8^+$ cells produce helper factors (lymphokines) and both can function as cytotoxic cells in certain circumstances (Luger et al., 1982). B Lymphocyte subpopulations express different B-cell-specific surface antigens, and the display of surface immunoglobulin classes (primarily IgD, IgM, and IgG) differs depending on the maturational state of the B cell (Cooper, 1983). Macrophage subpopulations have also been described as differing in membrane surface antigen expression (HLA/DR or ''Ia''), function, and buoyant density (Dougherty and McBride, 1984). No easily distinguishable subsets of PMNs have been described, but heterogeneity of mast cells has been recently reported (Schulman et al., 1983). Clearly, immunity results from many complex interacting cell types and subtypes, all of which have specialized functions and distinguishing characteristics.

Two general types of cellular defense mechanisms can be distinguished, depending on whether or not antigen stimulation is required for the particular defense mechanism to operate efficiently. Nonspecific or ''natural'' cellular defense mechanisms are those that do not require prior antigen exposure to be activated. One principle mechanism is phagocytosis and intracellular killing of infectious agents by PMNs, granulocytes, and macrophages. Phagocytic efficiency can be enhanced by the presence of antibody (opsonization) or activated complement components. Natural killer (NK) cells are a subpopulation of T cells that are able to lyse a variety of tumor cells and virus-infected cells without prior activation, although their cytotoxic activity can be augmented by certain stimuli such as interferon (Ortaldo and Herberman, 1984). Macrophages also express an antigen-nonspecific cytolytic activity.

Antigen-specific cellular immune reactions are those in which immune cells specifically recognize antigenic determinants through specific membrane receptors and evoke immunological memory that is responsible for the more rapid and vigorous nature of secondary (anamnestic) responses to the same antigen. These cellular immune reactions involve numerous cell types that interact with each other in a complex but ordered sequence ultimately resulting in cellular activation and the generation of effector cells or molecules capable of eliminating the invading agent (Fig. 1). The induction phase of T cell activation begins when antigen-presenting cells, such as macrophages and dendritic cells, initiate early events in T cell activation by ingesting and processing the antigen and presenting it in an immunogenic form to T lymphocytes bearing antigen receptors and the appropriate class II histocompatability antigens on their surface (Unanue, 1984). Macrophages also provide the necessary T helper factors, such as interleukin 1 (IL-1, formerly called lymphocyte-activating factor), to T cells (Dinarello, 1984). The regulatory phase of T cell activation is characterized by clonal expansion of subsets of antigen-reactive T cells and the generation of T helper or T suppressor cells that provides a system of checks and balances on the magnitude of the T cell response (Fathman, 1982). Finally, other subsets of antigen-reactive T cells, through a combined action of helper cells and soluble helper factors derived from these cells, differentiate into antigen-specific effector cells such as cytotoxic T cells (CTL) capable of lysing virus-infected cells, allogenic cells, tumor cells, and so on.

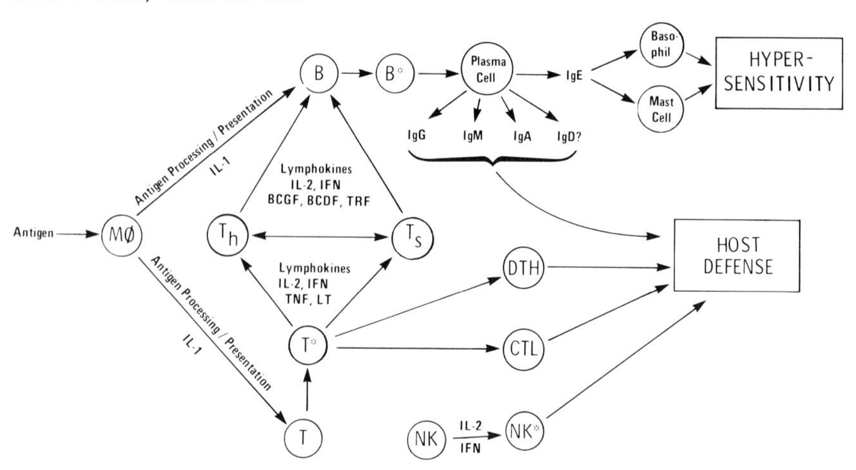

Fig. 1. Schematic representation of the major cell types involved in host defense. Abbreviations: MØ, macrophage; T_h, helper T cell; T_s, suppressor T cell; Ig, immunoglobulin; IL-1, interleukin 1; IL-2, interleukin 2; IFN, interferon; TNF, tumor necrosis factor; TRF, T cell replacing factor; BCGF, B cell growth factor; BCDF, B cell differentiation factor; NK, natural killer cell; DTH, delayed-type hypersensitivity effector cell; CTL, cytotoxic T lymphocyte; asterisk (*), activated cell type.

B lymphocytes are the mediators of humoral immunity and are are responsible for the production of antibody molecules (e. g., IgM, IgG, IgA, IgD, IgE, and their subclasses). When exposed to antigen, mature B cells become activated and undergo proliferative and differentiative processes that are regulated by macrophages and T cells and involve several soluble factors such as B cell growth factor (BCGF) and B cell differentiation factor (BCDF) (Cooper, 1983).

Circulating antibody can directly inactivate bacteria or viruses and, through the complement pathway, can cause tumor cell lysis. Alternatively, effector lymphoid cells bearing receptors for the Fc region of the immunoglobulin molecule (including NK cells, macrophages, B cells, and T cells) can be armed with specific antibody and lyse antigen-bearing cells in a process termed antibody-dependent cellular cytotoxicity (ADCC) (MacDonald et al., 1975).

Tissue and skin mast cells are responsible for allergic and hypersensitivity reactions to specific antigens, which in this case are termed "allergens." Antigen binding to IgE antibody, which is itself bound to the IgE receptor on the cell membrane, initiates a complex mast cell activation process culminating in the release of numerous biologically active mediators such as histamine, bradykinin, serotonin, and leukotrienes (Wasserman, 1983; Lewis et al., 1985). Under normal circumstances, these mediators are thought to regulate the microvascular tone, permeability, and blood flow, but under abnormal conditions (i. e., hypersensitivity), they can cause swelling, pain, edema, and eventual respiratory failure that may even result in death.

2.3. Role of Cytokines in Immunity

"Cytokine" is a collective term used to describe soluble factors derived from cells that mediate biological effects on other cell types. Cytokines can be produced by monocyte/macrophages (monokines) or lymphocytes (lymphokines). Over 100 cytokines have been described in the literature based upon activity of supernatant fluids from activated lymphoid cells in one or another bioassay procedure (Hansen et al., 1982). Two cytokines that play important roles in lymphocyte activation processes are IL-1 and interleukin 2 (IL-2, T cell growth factor). IL-1 is a 17,000-dalton glycoprotein that is produced by monocytes/macrophages (as well as a variety of other cell types), but not by lymphocytes (Dinarello, 1984; Duram et al., 1985). IL-1 plays a critical role in T cell activation since it provides a signal necessary for T cells to produce IL-2 (mw = 15,000 daltons in humans and rats, 30,000 daltons in mice) that acts as a direct growth factor to stimulate T cell proliferation and differentiation, completing the IL-1/IL-2 cascade system in T cell activation (Robb, 1984; Smith, 1984).

Originally, it was believed that IL-1 and IL-2 were involved in only T cell activation, but recently it has become clear that these cytokines play a role in B cell activation as well (Dinarello, 1984; Smith, 1984; Robb, 1984). Moreover, IL-1 has been shown to have numerous biological effects distinct from those involved in lymphocyte activation. For example, IL-1 mediates fever and acute phase reactant synthesis and induces chondrocytes and synovial fibroblasts to produce prostaglandin E_2 and a variety of proteases. For these reasons IL-1 is thought to be an important pathological mediator of some immunoinflammatory disorders such as rheumatoid arthritis (Dinarello, 1984).

Interferons comprise another important class of cytokines, and have a variety of effects including the ability to activate NK cells and macrophages, antiviral activity, and many have antiproliferative effects as well (Gresser, 1980). Other cytokines of immunological importance include transfer factor, tumor necrosis factor, lymphotoxin, colony stimulating factor(s), and macrophage-activating factor. A discussion of the nature and importance of these and the plethora of other cytokine mediators is beyond the scope of this chapter. The reader is therefore referred to several excellent recent reviews (Ihle et al., 1982; Gibson et al., 1983a; Hansen et al., 1982).

3. Potential Therapeutic Targets for Immunomodulatory Drugs

There are several potential therapeutic targets for immunomodulatory drugs that have evolved primarily from increasing theoretical and practical knowledge of the immune dysfunctions that accompany these disorders. The therapeutic approach in general has been to define the immunological defects associated with specific diseases and then identify therapeutic agents that would correct the functional immune status of the individual.

Some of the more important therapeutic targets are shown in Table 1. Operationally, these diseases are listed in distinct categories based primarily on the clinical manifestations and the current understanding of their underlying immunological mechanisms. However, it is important to note that these categories, although useful, are not rigid and many overlaps exist. Moreover, individual disorders within a given category vary widely with respect to their immune abnormalities, pathogenesis, clinical symptomology, laboratory findings, and so on. Table 1 also shows those lymphoid cells that could theoretically, if appropriately modulated by drug treatment, effect a beneficial clinical response. Since no immunomodulatory agents to date have effects on only one cell type (*see* section 4.3), and fewer still have undergone careful clinical evaluation, none of the indicated cellular targets are as yet definitively proven to be essential for a clinical response. It is clear that no one type of immunomodulatory agent is likely to be useful in all of these diseases. In fact, it is more likely that most diseases of this type will have to be treated on an individual basis, and therapeutic agents closely tailored to best suit particular diseases.

3.1. Immunodeficiency Diseases

Immunodeficiency disease represents a spectrum of immune defects that occur either spontaneously (primary immunodeficiency) because of an underlying inherited trait (metabolic disorders, enzyme deficiencies, and so on) or secondarily because of cancer or infection (acquired immune deficiency).

3.1.1. Primary Immunodeficiency

Primary immunodeficiencies are those disorders that occur naturally and have a primarily genetic rather than infectious basis. A spectrum of primary immunodeficiency disorders has been documented in the clinical literature. They are characterized by an inability to manifest normal cell-mediated and/or humoral immunity. In the general population, the incidence of primary immune deficiency is quite low and individual cases show marked heterogeneity in inheritance pattern and both laboratory and clinical findings. The most common clinical features of primary immune deficiencies include a failure to thrive, recurrent and severe bacterial and viral infection, and an early mortality.

Some primary immunodeficiencies, such as severe combined immunodeficiency disease (SCID), affect the function of both T and B cell lineages. Many SCID patients are hypogammaglobulinemic and show absent or low peripheral blood leukocyte (PBL) responses to phytohemagglutinin (PHA) (Gelfand and Dosch, 1983). On the other hand, some primary disorders predominantly affect either

Table 1
Potential Therapeutic Targets for Immunomodulatory Agents

Category	Example	Putative cellular targets for immunomodulatory drugs
Cancer	Leukemia, lymphoma, melanomas, and so on	NK, MØ, T
Autoimmune diseases	Rheumatoid arthritis	MØ, PMN, T, B
	Systemic lupus erythematosus	T, B, other
	Type I diabetes	T
	Pemphigus	B
Immunodeficiency		
Primary	SCID (several combined immunodeficiency disease)	T, B
	Purine nucleoside phosphorylase (PNP) deficiency	
	Adenosine deaminase deficiency	T
	Ataxia telangectasia	T
	DiGeorges syndrome	T
	X-linked agammaglobulinemia	B
	IgA deficiency	B
	Common variable immune deficiency	B
Acquired	Related to cancer (*see* Cancer above)	MØ, NK, T
	Viral infection	MØ, NK, T
	Bacterial infection	T, MØ, PMN
	Fungal infection	MØ, T
	Aging	T, B, MØ
	Miscellaneous	
	Post-surgical	T
	Burn patients	T
Allergy	Asthma	Mast cell, B
	Dermatitis	Mast cell, T
	Insect sting allergy	Mast cell
	Systemic anaphylaxis	Mast cell, B

the B or T lineage. Thus, several humoral immune deficiencies, such as X-linked agammaglobulinemia, have been described in which the primary defect is in the B cell lineage and, conversely, T cell-specific malfunctions have also been noted, for example, the genetic absence of purine nucleoside phosphorylase (Geha and Rosen, 1983; Hirschhorn, 1983; Seligmann and Ballet, 1983). In addition, congenital disorders primarily affecting phagocyte (PMN and macrophage) function, for example Chediak-Higashi syndrome, are also known (Twomey, 1982).

Clinical and laboratory studies have demonstrated at least two mechanisms through which primary immune deficiency can be manifested. The first is a genetic enzyme deficiency, as typified by adenosine deaminase deficiency (seen in many patients with SCID), purine nucleoside phosphorylase deficiency (resulting in cellular immune deficiency), and absence of transcobalamin II (which results in agammaglobulinemia) (Hirschhorn, 1983). The second possible mechanism is anomalous lymphoid cell maturation. Thus, lymphoid cell precursors could fail to differentiate appropriately into mature functional lymphocytes because of intrinsic stem cell defects of reduced thymus or bursal-equivalent function (Hong, 1982; Bach, 1983).

Based on these considerations, it is apparent that drugs capable of enhancing the reduced immune function or promoting lymphocyte differentiation provide a rational therapeutic approach to these disorders, especially when given in conjunction with standard antibiotic and antiviral therapy and/or immune globulin injections. However, the type of immunostimulatory agent must be carefully selected to suit the particular immune disorder. For example, agents that stimulate B cell differentiation and antibody secretion would be appropriate for patients with hypo- or agammaglobulinemia, but not necessarily for patients with T cell immunodeficiency.

3.1.2. Acquired Immune Deficiency Diseases

3.1.2.1. CANCER. The use of immunomodulatory drugs in cancer therapy has been a major driving force for development of these agents. Indeed, it was the limitations of conventional cancer treatments that provided the impetus for utilizing the immunostimulatory properties of a variety of bacterial products and other simple chemicals in the treatment of malignancy. Subsequently, this led to the application of this method of immune enhancement to autoimmune diseases and immunodeficiency. Although immunostimulation in cancer therapy is not a recent concept, interest in this approach has been rejuvenated following the development of the "biological response modifier (BRM)" program under the auspices of the National Institutes of Health (Oldham and Malley, 1983). The primary goal of this program is to identify, characterize, and ultimately evaluate clinically the antitumor effectiveness of natural and synthetic agents that modulate immune reactivity.

The role of nonspecific defense mechanisms appears to be one of importance in controlling tumor growth and reducing the probability of metastasis (Herberman, 1980). Thus, immunomodulators that have the ability to augment the cytotoxic activity of NK cells and macrophages (e. g., interferons) have the strongest rational basis for development in this area. However, even though interest in the role of specific T-cell-mediated antitumor immunity and antitumor antibody in controlling tumor growth has waned somewhat, in part because of the difficulty in demonstrating tumor-specific antigens on several tumor cells (Hewitt et al., 1976), agents that stimulate T and B cell reactivity may also be useful in tumor therapy.

Immunomodulatory drugs that stimulate antitumor immunity will not in general be effective first-line therapy in cancer patients. Evidence to date indicates that this type of drug will be most useful as an adjunct to cytoreductive therapy (surgical resection, irradiation, cytotoxic drug therapy). Thus, in animal tumor model systems, these agents rarely influence tumor growth or survival if the tumor burden is large, but do slow or stop tumor growth and reduce metastatic spread following cytoreductive therapy (Fidler et al., 1982; Kralovec, 1983). Examples of this class of agents include levamisole, aximexon, maleic anhydride divinylether copolymer (MVE), glucans, and several of the peptide cytokines, particularly the interferons and IL-2 (Fenichel and Chirigos, 1984; Kralovec, 1983; Talmadge et al., 1984).

3.1.2.2. INFECTIOUS DISEASES. The use of immunomodulatory drugs in the treatment of viral, bacterial, and fungal infections is a promising therapeutic arena. This is particularly true if these agents are designed to be used as adjunctive therapy in combination with antiviral, antibiotic, and antifungal agents. In principle, the antiviral/antibiotic would directly reduce the magnitude of the infection, whereas the immunomodulatory agents used would stimulate the functional activity of those particular lymphoid cells that attack the infectious agent in question, effecting a more rapid and complete elimination of the infectious agent and reducing the possibility of recurrence. The immunomodulatory agents must be selectively used, depending upon the type of infection. Since macrophages and PMNs are important antibacterial effector cells, agents that stimulate their functions (i. e., phagocytosis, intracellular killing, motility, and so on) would be of primary interest in bacterial infections. On the other hand, T or NK cell stimulators would be more appropriate for viral infections. The immunomodulatory therapy in infectious diseases would be particularly valuable in patients who are immunocompromised because of either the infection itself or surgical trauma, irradiation, cancer, severe burns, and so on (Easman and Gaya, 1983).

The idea that immunomodulatory agents, primarily those with immunopotentiating effects, may be useful in antiinfective therapy is the result of numerous studies in humans and experimental animals that clearly show that a transient blunting of immune responsiveness accompanies many viral, bacterial, and fungal infections (Easman and Gaya, 1983; Gilmore and Wainberg, 1985; Vivella and Fudenberg, 1982).

Perhaps the most extreme example of viral infection of lymphoid cells with resultant immunosuppression is the acquired immune deficiency syndrome (AIDS). The principle etiological factor in this disease appears to be infection of helper ($OKT4^+$) T cells by the human T cell leukemia virus type III [HTLV-III; also known as lymphadenopathy-associated virus (LAV) (Pinching, 1984)]. Both the absolute number and functional activity of helper T cells is dramatically reduced in AIDS patients and this leads to multiple clinical manifestations, the most prevalent and important of which are lymphadenopathy, Kaposi's sarcoma, and *Pneumocystis carinii* pneumonia (Weiss et al., 1985).

3.1.2.3. AGING. Immune function declines dramatically with advancing age in man and experimental animals. Age-related defects in the function of T cells, B cells, and macrophages have all been described. These alterations have been postulated to be responsible for the increased incidence of infections, malignancies, and certain autoimmune diseases in elderly populations (Weksler and Siskind, 1982; Gilman, 1984).

3.2. Autoimmune Diseases

Autoimmune diseases include rheumatoid arthritis (RA), systemic lupus erythematosis (SLE), type 1 (juvenile) diabetes, and autoimmune thyroiditis (Smolen and Steinberg, 1982). Although the primary pathological features of autoimmune disease are caused by an overactive immune reaction to ''self'' antigens, the hypothesis that these diseases result from abnormal immunoregulation has become quite attractive (Klareskog et al., 1982). Thus, in SLE and RA, in which hyperactive B cell function and elevated autoantibody secretion are characteristic features, regulatory T cell functions, such as lymphokine synthesis and suppressor T cell activity, are diminished (Smolen and Steinberg, 1982; Gilman et al., 1984; Miyasaka et al., 1984). This suggests that both immunosupressive and immunostimulatory agents could be therapeutically useful, the former by directly inhibiting those lymphocytes reacting against host components and the latter by stimulating regulatory cell function (e.g., suppressor cells). Indeed, both immunosuppressive agents, such as cyclophosphamide, and immunostimulators, such as levamisole and thymic peptides, do show clinical utility in RA patients (Gilman and Lewis, 1985).

3.3. Allergic Disorders

Immediate hypersensitivity is the immune response that has become synonymous with the term allergy, in part because it is the most common of the allergic reactions. However, there are four major types of allergic reactions and their mechanisms vary considerably (Table 2).

As indicated in Table 3, IgE production by antigens (i. e., allergens) is pivotal to immediate hypersensitivity reactions. This antibody binds to surfaces of mast cells that are abundant in the respiratory and gastrointestinal tracts, as well as the skin, and to basophils, which are circulating leukocytes.

Patients with allergic diseases often possess a constellation of immunological defects. For example, bronchial asthmatics and allergic rhinitics have demonstrated the involvement of both immediate (manifested by positive skin tests, antiallergen-specific IgE antibody, and increased levels of serum IgE) and delayed hypersensitivity (usually depressed cell-mediated responses). Patients with atopic dermatitis have elevated serum IgE, but also depressed cell-mediated immunity and abnormal leukocyte chemotaxis (Gupta and Good, 1979).

4. Immunomodulatory Agents

4.1. General Considerations

Immunomodulatory drugs can best be described as those agents, either natural or synthetic, that alter the immune system in such a way as to effect a therapeutic benefit. Considering the complexity of the immune system, it is not surprising that there are numerous potential mechanisms through which immunomodula-

Table 2
Types of Allergic Reactions

		Mechanisms	Examples
Type I	Immediate or anaphylactic hypersensitivity	IgE antibody on mast cell or basophil reacts with antigen resulting in release of mediators	Allergic asthma, allergic rhinitis, insect sting, some food and drug reactions, anaphylactic shock, atopic dermatitis
Type II	Cytotoxic	IgG antibody reacts with cell membranes or antigen associated with cell membrane	Transfusion of incompatible blood
Type III	Arthus	Antigen and antibody bind together to form immune complexes that deposit in the walls of blood vessels or kidneys	Serum sickness, some drug reactions
Type IV	Cell-mediated or delayed hypersensitivity	Interaction of T cells with antigen	Graft rejection, contact dermititis

Table 3
Potential Mechanism of Action of Antiallergics

Antigen distribution, processing, and presentation
Synthesis and modulation of IgE
Stabilization of mast cells/basophils to prevent mediator release
Synthesis of cellular mediators (e. g., phospholipase and lipoxygenase inhibitors)
Mediator-receptor interactions [e. g., antagonists of histamine, acetylcholine, leukotrienes, platelet activating factor (PAF), and so on]
Reflex bronchoconstriction
Movement, number, and function of neutrophils and platelets and their dependent mediators

tion can be achieved (Table 4). Moreover, it is now clear that ''immunomodulator'' is an appropriate term, since many factors including the time, dose, frequency, and route of administration greatly affect the response of the host qualitatively as well as quantitatively (Hadden, 1983). Moreover, several host characteristics also influence the nature and magnitude of drug response. Genetic factors, whether or not these are linked to the major histocompatability complex, are critical. Both the physiological (sex, age, diet, neuroendocrine) and pathological (intercurrent infections, spontaneous autoimmune disease, neoplasias, organ dysfunctions) can affect the immunopharmacological and pharmacokinetic properties of the drug. The characteristics of antigen challenge are also capable of modifying the end result; the dose, immunogenicity, and size of the antigen, be it tumor or pathogen, is critical. It may be necessary to reduce the tumor size (by surgery, x-irradiation of chemotherapy) or pathogen number (by concomitant antimicrobial therapy) in order to be successful with immunomodulatory therapy (Herberman, 1980; Fidler et al., 1982).

Table 4
Potential Mechanisms of Action of Immunomodulators

Antigen distribution, processing, and presentation
Production and differentiation of lymphoid cell precursors
Movement and number (relative or absolute) of mature effector or regulatory cells
Activation–inactivation thresholds of immune cells to antigen or differentiation factors (e. g., responsiveness to cytokines, expression of receptors)
Production, release, metabolism of cytokines that modulate immune functions
Movement, number, and function of neutrophils and platelets and their mediators
Lymphocyte recirculation/trafficking
Modulation of arachidonate metabolism
Other regulatory mechanisms (e. g., CNS, endocrinological homeostatis)

4.2. Classification

Categorizing agents that modulate the immune response has proven difficult and there is no concensus as to the appropriate classification. Four major categories of immunotherapy are described in Table 5. However, this classification is incomplete, and overlaps exist. Chemical classification of immunomodulators is extremely difficult in view of the plethora of agents with such activity

Table 5
Classification of Agents that Influence the Immune Response

Type	Activity
Immunostimulant, immunoenhancer, immunopotentiator, augmenting agent	Stimulates or enhances immune response
Immunorestorative	Restores or normalizes an abnormal immune response. Little or no effect on patients with a normal immune response
Immunosuppressant	Suppresses immune response
Adjuvant	Enhances immunogenicity of an antigen

and may be misleading since structurally different agents may have similar functions, such as interferon induction. Functional classification of immunomodulators is also a difficult task (Florentin et al., 1981, 1983; Spreafico et al., 1981; Hadden, 1983; Spreafico, 1985). It appears that immunomodulators act at different locations on the immune system, but the target cell for many agents has proved elusive. Indeed, many immunomodulators have several sites of action and the primary effect, if one exists, is not clearly apparent (Table 6). As a result, they are often capable of influencing both the humoral and cellular immune system. There is, nevertheless, a need to develop agents that can selectively enhance or inhibit specific classes or subclasses of immunocytes, e. g., increase suppressor T cells in systemic lupus erythematosus, increase NK cell activity in leukemia, and so on.

4.3. Biological Testing Procedures for Identifying Immunomodulatory Drugs

In view of the diverse mechanisms associated with immunomodulation, it is unlikely that a single in vitro or in vivo test system will be satisfactory for screening all types of immunomodulators. The use of a battery of appropriate tests is recommended, although

Table 6
Primary Cellular Targets of Selected Immunomodulators[a]

T Lymphocytes	Azathioprine, azimexon, BCG, bestatin, *C. parvum*, cyclosporin A, cyclophosphamide, glucans, IFN, IL-2, IL-3, isoprinosine, lentinan, levamisole, NPT 15392, MDP and analogs, MVE-2, picibanil, thymic hormones, Wy-18,251 (Tilomisole)
B Lymphocytes	Azathioprine, azimexon, cyclophosphamide, LPS, MVE-2, lentinan, Poly IC:LC, tuftsin
Macrophages	Azathioprine, BCG, *C. parvum*, glucans, lentinan, levamisole, MDP, MVE-2, picibanil, Poly IC:LC, pyrimidinones, tuftsin
NK Cells	BCG, bestatin, *C. parvum*, cyclophosphamide, glucans, IFN and inducers, IL-2, MVE-2, picibanil

[a]Modified from Spreafico (1985).

no consensus of specific tests has emerged (Talmadge et al., 1984). However, it is accepted that assays to be included involve a range of immunological tests of varying complexity, infectious models, autoimmune models, and neoplastic models (Hadden, 1981, 1983; Werner and Floc'h, 1981; Fidler et al., 1982; Florentin et al., 1983; Lagrange, 1983; Sedlacek, 1984; Talmadge et al., 1984). Unfortunately a comparative evaluation of the sensitivity, predictiveness, and limitation of such tests employing a diverse group of immunopharmacological agents is lacking.

Definitive studies using those models that allow for extrapolation to humans are plagued by numerous factors, including host reactivity and the type of antigen, as mentioned previously. Nevertheless, a pharmacological approach to a new immunomodulator often begins with an investigation of its effects in vitro on various cell populations constituting the immune system that are isolated from either normal animals or normal subjects. This allows the range of concentrations that modulate cell reactivity to be identified. However, these observations may not reflect events induced in vivo by the compound, effects on "abnormal" cells, or even effects on cells of the same lineage but obtained from different anatomical sites.

In vivo models involving animals with suboptimal resistance or tests involving suboptimal derangement are more appropriate than those involving optimal conditions since most immunomodulators restore rather than potentiate immune responses. Excessive challenge doses of pathogens and tumor cells should also be avoided; otherwise the result is overwhelming and may not be reversible by this class of drug.

The choice of animal is often critical. For example, some strains of mice are particularly susceptible to diseases such as collagen-induced arthritis (DBA/2) or lupus-like disease (MRL/1, NZB/W), whereas others (e. g., C57Bl/6-related strains) are highly resistant (Shirai, 1982; Stuart et al., 1984). The choice ot resistance model should also depend largely on the suspected primary mechanism of action of the immunomodulator. For example, immunomodulators in general will exert little or no effect in bacterial models of resistance unless macrophages are the primary targets of their activity. The basis for the selection of tumor models should include an evaluation of the immunogenicity of the tumor. Highly immunogenic tumors may not be optimal because of the strong immunoprotection in the absence of the immunomodulator, whereas use of a relatively nonimmunogenic tumor may also be counterproductive. Tumor cells should also be sensitive to killing by the effector cells stimulated by the immunomodulator under test. Tumor variants exist that are resistant to specific CTL or NK cell-mediated cytotoxicity and may thus be inappropriate.

Table 7 outlines one approach to the screening for immunomodulator activity. Many other tests could be included, but in practice fewer tests are used for the sake of expediency. It is apparent that a considerable amount of work is still necessary to make initial selections of agents with potential therapeutic efficacy in models of organ transplantation, autoimmunity (including inflammation), infection, and neoplasia. Drugs that have activity against selected cell populations should be considered for testing in disease models listed in Table 7.

A problem that commonly arises in the testing of immunomodulators with stimulatory activity is the bell-shaped dose–response curve. Increasing the dose will thus result in an increased response followed by a peak or plateau and finally a reversal of the effect. The precise interval between the maximally effective dose and the paradoxical effect varies with the agent tested. The highest tolerated dose is, therefore, not always the most effective.

Some immunomodulators may exert a degree of selectivity toward specific cell types and exert uniform in vivo responses; however, absolute selectivity does not yet exist and the possibility

Table 7
Commonly Used Assays To Identify and Characterize Immunomodulatory Agents[a]

	Immune function, in vitro/in vivo	Disease models
T cells	T-cell cytotoxicity	Organ transplantation Skin graft survival
	Blastogenesis (LPS, Con A, PHA, alloantigen)	Graft vs host disease (popliteal lymph node assay)
	Lymphokine production	
	Th/Ts ratios	Autoimmune diseases Experimental allergic encephalomyelitis (EAE)
	Delayed type hypersensitivity	Murine lupus (MRL/1, NZB)
	DTH (SRBC, MBSA, oxazolone)	Murine immune complex nephritis [DBA/2 into (C57B1/6 × DBA/2) F_1] Rat adjuvant arthritis Rat collagen arthritis
		Inflammation Rat carrageenan edema/pleurisy
B cells	Blastogenesis (PWM, LPS)	Infection Viral prophylaxis (herpes simples)
	Plaque-forming cells (PFC) to sheep red blood cells (SRBC)	Bacterial and fungal (*Salmonella typhimurium, E. coli, Candida albicans, Listeria monocytogenes,* and so on)
Nonspecific	NK cytotoxicity	Tumor Primary and metastatic (Lewis lung carcinoma; lymphoma, Madison lung M109,
	Macrophages (phagocytosis, chemotaxis, cytotoxicity; release of O_2 radicals, ly-	

Table 7 *(Continued)*
Commonly Used Assays To Identify and Characterize Immunomodulatory Agents[a]

	Immune function, in vitro/in vivo	Disease models
	sosomal enzyme release, arachidonate metabolism, and monokine production; Ia and C3 expression)	mammary adenocarcinoma fibrosarcoma, and so on)
	Antibody-dependent cellular cytotoxicity (ADCC)	
	Neutrophils (chemotaxis, release of O_2 radicals and lysosomal enzymes)	

[a]Modified from Sedlacek (1984).

of finely targeted immunomodulation remains a future goal. As a consequence, all of the current immunomodulators have inherent limitations for the treatment of disease. The transition from laboratory to clinic is also complicated by ignorance of the etiological, immunological, and physiological bases of many human diseases, and thus the relevance of animal models frequently cannot be properly validated. In some diseases it may also be necessary to combine immunomodulator administration with more established therapy in order to demonstrate maximum benefit. An example is the combination of immunomodulators with macrophage activating, T cell restoring, and/or NK cell-enhancing activity with cancer cytoreductive therapy (chemotherapy, surgery, or irradiation).

4.4. Immunomodulatory Drugs in Current Use or Under Development

Table 8 lists some of the more established biologically and synthetically derived immunomodulators with clinical efficacy or under clinical investigation. This list is by no means complete and more extensive information concerning immunomodulators can be found elsewhere (Hersh, 1982, 1983; Fenichel and Chirigos, 1983; Lewis et al., 1982). The biologically derived immunomodulators are largely isolated from microorganisms such as bacteria and fungi. The synthetic immunomodulators have been divided into stimulators and suppressors according to the primary direction of modulation. Although not listed here, agents capable of modulating the immune

Table 8
Immunomodulatory Drugs in Current Use or Under Development

Biologically derived	Synthetic
Bacillus calmuette guerin (BCG) and extracts	Stimulators
Bestatin	Azimexon
Brucella abortus	Diethyldithiocarbamate
Corynebacterium parvum	Isoprinosine
Glucans	Levamisole
Interferons (α, β, γ)	Lipoidalamines
Krestin (PSK)	Maleic anhydride divinylether copolymer (MVE)
Lentinan	MDP analogs
Lymphokines (IL-1, IL-2, CSF, TNF, and so on)	NED-137
Muramyldipeptide (MDP)	NPT-15392
OK432 (Picibanil)	Polyinosinic-polycytidylic-poly-*L*-lysine (Poly IC:LC)
Thymic factors	Pyrimidinones
Thymosins (α_1,α_5,α_7,β_3,β_4)	Tilomisole (Wy-18,251)
Thymopentin (thymopoietin)	
Thymostimulin	Suppressors
Thymic humoral factor	Corticosteroids
Thymulin (factor thymique serique)	Cyclophosphamide
Trehalose dimycolate (cord factor)	Cyclosporin A
Tuftsin	6-mercaptopurine and azathioprine
	Methotrexate

response exist in each of the major nonimmunotherapeutic categories and include calmodulin antagonsits, cannabinoids, bromocriptine, diazepam, insulin, thyroxine, and tricyclic antidepressants. The immunological activities of such a diverse group of agents are better understood when it is realized that lymphocytes possess a variety of surface receptors including β-adrenergic, dopamine, cholinergic, H_2-histamine, opiate, and benzodiazepine receptors (Sorkin et al., 1981).

5. New Approaches Under Development

Several recent approaches to immunomodulation deserve mention, although some of these are in their infancy and are based, at the present time, more on theory than solid experimental data.

5.1. Monoclonal Antibodies

Monoclonal antibodies are immunoglobulins derived from cell lines that are constructed by fusing together antibody-producing normal B cells and an immortal mouse myeloma cell line. The resultant hybrid cell lines (hydridomas) possess the ability to secrete antibody and can grow permanently in culture (Milstein, 1982). Since these hydridomas arise from a single original B cell, they continuously produce a single class of immunoglobulin molecules with a single, defined antigenic specificity. Thus, large quantities of highly specific homogenous antibody can be readily obtained. The development of hybridoma technology and the availability of an array of monoclonal antibodies directed against a multitude of antigens has had a remarkable impact on many scientific disciplines. This impact has been particularly dramatic in the diagnostic and immunotherapeutic areas.

The use of monoclonal antibodies in the diagnosis of viral and bacterial infections has led to more rapid, accurate, and sensitive diagnostic procedures. Noteworthy examples of this include detection of hepatitis surface antigen, herpes simplex virus, *Chlamydia trachomatis*, and *Niserris gonnorhea* (Nowinski et al., 1983). Cancer diagnosis is another exciting application for monoclonal antibody technology. Monoclonals have already been described that react with a variety of tumor cell types, including neuroblastomas, colorectal and breast carcinomas, melanomas, and leukemias (Dippold et al., 1982; Marx, 1982; Cheung et al., 1985; Iacobelli et al., 1985; Lindmo et al., 1985).

Monoclonal antibodies have also led to rapid advances in the understanding of immune regulation and how the normal regulatory balance is perturbed in patients with autoimmune and immunodeficiency diseases. For example, monoclonal antibodies to helper and suppressor T cells (OKT4$^+$ and OKT8$^+$ cells, respectively) have been used to assess alterations in the number and ratio of these cell types in such diseases as rheumatoid arthritis, systemic lupus erythematosus, multiple sclerosis, myasthenia gravis, and AIDS (Bach and Chatenoud, 1982; Olsson, 1983; Pinching, 1984).

Immunotherapy with monoclonal antibodies directed against tumor antigens has been proposed as a viable anticancer modality (August, 1982). These antibodies can be used alone, or can first be coupled to a toxin (i. e., ricin A) or other drug to "target" such chemicals directly to the cancerous tissue, thereby reducing the total dose of drug and diminishing the probability and/or severity of side effects (Vitetta et al., 1983). Attaching immunostimulants or immunosuppressants could be used to target such drugs to specific lymphoid cell populations. In fact, some monoclonal antibodies, such as antibody to the T3 molecule on human T cells, directly stimulate T cell proliferation (at least in vitro) and could be used essentially as immunostimulatory "drugs" themselves. Similarly, antibodies that suppress lymphocyte activation, such as antibodies to class II major histocompatability antigen (Ia or HLA/DR) could be used directly as immunosuppressive agents (August, 1982; Olsson, 1983; Kolb and Toyka, 1984).

Monoclonal antibodies have shown therapeutic antitumor efficacy in animal models such as murine leukemia (Bernstein and Nowinski, 1982; Vitetta et al., 1983), and several studies in humans have also shown positive effects. For example, three patients with leukemia showed a transient clinical response to therapy with a monoclonal antibody specific for the leu-1 antigen on mature T cells (Miller et al., 1982). In another study, treatment of patients with chronic lymphocytic leukemia with monoclonal antibody to the T101 antigen (present on mature T cells) caused a dramatic drop in circulating leukemic cells, although no positive clinical responses were noted (Dillman et al., 1982).

Treatment of autoimmune disease is another therapeutic target for monoclonal antibodies. Treatment of mice with monoclonal anti-L3T4 antibody (a rat antibody to a mouse T cell antigen similar to the human OKT4 antigen) has been shown to prevent the development of collagen-induced arthritis (Ranges et al., 1985), lupus disease (Wofsy and Seaman, 1985), and experimental allergic encephalomyelitis (Waldor et al., 1985). Analogous studies in humans have not been reported to date.

Although the use of monoclonal antibodies as therapeutic agents is an exciting possibility, several theoretical and practical problems must be addressed before the utility of this approach can be fully realized. Since most of the monoclonals available today are of murine origin, antibody responses to the monoclonal can develop following administration to humans. Even if human monoclonal antibodies are used, antiidiotype antibody responses (antibodies to the unique antigen-binding regions on the injected antibody) could ensue (Geha, 1984; Kaprowski et al., 1984). Thus on subsequent administration, the monoclonal antibody would be rapidly cleared and inactivated by host antibody, limiting therapeutic effectiveness at best and, at worst, resulting in anaphylactoid reactions (Dillman et al., 1982). Clearance rates and in vivo localization of therapeutically administered monoclonal antibodies are also problematic.

5.2. Cytokines

As discussed earlier, cytokines are important mediators of a variety of effector functions of immune cells. Cytokines that stimulate T cells (IL-1, IL-2), macrophages (macrophage-activating factor, interferon), NK cells (IL-1, IL-2, interferon), and B cells (IL-1, IL-2, T cell replacing factor; TRF), as well as cytokines that suppress lymphoid cell reactivity, have all been described (Hansen et al., 1982). In addition, cytotoxic cytokines such as lymphotoxin and tumor necrosis factor are known. In vitro data utilizing these and other lymphokines suggest that they could be used therapeutically to augment or suppress immune function (Kleinerman et al., 1983). Only interferons, primarily α and γ, have been extensively assessed in the clinic so far, with disappointing results overall (with the exception of some specific tumor types) (Strander and Einhorn, 1982). Early clinical trials with IL-2 have shown some promising results (Merluzzi and Last-Barney, 1985). However, further information on the biology, pharmacokinetics, and mode of action of lymphokines is needed before the means for successful therapeutic use for these agents can be developed.

6. Summary

The field of immunopharmacology is blossoming from an ever-increasing knowledge of the diversity and complexity of the immune system. Our understanding of the many ways in which the immune function can malfunction in disease processes has led to a multifaceted approach to immunopharmacological drug development. Although many chemical and synthetic agents can modify immune function, few have selective, specific effects and their mechanism(s) of action are as yet ill-defined. Clinical trials continue to show promise for immunopharmacological agents in the treatment of cancer, autoimmunity, immunodeficiency, and hypersensitivity, but definitive proof of their comparable efficacy relative to the classes of therapeutic agents awaits further study.

References

August, J. T. (1982) *Monoclonal Antibodies in Drug Development*. Soc. Pharmacol. Exp. Therapeutics, Bethesda, Maryland, pp. 1–237.

Bach, J.-F. (1983) The thymus in immunodeficiency diseases: New therapeutic approaches. *Birth Defects: Original Article Series* **1**, 245–253.

Bach, J.-F. and Chatenoud, L. (1982) The significance of T-cell subsets defined by monoclonal antibodies in human diseases. *Ann. Immunol. Inst. Pasteur* **183D**, 131–136.

Bernstein, I. D. and Nowinski, R. C. (1982) Monoclonal Antibody Treatment of Transplanted and Spontaneous Murine Leukemia, in *Hybridomas in Cancer Diagnosis and Treatment* (Mitchell, M. S. and Oettgen, H. F., eds.) Raven, New York.

Cheung, N.-K. V., Saarinen, U. M., Neely, J. E., Landmeier, B., Donovan, D., and Coccia, P. F. (1985) Monoclonal antibodies to a glycolipid antigen on human neuroblastoma cells. *Cancer Res.* **45**, 2642–2649.

Cooper, M. D. (1983) B cell differentiation. *Birth Defects: Original Article Series* **19**, 25–29.

Dillman, R. O., Sobol, R. E., Collins, H., Beauregard, J., and Royston, I. (1982) T101 Monoclonal Antibody Therapy in Chronic Lymphocytic Leukemia, in *Hybridomas in Cancer Diagnosis and Treatment* (Mitchell, M. S. and Oettgen, H. F., eds.) Raven, New York.

Dinarello, C. A. (1984) Interleukin 1. *Rev. Infect. Dis.* **6**, 51–95.

Dippold, W. G., Lloyd, K. O., Houghton, A. N., Li, L. T. C., Ikeda, H., Oettgen, H. F., and Old, L. J. (1982) Human Melanoma Antigens Defined by Monoclonal Antibodies, in *Hybridomas in Cancer Diagnosis and Treatment* (Mitchell, M. S. and Oettgen, H. F., eds.) Raven, New York.

Dougherty, G. J. and McBride, W. H. (1984) Macrophage heterogenicity. *J. Clin. Lab. Immunol.* **14**, 1–11.

Duram, S. K., Schmidt, J. A., and Oppenheim, J. J. (1985) Interleukins 1: An immunological perspective. *Ann. Rev. Immunol.* **3**, 263–287.

Easman, C. S. F. and Gaya, H., eds. (1983) *Second International Symposium on Infections in the Immunocompromised Host*, Academic, New York.

Fathman, D. G. (1982) Regulation of the Immune Response, in *Clinical Cellular Immunology. Molecular and Therapeutic Reviews* (Luderer, A. A. and Weetall, H. H., eds.) Humana, New Jersey.

Fenichel, R. L. and Chirigos, M. A., eds. (1983) *Immune Modulation Agents and Their Mechanisms* Marcel Dekker, New York.

Fidler, I. J., Berendt, M., and Oldham, R. K. (1982) The rationale for and design of a screening procedure for the assessment of biological response modifiers for cancer treatment. *J. Biol. Response Modif.* **1**, 15–26.

Florentin, I. M., Bruley-Rosset, M., Schulz, J., Davigny, M., Kiger, N., and Mathe, G. (1981) Attempt at Functional Classification of Chemically-Defined Immunomodulators, in *Advances in Immunopharmacology* (Hadden, J. W., Chedid, L., Mullen, P., and Spreafico, F., eds.) Pergamon, Oxford.

Florentin, I., Kraus, L., Bruley-Rossett, M., and Mathe, G. (1983) In Vivo Functional Characterization of Immunomodulators, in *Advances in Immunopharmacology* vol. 2 (Hadden, J. W., Chedid, L., Dukor, P., Spreafico, F., and Willoughby, D., eds.) Pergamon, Oxford.

Ford, W. L. (1975) Lymphocyte migration and immune responses. *Prog. Allergy* **19**, 1–59.

Geha, R. S. (1984) Idiotypic–antiidiotypic interactions in humans. *J. Biol. Resp. Modif.* **3**, 573–579.

Geha, R. S. and Rosen, F. S. (1983) Immunoregulatory T cell defects. *Immunol. Today* **4**, 233–237.

Gelfand, E. W. and Dosch, H.-M. (1983) Diagnosis and classification of severe combined immunodeficiency disease. *Birth Defects: Original Article Series* **19**, 65–72.

Gibson, J., Basten, A., and Van Der Brink, C. (1983a) Clinical use of transfer factor 25 years on. *Clinic Immunol. Allergy* **3**, 331–357.

Gibson, G. G., Hubbard, R., and Parke, D. V., eds. (1983b) *Immunotoxicology* Academic, London.

Gilman, S. C. (1984) Lymphokines in immunological aging. *Lymphokine Res.* **3**, 119–123.

Gilman, S. C. and Lewis, A. J. (1985) Immunomodulatory Drugs in the Treatment of Rheumatoid Arthritis, in *Antiinflammatory and Antirheumatic Drugs* vol. III (Rainsford, K. D., ed.) CRC, Boca Raton, Florida.

Gilman, S. C., Daniels, J. F., Wilson, R. E., Carlson, R. P., and Lewis, A. J. (1984) Lymphoid abnormalities in rats with adjuvant-induced arthritis. I. Mitogen responsiveness and lymphokine synthesis. *Ann. Rheum. Dis.* **43**, 847–855.

Gilmore, N. and Wainberg, M. A. (1985) Viral Mechanisms of Immunosuppression, in *Progress in Leukocyte Biology* vol. 1, Alan R. Liss, New York.

Gresser, I. (1980) Interferon and the Immune System, in *Progress in Immunology* vol. 4 (Fougereau, P. and Nausset, J., eds.) Academic Press Congress of Immunology, New York.

Gupta, S. and Good, R. A. (1979) *Cellular Molecular and Clinical Aspects of Allergic Disorders* Plenum, New York.

Hadden, J. W. (1981) The Immunopharmacology of Immunotherapy: An Update, in *Advances in Immunopharmacology* (Hadden, J. W., Chedid, L., Mullen, P., and Spreafico, F., eds.) Pergamon, Oxford.

Hadden, J. W. (1983) Characterization of Immunotherapeutic Agents: An Overview, in *Advances in Immunopharmacology* vol. 2 (Hadden, J. W., Chedid, L., Dukor, P., Spreafico, F., and Willoughby, D., eds.) Pergamon, Oxford.

Hadden, J. W., Coffey, R. G., and Spreafico, F., eds. (1977) *Immunopharmacology* Plenum, New York.

Hansen, J. M., Rumjanek, V. M., and Morley, J. (1982) Mediators of cellular immune reactions. *Pharmacol. Ther.* **17**, 165–198.

Herberman, R. B., ed. (1980) *Natural Cell Mediated Immunity Against Tumors*. Academic, New York.

Hersh, E. (1982) Perspectives for Immunological and Biological Therapeutic Intervention in Human Cancer, in *Immunological Approaches to Cancer Therapeutics* (Mihich, E., ed.) Wiley, New York.

Hersh, E. M. (1983) Immunotherapy of Human Cancer: Current Status and Prospects for Future Development, in *Advances in Immunopharmacology* (Hadden, J. W., Chedid, L., Dukor, P., Spreafico, F., and Willoughby, D., eds.) Pergamon, Oxford.

Hewitt, H. B., Blake, E. R., and Walder, A. S. (1976) A critique of the evidence for active host defense against cancer; based on personal studies of 27 murine tumours of spontaneous origin. *Br. J. Cancer* **33**, 241–259.

Hirschhorn, R. (1983) Genetic deficiencies of adenosine deaminase and purine nucleoside phosphorylase: Overview, genetic heterogeneity and therapy. *Birth Defects: Original Article Series* **19**, 73–81.

Hong, R. (1982) Congenital Immunodeficiencies, in *The Pathophysiology of Human Immunologic Disorders* (Twomey, J. J., ed.) Urban and Schwarzenberg, Baltimore, Maryland.

Iacobelli, S., Natoli, V., Scambia, G., Sanetusanio, G., Negrini, R., and Natoli, C. (1985) A monoclonal antibody (AB/3) reactive with human breast cancer. *Cancer Res.* **45**, 4334–4338.

Ihle, J. N., Rebar, L., Keller, J., Lee, J. C., and Hapel, A. J. (1982) Interleukin 3: Possible roles in the regulation of lymphocyte differentiation and growth. *Immunol. Rev.* **63**, 5–32.

Kaprowski, H., Herlyn, O., Lubeck, M., DeFreitas, H., and Sears, H. F. (1984) Human anti-idiotype antibodies in cancer patients: Is the modulation of the immune response beneficial to the patient? *Proc. Natl. Acad. Sci. USA* **81**, 216–219.

Kay, N. E., Ackerman, S. K., and Douglas, S. D. (1979) Anatomy of the immune system. *Seminars Hematol.* **16**, 252–282.

Klareskog, L., Forsum, U., Sheynius, A., Kabelitz, D., and Wigzell, H. (1982) Evidence for a self-perpetuating HLA-DR-dependent delayed-type hypersensitivity reaction in rheumatoid arthritis. *Proc. Natl. Acad. Sci. USA* **79**, 3632–3636.

Kleinerman, E. S., Schroit, A. J., Fogler, W. E., and Fidler, I. J. (1983) Tumoricidal activity of human monocytes activated in vitro by free and liposome-encapsulated human lymphokines. *J. Clin. Invest.* **72**, 304–315.

Kolb, H. and Toyka, K. V. (1984) New concepts in immunotherapy. *Immunol. Today* **5**, 307–308.

Kralovec, J. (1983) Synthetic immunostimulants in antitumor therapy. *Drugs of the Future* **8**, 615–638.

Langrange, P. H. (1983) Clinical Immunomodulation of Bacterial Infection, in *Advances in Immunopharmacology* vol. 2 (Hadden, J. W., Chedid, L., Dukor, P., Spreafico, F., and Willoughby, D., eds.) Pergamon, Oxford.

Lewis, A. J., Carlson, R. P., and Chang, J. (1982) Therapeutic modulation of cellular mediated immunity. *Annu. Reports. Med. Chem.* **17**, 191–202.

Lewis, A. J., Musser, J. H., Chang, J., and Silver, P. J. (1985) New Approaches to Bronchodilator and Antiallergic Drug Therapy, in *Progress in Medicinal Chemistry* vol. 22 (Ellis, G. P. and West, G. B., eds.) Elsevier Science, Amsterdam.

Lindmo, T., Boven, E., Mitchell, J. B., Morstyn, G., and Bunn, P. A. (1985) Specific killing of human melancoma cells by ^{125}I-labeled 9.2.27 monoclonal antibody. *Cancer Res.* **45**, 5080–5087.

Luger, T. A., Smolen, J. S., Chused, T-M., Steinberg, A. D., and Oppenheim, J. J. (1982) Human lymphocytes with either the OKT4 or OKT8 phenotype produce interleukin 2 in culture. *J. Clin. Invest.* **70**, 470–473.

MacDonald, H-R., Bonnard, G. D., Sordat, B., and Zawodnik, S. A. (1975) Antibody-dependent cell-mediated cytotoxicity: Heterogeneity of effector cells in human peripheral blood. *Scand. J. Immunol.* **4**, 487–497.

Marx, J. L. (1982) Monoclonal antibodies in cancer. *Science* **216**, 213–285.

Merluzzi, V. J. and Last-Barney, K. (1985) Potential use of human interleukin 2 as an adjunct for the therapy of neoplasia, immunodeficiency and infectious disease. *J. Bio.. Resp. Modif.* **7**, 31–39.

Miller, R. A., Maloney, D., Warnke, R., McDougall, R., Wood, G., Kawakami, T., Dilley, J., Goris, M. L., and Levy, R. (1982) Considerations for Treatment with Hybridoma Antibodies, in *Hybridomas in Cancer Diagnosis and Treatment* (Mitchell, M. S. and Oettgen, H. F., eds.) Raven, New York.

Milstein, C. (1982) Monoclonal antibodies. *Cancer* **49**, 1953–1957.

Miyasaka, N., Nakamura, T., Russell, I. J., and Talal, N. (1984) Interleukin 2 deficiencies in rheumatoid arthritis and systemic lupus erythematosis. *Clin. Immunol. Immunopathol.* **31**, 109–117.

Mullen, P. W. (1979) An immunopharmacology journal: Reflections on its interdisciplinary and historical context. *Int. J. Immunopharmacol.* **1**, 1–4.

Nowinski, R. C., Tarn, M. R., Goldstien, L. C., Stong, L., Kuo, C.-C., Corey, L., Stamm, W. E., Handsfield, H. H., Knapp, J. S., and Holmes, K. K. (1983) Monoclonal antibodies for diagnosis of infectious diseases in humans. *Science* **219**, 637–644.

Oldham, R. K. and Malley, R. V. S. (1983) Immunotherapy: The old and the new. *J. Biol. Resp. Modif.* **2**, 1–37.

Olsson, L. (1983) Monoclonal antibodies in clinical immunobiology. *Allergy* **38**, 145–154.

Ortaldo, J. R. and Herberman, R. B. (1984) Heterogeneity of natural killer cells. *Ann. Rev. Immunol.* **2**, 359–394.

Pinching, A. (1984) The probable cause of AIDS. *Immunol. Today* **5**, 196–199.

Ranges, G. E., Sriram, S., and Cooper, S. M. (1985) Prevention of type II collagen-induced arthritis by in vivo treatment with anti-L3T4. *J. Exp. Med.* **162**, 1105–1110.

Reinherz, E. L. and Schlossman, S. F. (1980) The differentiation and function of human T lymphocytes. *Cell* **19**, 312–326.

Robb, R. J. (1984) Interleukin 2: The molecule and its function. *Immunol. Today* **5**, 203–209.

Rosenthale, M. E. and Mansmann, H. C., eds. (1975) *Immunopharmacology* Spectrum, New York.

Schulman, E. S., Kagey-Sobotka, A., Macglashan, Jr., D. W., Adkinson, Jr., N. F., Peters, S. P., Schleimer, R. P., and Lichtenstein, L. M. (1983) Heterogeneity of human mast cells. *J. Immunol.* **131**, 1936–1941.

Sedlacek, H. H. (1984) Test systems for immunomodulators—how to find out immunomodulators. *Behring Inst. Res. Comm.* **74**, 122–131.

Seligmann, M. and Ballet, J.-J. (1983) Diagnosis criteria and classification of human primary defects of humoral immunity. *Birth Defects: Original Article Series* **19**, 153–160.

Shirai, T. (1982) The genetic basis of autoimmunity in murine lupus. *Immunol. Today* **3**, 167–174.

Smith, K. A. (1984) Interleukin 2. *Ann. Rev. Immunol.* **2**, 319–334.

Smolen, J. S. and Steinberg, A. D. (1982) Disorders of Immune Regulation, in *The Pathophysiology of Human Immunologic Disorders* (Twomey, J. J., ed.) Urban and Schwarzenberg, Baltimore, Maryland.

Sorkin, E., Del Rey, A., and Besedovsky, H. O. (1981) Neuroendocrine Control of the Immune Response, in *The Immune System* vol. 1 (Steinberg, C. M. and Lefkovits, I., eds.) Karger, Basel.

Spreafico, F. (1985) Problems and challenges in the use of immunomodulating agents. *Int. Archs. Allergy Appl. Immunol.* **76**, 108–118.

Spreafico, F., Vecchi, A., Conti, G., and Sironi, M. (1981) On the Heterogeneity of Immunotherapeutic Agents, in *Advances in Immunopharmacology* (Hadden, J. W., Chedid, L., Mullen, P., and Spreafico, F., eds.) Pergamon, Oxford.

Strander, H. and Einhorn, S. (1982) Interferon and cancer: Faith, hope and reality. *Amer. J. Clin. Oncol.* **5**, 297–301.

Stuart, J. M., Townes, A. S., and Kang, A. H. (1984) Collagen immune arthritis. *Ann. Rev. Immunol.* **2**, 199–218.

Talmadge, J. E., Oldham, R. K., and Fidler, I. J. (1984) Practical considerations for the establishment of a screening procedure for the assessment of biological response modifiers. *J. Biol. Resp. Modif.* **3**, 88–109.

Twomey, J. J. (1982) Disorders of Macrophages, in *ThePathophysiology of Human Immunological Disorders* (Twomey, J. J., ed.) Urban and Schwarzenberg, Baltimore, Maryland.

Unanue, E. R. (1984) Antigen-presenting function of the macrophage. *Ann. Rev. Immunol.* **2**, 395–428.

Vitetta, E. S., Krolick, K. A., Miyama-Inaba, M., Cushley, W., and Uhr, J. W. (1983) Immunotoxins: A new approach to cancer therapy. *Science* **219**, 644–647.

Vivella, G. and Fudenberg, H. H. (1982) Secondary Immunodeficiencies, in *The Pathophysiology of Human Immunologic Disorders* (Twomey, J. J., ed.) Urban and Schwarzenberg, Baltimore, Maryland.

Waldor, M. K., Sriram, S., Hardy, R., Herzenberg, L. A., Herzenberg, J. A., Lancer, L., Tim, M., and Steinman, L. (1985) Reversal of experimental allergic encephalomyelitis with monoclonal antibody to a T-cell subset marker. *Science* **227**, 415–417.

Wasserman, S. I. (1983) Mediators of immediate hypersensitivity. *J. Allergy Clin. Immunol.* **72**, 101–115.

Weiss, A., Hollander, H., and Stobo, J. (1985) Acquired immunodeficiency syndrome: Epidemiology, virology and immunology. *Ann. Rev. Med.* **36**, 545–562.

Weksler, M. E. and Siskind, G. W. (1982) Age and the Immune System, in *The Pathophysiology of Human Immunologic Disorders* (Twomey, J. J., ed.) Urban and Schwarzenberg, Baltimore, Maryland.

Werner, G. H. and Floc'h, F. (1981) Persistence Models for the Testing of Immunopotentiating Agents, in *Advances in Immunopharmacology* (Hadden, J. W., Chedid, L., Mullen, P., and Spreafico, F., eds.) Pergamon, Oxford.

Wofsy, D. and Seaman, W. E. (1985) Successful treatment of autoimmunity in NZB/NZW F_1 mice with monoclonal antibody to L3T4. *J. Exp. Med.* **161**, 378–385.

Toxicological Evaluation and Clinical Aspects

Toxicological Evaluation of Drugs

JOY CAVAGNARO AND RICHARD M. LEWIS

1. Introduction

All drugs have the potential to produce toxicity. The degree to which such toxicity occurs in a compound as balanced against its therapeutic activity is termed the therapeutic index. Toxicity should be clearly dissociable from efficacy. The goal of toxicity testing is therefore to determine the adverse effects of drugs in well-defined biological systems, which have their own unique susceptibility to toxicity. Once determined, these effect(s) of a drug and its metabolites are extrapolated to humans so that potential risks can be weighed against potential benefits. The ultimate objective is to evaluate the probability that a drug will not produce significant damage under specific conditions of use (Health and Welfare, Canada, 1981).

Although numerous regulatory agencies have different criteria for measuring drug toxicity, there is general agreement with respect to the major testing regimen. The principal requirement for all agencies is in vivo mammalian toxicity data (Kroes and Feron, 1984).

In vivo mammalian toxicity testing has traditionally consisted of acute, subacute or subchronic, and chronic studies designed to determine the general effects of compounds on animal systems. Sufficient numbers of animals are included per treatment (including controls) to permit statistically valid dose-related estimates of the incidence of toxic effects.

Special studies designed to estimate the potential of drugs to influence reproduction, development, or behavior, or induce tumors or genetic damage, or specifically alter the immune system have not only been utilized, but are increasingly becoming a prerequisite before new drugs are tested in humans.

1.1. The Drug

Before toxicological studies are conducted, the drug should be well characterized, especially in terms of its efficacy, specificity, and bioavailability. Satisfactory stability of the drug should also be demonstrated for the required conditions and duration of storage, as should its solubility characteristics. Toxicologic studies are greatly facilitated if qualitative and quantitative analytic methods for determination of the drug and its metabolites in biologic material are available.

The drug should be administered in the vehicle intended for therapeutic application, especially if the vehicle influences delivery or activity of the drug. Control studies should be performed with the vehicle alone if pertinent toxicologic data are unavailable. Dosing schedules and patterns are selected after the in vivo pharmacodynamics of the drug are known and must be consistent with them. Establishment of the routes and rates of absorption and excretion are important in identifying biological steady states, or degrees of bioretention, at the dose levels ultimately selected for chronic studies.

1.2. The Test System

Selection of the appropriate animal species for in vivo toxicity studies is generally based upon the results of concurrent pharmacologic and metabolic studies, as well as clinical observations. The sensitivity of possible target organs is also considered when deciding upon the most appropriate test system. The laboratory animals generally available for toxicity testing of drugs include: the rodent (rat, mouse, guinea pig, hamster); lagomorph (rabbit); carnivore (dog, cat); or primate (monkey, baboon). Two or more of these different orders of animals should be used. Similar toxicity in more than one order of animal may increase the predictability of toxicity in man.

1.3. The Regimens

The basic design of acute, subchronic, and chronic toxicity testing is outlined in the Appendix.

2. Toxicity Data Requirements

A comprehensive approach to toxicity testing proceeds in a stepwise fashion through a battery of metabolism, acute, and subacute and/or subchronic studies, and decision points, leading ultimately to chronic tests, if necessary. At the present time, short-term in vivo and in vitro tests are considered useful supplements to chronic studies in identifying and assessing potential human cancer risks. These may be used as indicators of the need for a chronic bioassay and to set priorities for long-term testing. Such tests may also be helpful in studying possible mechanisms and sites of activity of carcinogens identified in the bioassay (report of the National Toxicology Program Ad Hoc Panel on Chemical Carcinogenesis Testing and Evaluation, 1984).

A summary of those studies generally performed in the safety evaluation of single ingredient drugs for oral or parenteral administration is shown in Table 1 and includes acute toxicity, subchronic, chronic, carcinogenicity, mutagenicity, reproductive, and special toxicity studies. All such studies should be carried out according to Good Laboratory Practices (GLP) guidelines (*see* section 4).

2.1. Acute Toxicity Studies

Acute toxicity comprises the adverse effects occurring within a short time of administration of a single dose of a test compound. The route of administration is either oral or parenteral. Acute toxicity studies are performed in rodent and nonrodent species. Unless there are contraindications, the rat is the preferred species. Both sexes are used unless the proposed clinical use of the drug precludes its use in both sexes in humans. These studies are required for all drugs before an initial (Phase I) human tolerance study can be done.

The observation and recording of any signs characteristic of toxicity, both at lethal and sublethal dosage levels, is an essential part of an acute toxicity study. Animals that die during the test, as well as surviving animals at the conclusion of the test, are necropsied.

At least two animal species, one a nonrodent, are used in toxicology studies. Ideally these should include those predominating in the pharmacological testing of the drug and those likely to be used in the multidose toxicity and reproduction studies.

To determine appropriate doses, an acute toxicity study may be preceded by a preliminary range-finding test on a small number

Table 1
Summary of Studies Generally Performed in Safety Evaluation of Single Ingredient Drugs

Type of test	Species used	Number of animals	Number of dose levels	Duration
Acute toxicity	At least three; one should be nonrodent	Not specified	Not specified	2–4 wk
Subchronic toxicity	Rodent (rat); nonrodent	10–20/sex (rodent); 2–4/sex (nonrodent)	Sufficient to ensure that at least one level has no effect and that doses are provided that show detailed toxic effects	90-d rat (max); 12-mo nonrodent (max)
Chronic toxicity	Rat	25/sex	Three + control (min)	18–24 mo
Reproduction study	Two for embryotoxicity studies at least in fertility and perinatal studies; one of the species (if possible) that is used in long-term studies. Third specifies if conflicting results are obtained.	Embryotoxicity: 20 pregnant females (rodent); 12 pregnant females (nonrodent)	Three + control	Not specified
Carcinogenicity	At least two, preferably those with metabolism of drug similar to humans.	50/sex	Three + control	Majority of lifetime

of animals. The purpose of dose selection is to achieve a range of toxic responses. As a rule, at least three treatment levels are required to define the slope of a dose–response curve that describes toxicity. These are prepared to induce severe toxicity or lethality at the highest level and to be without adverse effect at the lowest level. An intermediate level should cause tolerated adverse effects without causing drug-induced death. Animals are generally fasted prior to dosing and should be observed for at least 1 wk after dosing or longer if either overt signs persist or delayed deaths occur. Sometimes in the case of nonrodent animals, a range-finding study may be more productive in obtaining useful information and may be substituted for the more classical acute toxicity study. Animals that die after acute drug administration should be necropsied, when feasible.

Acute toxicity studies contribute to the overall profile of a drug's activity, and provide information about the magnitude of its toxicity and overt effects. They may also provide information on sex differences in toxic responses; the mode of action of the drug, the tissues and organs affected, and the speed and completeness of recovery in relationship to dosage and time after dosing. In addition they provide apparent dosing information for subsequent subacute or subchronic studies (Table 2).

Table 2
Objectives of Acute Toxicity Testing

Determine maximum no-effect dose
Detect species, sex, and age differences
Detect differences in toxicity related to different routes of administration
Determine the occurrence of specific effects on, e. g., locomotion and behavior, respiration, circulation
Detect existence of delayed toxicity
Determine ratio between doses producing pharmacological effects and those producing significant toxic effects
Determine doses for repeat-dose toxicity studies
Predict toxic reactions after acute overdosage in humans

2.2. Subacute or Subchronic Toxicity (Short-Term Studies)

The purpose of investigating subacute or subchronic toxicity is to establish the effects of repeated exposure to the test compound by administering it to animals regularly, preferably daily, for at least

14 d. Short-term repeated dose studies are often used as dose-range-finding studies for subsequent testing and are required for all drugs before the initial (Phase I) human tolerance study. They provide additional information, including sex-related differences, target organs, and dose levels in relation to toxicological responses, necessary for planning chronic studies and in helping to define the toxicity of the test compound.

The route of compound administration is determined by intended use, its physical and chemical properties, and its toxicological properties as revealed by acute toxicity studies. Short term repeated dose studies are performed on rodent and nonrodent species. At least three dose levels and a control are used.

During the period of drug administration, animals are observed daily to detect signs of toxicity, including behavioral and neurological effects. Periodic recordings of the effects of the drug on body weight, food and water consumption, and the eyes should be done in all studies. Cardiovascular as well as other physiological and pharmacological manifestations may also be monitored in the nonrodent species. Animals that die during the test are necopsied and, at the conclusion of the test, the surviving animals are necropsied. Histopathological examinations are performed on the high-dose group and the control group. Target organs and tissues showing defects attributable to the test compound at the highest dose are examined at all lower doses. Hematological tests (e.g., hemoglobin, hematocrit, white blood cell and differential counts, coagulation test) and clinical biochemistry in blood that assess carbohydrate metabolism (e.g., blood glucose), liver function (e.g., AST, ALT), kidney function (BUN), and other tests that are pertinent to the drug's effect, e.g., electrolytes, lipids, hormones, are also evaluated. Urinalysis may be limited to nonrodents but should include microscopic examination of the sediment.

Subchronic studies are usually carried out with rodents in the period of growth for about one-tenth of their lifetime. Subchronic testing is indicated in cases in which toxic effects may become apparent only after more prolonged exposure to the compounds, e. g., compounds that have a long biological half-life.

A subchronic study that may be up to 13 wk in duration is necessary if there is a potential for prolonged or repeated exposure via a route that is likely to lead to significant absorption; if acute or subacute experiments suggest that the compound is slowly excreted and accumulated in tissues; or if serious, irreversible, or unusual toxic effects were observed in the 2-wk study; and/or a no-effect level is not identified. These tests are performed in rodent and nonrodent species and are required before early human safety and efficacy studies (Phase II) for limited use and unlimited use of drugs. A typical subchronic study should include 10–20 rodents per sex per group, and 2–4 nonrodents per sex per group.

The guidelines for subchronic studies reflect a change in attitude toward the importance of such testing. Previously, the primary purpose of a subchronic study was to establish a dose (e. g., maximum tolerated dose) for use in subsequent chronic studies. Subchronic studies are now used to study a broad system of toxicological information, such as reproduction, teratology, immunology, neurobehavior, clinical pathology, and pharmacokinetics (de Serres and Matsushima, 1984) (Table 3).

Table 3
Objects of Short-Term Toxicity Testing

Determine the appropriate exposure levels for chronic or long-term and carcinogenicity studies
Establish dose-response and maximum no-effect dose for each species
Identify organs and tissues injured by exposure to detect possible cumulative toxicity
Detect species, age, and sex differences
Identify special monitoring techniques, i. e., which can be included in chronic studies
Assess possible health hazards to humans associated with repeated or continuous exposures

2.3. Chronic Toxicity (Long-Term Studies)

Chronic toxicity relates the deleterious effects produced by a drug when administered repeatedly for from 3 mo to several years. These studies must be ongoing before extended safety and efficacy studies (Phase III) are done and must be completed before the drug is marketed. Chronic testing is performed on compounds that have an anticipated use leading to a relatively long or continuous exposure in humans (Table 4).

The duration of chronic toxicity tests performed on rodent and nonrodent species has been a matter of discussion. In the case of rodents, except for carcinogenicity, the detection of chronic symptoms of toxicity may not require exposure of longer than 6–12 mo.

Table 4
Objectives of Long-Term Toxicity Testing

Determine therapeutic margins in the species studied
Determine dose-response relationship of behavioral, functional, and histopathological effects of a drug when administered repeatedly
Determine the target organs for potential toxicity
Determine the existence of species and sex differences
Determine whether the toxicity is reversible

Administration of drug includes the route that will be used in clinical studies. If it is suspected that the pharmaceutical dosage form may affect the toxicity of the drug, this dosage form should be administered as well. When the drug is administered orally, evidence for absorption (blood levels, pharmacologic effects, and so on) should be provided. Oral administration may be substituted for parenteral administration if studies show that the metabolic patterns obtained after administration by each route are comparable.

Chronic toxicity studies are generally conducted with at least three different dose levels. The highest dose should cause toxic effects, but should also allow for survival of the majority of animals. The high dose should not compromise the physical condition or appearance of the animal during the first 25% of the dosing period. The lowest dose should exert appropriate pharmacological or therapeutic effects in the species concerned. To permit the determination of dose–response relationships, one or more intermediate doses should also be included.

Control groups generally include untreated, vehicle-treated, and positive-control animals as required for various comparisons.

When deciding on the number of animals per group, consideration is given to the understanding that at least some animals for the highest dosage level and some in the control group will be used for periodic laboratory investigations and histopathology, whereas others will be retained for observation after the end of the compound administration period. The selection of group size for chronic studies will depend upon the toxicity mortality findings found in the preceding subacute/subchronic studies.

Throughout the duration of the study, all animals are observed for changes in appearance and behavior, neurologic effects, body weight, rate of weight gain, and food and water consumption. Additional observations include hematology, clinical chemistry, urinalysis, and histopathology.

Chronic toxicity studies may be designed to investigate carcinogenic potential and general toxicity within the same study. This requires careful consideration of the number of animals per sex and per group used; interim observations of clinical chemistry, hematology, urinalysis, and pathology (gross and microscopic) must be frequent; and the number of dose levels, duration of the study, and usage of recovery observation periods must be properly determined. Alternatively, chronic bioassays for carcinogenicity can be performed.

2.4. Carcinogenicity

The induction of cancer by drugs is a stepwise process involving conversion of normal cells to neoplastic cells and the progression of the neoplastic cell to formation of a tumor. The interaction of a chemical with DNA is generally accepted to be a requisite step for the initiation of chemical carcinogenesis. A carcinogen can be described as an agent capable of inducing benign and/or malignant neoplasms in animals or humans. The term refers to initiators, tumor promoters, potentiators, modifiers, (e. g., nutritional, hormonal, immune, and genetic), and toxicity factors (e. g., disruption of protective enzymes or membranes and depletion of free radical scavengers) (Farber, 1982, Miller and Miller, 1981).

A carcinogenicity study or chronic bioassay in animals is required when there is a potential for continuous or repeated exposure of the drug over long periods of time or when the drug or any of its metabolites has a chemical structure similar to a known carcinogen. Carcinogenicity tests are performed in rodent and nonrodent species. They must be ongoing before extended safety and efficacy studies (Phase III) are done and must be completed before marketing. They are usually initiated as soon as efficacy is confirmed in the Phase II clinical trials. Nonrodent (i. e., primates and dogs) tests are required for contraceptive products (Squire, 1984).

Administration of test compound begins at 6 wk of age and continues until approximately 2 yr, unless there is excessive mortality prior to that time. A multiple of 25–50 times the maximum daily human dose is often considered an acceptable high dose for rodent carcinogenicity studies. This estimate may change with different species. The protocol is designed to focus on a single toxic or carcinogenic effect, although other toxicities are likely to occur in such studies.

Because of the usually long latent period required for the production of tumors, the mose useful species are the rat, mouse, and hamster. Susceptibility to a particular carcinogen or class of carcinogens is a major consideration in selecting appropriate species.

Chronological tables are prepared for animals that die or are sacrificed during an experiment. These list individual times of death, the cause of death when known, and the occurrence of preneoplastic and neoplastic lesions. Well-constructed tables are fundamental to the reporting and evaluation of toxicological data. Composite tables for all experimental animals examined indicate the incidences of pathogenetically/histopathologically correlated preneoplastic and neoplastic lesions for each affected organ or tissue for each group examined. The database should be capable of displaying all events of interest for each animal (toxic signs, tumor palpation, clinical pathology, necropsy results, sacrifice or deaths, and so on), and the dates of these events. Data obtained from lifetime exposure of laboratory animals, however, do not differentiate between the carcinogen as initiator, potentiator, or modifier of cancer.

All biological tests, laboratory determinations, and statistical methods should be described by protocol or reference. Consideration of the structure of the test compound is important not only in predicting potential carcinogenicity, but in anticipating the likelihood of positive results in short-term and chronic tests. The next step is the application of a battery of short-term tests.

2.5. Mutagenicity Studies

The accuracy of short-term in vitro or in vivo tests in predicting the cancer-causing ability of drugs or chemicals is described by two parameters: the ''sensitivity'' of the assay as determined by the proportion of carcinogens correctly identified, and the ''specificity'' of the assay as determined by the proportion of noncarcinogens correctly ruled out (Cooper et al., 1979; Hollstein et al., 1979).

In the actual application of in vitro test methods, the significance of endpoints resides mainly in the correlations between the results obtained in well-conducted and definitive, long-term carcinogenicity bioassays (Dunkel et al., 1981).

No individual test, including many widely used assays, has detected all carcinogens tested, and thus, from this practical consideration alone, a battery of tests is essential. However, the importance of a battery becomes even more obvious upon consideration of the complexities of biotransformation and mechanisms of action of chemical carcinogens.

Mutagenicity testing emerged as radiation biologists began to implicate DNA damage in somatic cells as a mechanism potentially involved in the etiology of cancer (Kolbye, 1980). An appreciation of the mechanisms of possible DNA damage developed along with knowledge relating to the chemical composition of DNA and its structure and nucleotide sequence. As a means to identify potential carcinogens, efforts have been directed to detecting and identifying chemicals with the potential to damage DNA.

Studies over the past decade have shown that there is a relatively high empirical correlation between the results of carcinogenicity and mutagenicity testing. Because of this correlation and the pervading theory that carcinogen transformation or ''initiation'' involves an alteration of genetic material in cells, genotoxicity assays have been used as prescreens or as supportive evidence for carcinogenicity testing in animals (Table 5).

Table 5
Objectives of Mutagenicity Testing

Objectives
Evaluate drugs under development for possible adverse genetic or carcinogenic effects before costly product development is attempted
Screen natural or synthetic compounds for genotoxic or carcinogenic potential
Screen body fluids (e. g., blood, urine) or excreta for genotoxic agents that may indicate exposure to noxious agents
Aid in understanding the mechanism of cancer or mutation induction

A mutagenic compound is one capable of inducing transmissible genetic changes. There is no practical method for directly assessing the mutagenic potency of a compound in mammals. Apart from the mouse-specific locus test, all in vivo tests currently available for heritable effects in mammals are limited to the detection of effects caused by gross chromosomal damage.

Over 100 short-term tests are available for determining whether a compound has the potential to cause genetic damage, i. e., whether it can interact directly with the DNA itself or have an effect on the chromosome to produce either an alteration in structure or a change in chromosome number. Over the last 10 yr, studies of short-term test systems designed to detect possible carcinogens have been extensive.

Genetic toxicology assays belong to two main categories: tests that use specific endpoints and tests that respond to a wide spectrum of mutagens or carcinogen-induced lesions. In general, genetic toxicology batteries are designed to measure three endpoints: point mutation, DNA damage, and chromosomal aberration (Hollstein et al. 1979). Figure 1 outlines a proposed testing battery with various decision points.

As in the case for most toxicological studies, negative in vitro results do not necessarily lead to an affirmation of lack of toxicity, nor do the existence of safety and positive findings necessarily result in condemnation of the drug. Evidence that might invalidate positive in vitro results could include data that prove that the test chemical does not reach the germinal tissue, or other pertinent metabolic or pharmacokinetic considerations.

2.5.1. Specific Gene Mutations in Nonmammalian Cells

2.5.1.1. *SALMONELLA TYPHIMURIUM*/MICROSOME PLATE INCORPORATION ASSAY (AMES TEST). The Ames test is a microbial assay that measures histidine (his^- to his^+) reversion by chemicals that cause base substitutions or frameshift mutations in the genome of this organism.

Bacteria are exposed to test compounds with and without metabolic activation and plated on minimal medium. After a period of incubation, revertant colonies are counted and compared to the number of spontaneous revertants in an untreated and/or solvent control culture.

2.5.2. Specific Gene Mutation in Mammalian Cells

2.5.2.1. MOUSE LYMPHOMA FORWARD MUTATION ASSAY (MLFA). The MLFA utilizes a mammalian cell line as a target cell to measure forward mutations. The L5178Y mouse lymphoma cell line originated with Dr. Donald Clive, Burroughs Wellcome Company, Research Triangle Park, NC. These cells are presumed to be diploid, and three thymidine kinase (TK) phenotypes have been recognized: $TK^{+/+}$, $TK^{+/-}$, and $TK^{-/-}$. The $TK^{+/+}$ and $TK^{+/-}$ cells are sensitive to trifluorothymidine (TFT) and resistant to methotrexate. The $TK^{-/-}$ phenotype exhibits reverse sensitivity and resistance. The heterozygous $TK^{+/-}$ phenotype is used as the target cell in this test system.

When $TK^{+/-}$ cells are exposed to agents that can alter DNA, one of the possible consequences is the induction of forward mutations from $TK^{+/-}$ to the $TK^{-/-}$. This assay measures the induction of the $TK^{-/-}$ phenotypes as its endpoint.

2.5.2.2. CHO/HGPRT FORWARD MUTATION ASSAY. The CHO/HGPRT forward mutation assay also utilizes a mammalian cell line as a target cell to measure forward mutations. The Chinese hamster ovary cells used in this assay are a subclone of Kao and Puck's Chinese hamster ovary cells, clone K_1 (Dr. A. Hsie at the Oak Ridge National Laboratory, Oak Ridge, TN.).

This assay utilizes the X-linked hypoxanthine-guanine phosphoribosyl transferase locus in Chinese hamster ovary cells. Hypoxanthine-guanine phosphoribosyl transferase is a ''salvage enzyme'' that permits cells to use exogenous hypoxanthine and guanine in their DNA synthesis. The assay can be used to measure the induction, by mutagenic agents, of a forward mutation in the HGPRT locus from $HPGRT^{+}$ to $HGPRT^{-}$. Since the $HGPRT^{-}$ mutant has little or no HGPRT enzyme activity, the mutant can be selected by growing the cells in medium containing a purine analog such as 6-thioguanine (TG). Incorporation by the $HGPRT^{+}$ cells of TG into their DNA is lethal. The mutant $HGPRT^{-}$ cells, which lack HGPRT enzyme activity, are not capable of utilizing the TG in their DNA synthesis and, therefore, survive and grow in the selection medium containing TG.

2.5.3. Primary DNA Effects

2.5.3.1. SISTER CHROMATID EXCHANGE (SCE). Sister chromatid exchange in CHO cells is a short-term test for screening compounds for potential genotoxic activity. This test utilizes a mammalian cell line as a target cell.

CHO cells are a permanent cell line with an average modal number of 20, and an average cycling time of 10–14 h. Cells are harvested at appropriate times and chromosome preparations are made. Preparations are stained and metaphase cells are analyzed for chromosomal aberrations.

SCEs that are likely to be produced by a variety of lesions occur during both the first and second cell cycles and are detectable in second-division cells. In vitro assays generally incorporate BUdR during both periods of DNA synthesis.

2.5.3.2. UNSCHEDULED DNA SYNTHESIS IN PRIMARY RAT HEPATOCYTES. The autoradiographic identification of unscheduled DNA synthesis in primary cultures of adult rat hepatocytes is used as a predictive test for mutagens and/or carcinogens. In addition, the autoradiographic identification of chemically induced unscheduled

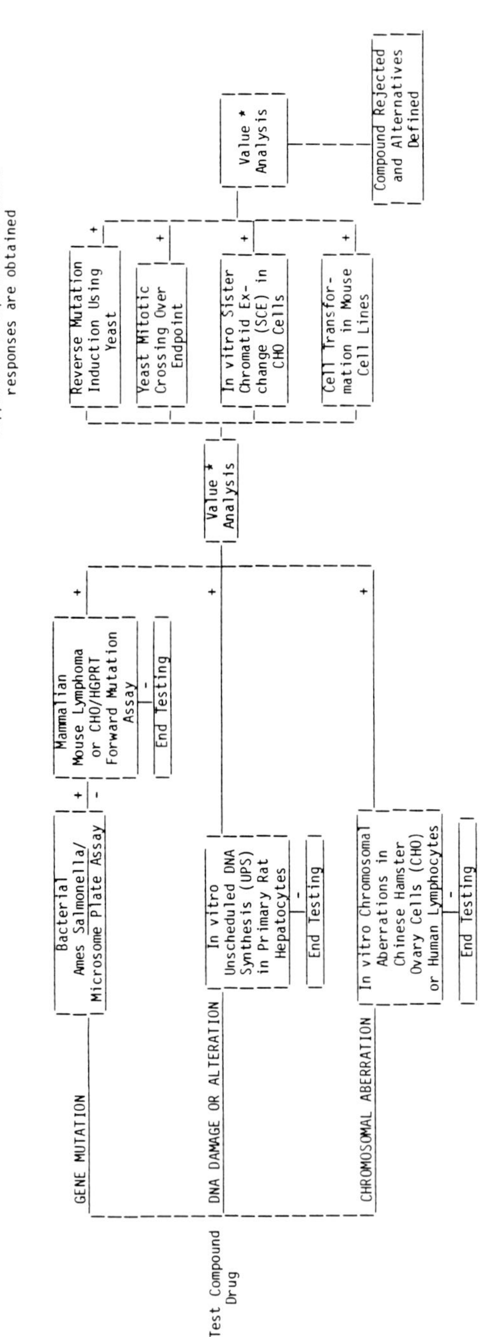

*At each decision point, care should be taken to plan studies and control for extraneous factors which might introduce artifacts (problem compounds, physical form, solvent requirements, etc.).

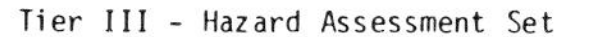

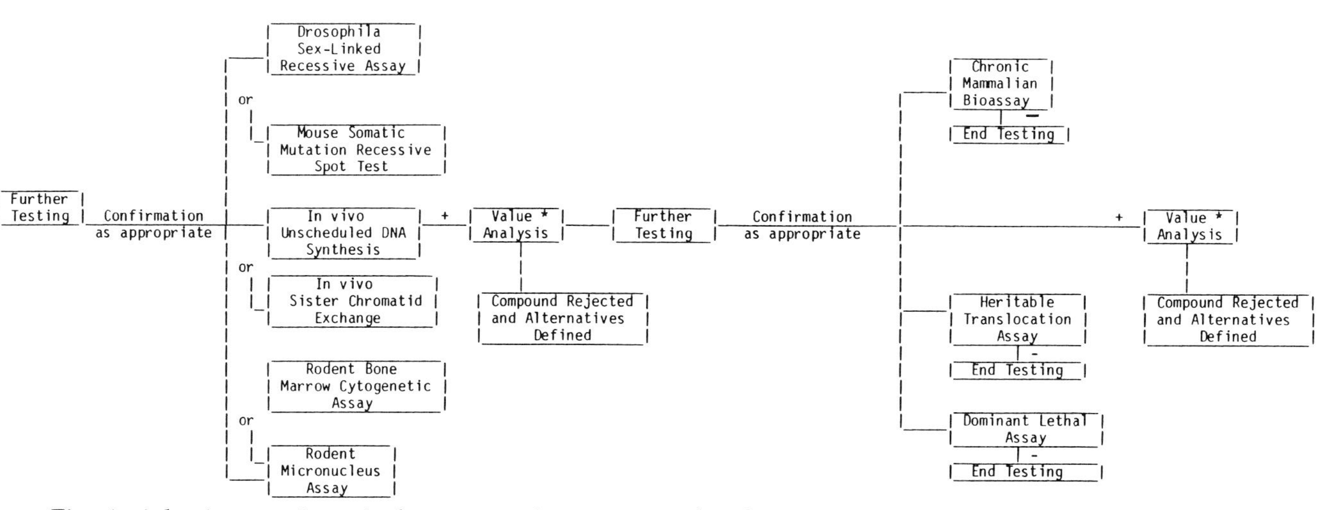

Fig. 1. A basic genetic toxicology screening program for drugs. For general screening, a battery of three to five in vivo tests is recommended, which includes assays for detecting gene mutation, chromosome alterations, and primary DNA effects. The profile of response in this battery can then be elevated and used to classify the drug.

DNA synthesis (UDS) provides a relevant, distinct, and quantifiable endpoint that is not confounded by DNA replicative synthesis in nondividing cells.

Freshly prepared hepatocytes contain the enzymes required for the activation of many promutagens and therefore provide a model possessing both the metabolic capability for activation, as well as the genetic target within the same cell.

2.5.4. Chromosome Aberrations in Mammalian Cells

2.5.4.1. IN VITRO MAMMALIAN CYTOGENETIC ASSAY. The in vitro cytogenetic test is a short-term mutagenicity test for the detection of structural chromosomal aberrations in cultured mammalian cells. Cultures of established cell lines, as well as primary cell cultures, may be used. After exposure to test compounds with and without a liver enzyme activation mixture, cell cultures are treated with a spindle inhibitor to accumulate cells in a metaphase-like stage of mitosis. Chromosome preparations, arrested in metaphase, are fixed and stained and analyzed microscopically for chromosomal aberrations.

2.5.4.2. IN VIVO CYTOGENETIC ASSAY. The in vivo cytogenetic test is a short-term mutagenicity test for the detection of structural chromosomal aberrations. Chromosomal aberrations are generally evaluated in first posttreatment mitoses. With chemical mutagens, the majority of the induced aberrations are of the chromatid type.

The method employs bone marrow cells of mammals that are exposed to test compounds by appropriate routes and are sacrificed at sequential intervals. Test compounds are generally administered only once. Based on toxicological information, a repeated treatment schedule can be employed. Animals are further treated, prior to sacrifice, with a spindle inhibitor to accumulate cells in a metaphase-like stage of mitosis. Chromosome preparations from the cells are made and stained, and metaphases are analyzed microscopically for chromosomal aberrations.

2.5.5. Morphological Cell Transformation

2.5.5.1. BALB/C 3T3 AND C_3H/10T1/2. A variety of cell culture systems have been developed with applications for the study and detection of chemical carcinogens. The majority utilize fibroblast cultures derived from embryonic tissues. The systems commonly used employ the normal cell lines Balb/c 3T3 or C_3H/10T1/2. Single treatments with carcinogens are frequently sufficient to produce transformed cells that are easily recognized on the basis of distinct changes in morphology and growth control.

The cellular mechanisms responsible for the morphological transformation of cells in culture are not understood. Since such systems appear to assay the ability of chemicals to convert non-tumorigenic cells to a tumorigenic state or normal cells to a preneoplastic state, it is generally assumed that these systems mimic central aspects of the carcinogenic process. For this reason, cell transformation systems may provide a biologically relevant means of detecting chemical carcinogens and may be sensitive to substances that pose a carcinogen risk by mechanisms that will not elicit a response in other assays for genotoxicity (National Toxicology Program, 1984).

2.6. Reproduction and Teratology Studies

Reproductive studies are aimed primarily at evaluating possible drug effects on fertility; on the zygote, its transport, implantation, and development; on parturition and the newborn; on lactation, weaning, and care of the young; and on the teratogenic potential of the drug.

General reproductive and perinatal/postnatal tests are conducted in at least one rodent species. The rat has been the species of choice for the study of the effects of drugs on reproduction. These tests are required for all drugs before including women of child-bearing potential in human studies.

In the case of general reproductive, prenatal, and postnatal studies, three dose levels are usually sufficient.

2.6.1. Fertility and General Reproductive Performance

For an adequate study of fertility, both male and female animals are studied. The rat has been used most often. Emphasis is placed on the effects of a given drug on gonadal function, estrous cycles, mating behavior, conception rates, and the early stages of gestation. Information concerning drug action on these functions is usually available in a relatively short time. In the study of a drug's effects on male fertility, male rats should have attained a minimum age of 40 d before drug administration begins and should have been treated for 9–10 wk prior to mating to assure the absence of a potential effect on spermatogenesis. Male animals from subchronic or chronic studies may be used. These pretreated males can be mated with either treated or untreated females.

For the study of drug action on female fertility, adult or at least sexually mature animals are used. Estrous cycles should be established by daily vaginal smears prior to and during dosing period.

After 14 d of drug administration, female animals are exposed to males. The results from these studies should serve as a guide for subsequent in-depth studies.

2.6.2. Teratological Studies

These studies concentrate on determining whether a drug has a potential for embryotoxicity, fetal toxicity, and/or teratogenicity. Drug administration is, therefore, restricted to the period of organogenesis. Teratological studies are conducted in two species, routinely rat and rabbit, and are required for all drugs before including women of childbearing potential in human studies. Rats are usually 3.5 mo old and dosed on d 6–15 of gestation (d 6 being the day sperm are found). Rabbits are usually 6–8 mo and the corresponding period of dosing is 7–9 d. Drug treatment is confined to females and started early enough and continued long enough to cover organogenesis of the species.

In determining what is teratogenic, it is generally accepted that any substance that will produce embryonic death will also produce deformities when administered in lesser dosages at the appropriate time during development of the fetus. Experiments, however, are often repeated with gradually decreasing doses in order to confirm that embryotoxicity is not obscuring the discovery of possible teratogenic effects.

Fetuses should be delivered by Cesarean section 1 or 2 d prior to the anticipated date of partuition. The number of fetuses, their placement in the uterine horn, the number of live and dead fetuses, and the number of early and late resorptions should be determined. Each fetus should be weighed and examined immediately for any extended anomalies. The order of performance depends on the drug being studied. All three segments may not have to be evaluated in certain instances, whereas in other instances specially designed studies may replace or supplement existing procedures. In addition, exhaustive examinations are made of the products of conceptions; careful examinations are made for visceral and skeletal abnormalities.

2.6.3. Perinatal and Postnatal Studies

The purpose of perinatal and postnatal studies is to study the effects of a drug administered during the last trimester of pregnancy and the period of lactation. They delineate the effects of the drug on late fetal development, labor and delivery, lactation, neonatal viability, and growth of the newborn. Properly extended, these studies may answer questions pertaining to chronic pediatric use and transplacental carcinogenesis.

2.7. Metabolism Studies

Toxicokinetics is the study of the rate of absorption, distribution, metabolism, and excretion of compounds. It involves the determination of the concentration of a substance or its metabolites in organs, tissues, and excreta at appropriate time intervals after compound administration, thereby providing some insight into the mode of action of the compound or its metabolism at the target site. This test is variable in duration and is performed on rodent and nonrodent species. It is required for all drugs before the initial (Phase I) human tolerance study.

The determination of toxicokinetics is especially relevant for the detection of possible accumulation in the body, since it is likely that accumulation may lead to toxic effects after long-term exposure. Information about the concentration of a compound in different body compartments may aid in elucidating mechanisms of action or species-specific differences in compound effects. The studies of species-specific differences may indicate the relevance for the extrapolation of data gathered from the use of a single species to humans (Kroes and Feron, 1984).

2.8. Immunotoxicity Studies

Over the past 10 years, there has been an increasing awareness of the capability of drugs to alter and modulate the immune response. This awareness has led the government, industry, and academia, individually and collectively, to become actively involved in the emerging discipline of immunotoxicology. As a scientific discipline, immunotoxicology can be viewed as both the study of the direct toxic effects of compounds on the immune system and the indirect adverse immunological consequences of current systems of toxicity.

The potential significance of agents affecting the immune system include: the possible indication of more general effects on metabolism or other functions of host cells; an altered susceptibility to infectious diseases (e. g., AIDS, which shows change in immune parameters and, as a consequence, a predisposition to opportunistic infections), autoimmune diseases, and tumors; and an altered response to immunization.

Adverse effects on the immune system may be mediated by nonspecific mechanisms. Stress, nutritional deficiencies, debilitation, endocrine changes, and age (chronic changes encountered in older animals as a result of the normal aging process) all have been implicated in indirectly affecting the immune response. Immune function studies should be performed when the effect of a compound on the lymphoid system cannot be attributed to an indirect influence of known functional significance.

The value in elucidating mechanisms of immune response may include: (i) a more accurate assessment of the nature and severity of the immunologic defect, (ii) classification of immunotoxicants by biological response rather than by chemical structure, (iii) prediction of immunotoxic reactions of new chemicals, (iv) selection or construction of nontoxic alternatives, and (v) treatment of host.

Types of immune alterations may include: depletion of immunocompetent cells by direct cytolysis (e. g., loss of responding or loss of regulatory cells); impairment of cell metabolism (e. g., block of maturation or differentiation); chemical interference with cell metabolism (e. g., selective destruction of those cells that divide in the course of the immune response by mitotic poisons); or induction of suppressor cells.

Many variables complicate analyses of an adverse immune response. Such parameters as dosage level, frequency, regimen, and route of administration, age and species of test animal, and environmental factors will determine the test compound's direct effect on immune competence (Table 6).

Since the immune system is extremely complex and has numerous internal negative feedback controls, it would be rare to find a situation in which all components of the immune system are defective, with the exception of congenital immune deficiency syndromes (Loose, 1985). It is precisely because of this complexity that it is not only important to qualify the uses of the term immunotoxic as it pertains to a given compound and to a given series of experimental assays, but also against a background of current toxicological parameters.

2.9. Behavioral Toxicity Studies

Behavioral toxicity testing relates to the determination of the potential adverse effect of a compound on a wide range of neurological and motor responses. As in other toxicity tests, behavioral toxicology approaches testing from two perspectives. One deter-

Table 6
Approaches Toward Immunotoxicological Assessment

Studies that are capable of detecting major alterations in immune response may include:
Background toxicological parameters including organ or body weight ratios and clinical parameters, e. g., hematology, clinical chemistry, as well as measurement of specific steroids and prostaglandins
Gross initial assessment of an immune response, e. g., lesions or alterations in lymphatic organs, cell viabilities, and peripheral lymphocyte counts
Studies that focus on the three basic cell types of the immune system, namely, assessment of humoral immunity (B cells), cell-mediated immunity (T cells), and macrophage function assays. Such assays might include measurement of specific antibody production, delayed type hypersensitivity, natural killer cell activity, and phagocytosis by macrophages. Attempts to discern possible common mechanisms for multiple responses can be made. Information regarding cellular populations, cellular interactions, and factor–cell interactions can be elicited from these studies.
Studies using model system (in vitro primary immunization, host-resistance), may provide additional information for discerning mechanism(s), and since these assays require the interaction of a variety of cells and cellular products.

mines the direct behavioral effects of the sample drug and the other approach determines the effect on development of the behavior-controlling system. Major differences between the approaches are the method and frequency of dosage and the statistical treatment of the results (e. g., individuals vs litters). An expectation is that behavioral changes may be an early manifestation of toxicity and, therefore, will provide very sensitive indices of toxicity (Dews and Wengers, 1979).

An inclusive battery of tests for neurobehavioral toxicity suggested by Tilson et al. (1979) measures effects on sensory, motor, arousal, and physiological functions. Behavioral studies in general require large numbers of subjects because of the wide range of expected normal values. This, therefore, compounds the multiple number of tests necessary for complete evaluation and emphasizes the importance of the subchronic testing protocols. This demands an awareness by the evaluator of signs of possible behavioral toxic effects.

Objective evidence that indicates a need for specific neurobehavioral toxicity testing includes findings of CNS toxicity at high doses. In addition, such testing should include developmental behavioral studies when the drug has been shown to cross the placenta and/or cause teratogenic effects.

3. Alternatives to Animal Testing

Russell and Burch (1959) defined an alternative test as "any technique that replaces the use of animals, that reduces the need for animals in a particular test, or that refines a technique in order to reduce the amount of suffering endured by the animal." The impetus for the development of in vitro systems to reduce the use of living animals in research and testing continues for ethical, moral, scientific, and monetary reasons. The alternatives most commonly considered include cell and organ culture, computer modeling, the use of minimally invasive procedures, and endpoints that produce little stress or suffering (Rowan, 1985).

3.1. Short-Term In Vivo Studies

Short-term studies may be carried out with a limited number of animals. In the case of range-finding studies, five animals/sex/group may be sufficient, whereas for short-term studies serving as a final study in toxicity, 10 animals/sex/group are needed. In long-term studies, a considerable reduction in the use of animals may be obtained when a combined chronic toxicity/carcinogenicity assay is conducted.

The likelihood of replacement of short-term and chronic toxicity tests by nonanimal experiments is not promising. The main reason for this is that there is no conceivable way at present by which the toxicokinetics of the whole animal can be mimicked in tissues in vitro.

3.2. In Vitro Studies

Significant advances in tissue culture methodology and knowledge of the behavior of specific cell types in vitro have occurred during the past decade. In vitro cell and tissue culture systems can be used as screening tests to perform studies on mechanism of action for personnel monitoring, and for risk assessment to supplement and expand whole animal studies (Nardonne and Bradlaw, 1983).

The biological endpoints that can be investigated and quantitated may be associated with functions common to all cells (e. g., cell viability, cell respiration, protein synthesis, cell membrane selective permeability), or with specialized or differential functions, such as those that are characteristic of the target cells or tissues (e. g., keratin production by corneal epithelium, lens-specific protein produced by lens tissue). These options must be carefully considered in developing a strategy for the evaluation of a new compound.

A comprehensive testing program would use a variety of biological systems that are useful for the quantitation of effects in unspecialized as well as differentiated functions.

3.3. Structure–Activity Models

It is generally agreed that, with the present state of knowledge, it is not yet possible to determine with any certainty the pharmacological and toxicological action of a drug from consideration of its chemical structure. In many anecdotal instances compounds that as a chemical class have been free of overt toxicity can unexpectedly be found to possess such properties. Yet in many instances, substances with closely related structures do have similar actions—a phenomenon that has been of great importance in the discovery of drugs.

The purpose of structure–activity models is not to replace standard animal bioassays, but rather to provide information that can aid in decision-making with or without standard bioassays (Enslein, 1984). A more complete analysis, based upon the molecular structure of each drug and related compounds, will aid in the establishment of a structure–activity relationship database that will be important in predicting the potential risks of the drug and will also facilitate the design of new and less toxic drugs.

4. Good Laboratory Practices

In 1978, the Food and Drug Administration (FDA) released a series of guidelines that took effect in June, 1979 (Federal Register, 1978). These guidelines, the Good Laboratory Practices (GLP), defined the methods of experimentation to be used when determining the safety of products that are controlled by the FDA. The intention of these guidelines was to assure that safety testing in all designated laboratories was performed with consistently high standards and that experimentation would be conducted by methods approved by the FDA.

The GLPs narrowly define all aspects of experimentation, including the proper characteristics of the laboratories and animal rooms, method for recording and developing raw data, and qualifications of the individuals who perform the testing.

The central individual in a safety testing program is the study director (SD). The SD has responsibility throughout the testing for the proper execution of the protocol, which includes daily observation, bleeding, necropsy, safety procedures, and equipment operation. In addition, the SD must verify that data collection is done properly and that the statistical methods are properly applied.

The GLP protocol requires that the staff of the testing laboratory include a quality assurance (QA) unit. This division, as its name implies, reviews all completed studies to verify that GLP and appropriate scientific methods were applied during the experimental phase of the testing. The QA unit, in order to function properly, must be completely independent of the study director and the management personnel of the study area, and must be composed of qualified scientists who can properly determine that the conclusions of the study have been reached by means of strict adherence to the testing protocol and have been drawn on the basis of appropriate scientific principles.

Although its role is not defined by the FDA, the management of a testing laboratory takes ultimate responsibility that a study is of high quality and that all FDA regulations are followed (Traina, 1983). With this responsibility is the burden of providing approved animal care facilities, proper laboratory space and equipment, and proper methods for archiving records of completed studies. In addition, standard operating procedures (SOP) must be instituted, recorded, and followed. SOPs must be regularly updated and made available to all individuals who carry out tasks that are in any way associated with proper completion of study protocol. Deviation from the SOP can seriously jeopardize the validity of a study.

The development of a study protocol is one of the most important responsibilities of toxicological management. This process requires cooperation between drug manufacturer, FDA, and the testing laboratory. The protocol must incorporate the requirements of the FDA, the known possible effects of the drug to be tested, and the capabilities of the testing laboratory.

5. Future Trends in Toxicity Studies— The Age of "Biotoxicology"

For the past 25 years, the discipline of toxicology has effectively established itself as a science. Short-term test systems, including animal in vivo and in vitro and human in vitro, have been validated, along with long-term tests that include animal in vivo bioassays and human epidemiology.

Basic research in the fields of immunology, teratology, neurobiology, and gerontology has provided and is continuing to provide the toxicologist with assays that are feasible for integration as part of a routine toxicologic/safety study. Current research is focusing more and more on fundamental mechanisms of toxicity and how they impact on in vitro models.

It has been demonstrated that a battery of short-term tests can now be used as a qualitative prediction of human responses to a new compound because of the high positive correlation between the carcinogenic and mutagenic activities of chemical carcinogens.

Short-term assays are being developed to detect carcinogen–oncogene interaction, and monoclonal antibodies have proven useful in achieving this goal. Genetic toxicology tests that are being developed include mammalian cell assay systems that use cocultivation of hepatocytes as the activation system to replace conventional S9 mixtures. In addition, tests for aneuploidy have been described for yeast and *Drosophila*, and assays for gene transposition are currently being performed in *Drosophila* and in rodent cells. Genetic toxicity endpoints are also beginning to be evaluated using human cells in culture.

In vitro assessment of potential teratogens has been determined based upon validation of the mouse embryo limb bud system (Hassell and Horigan, 1982). In the future, research on the use of human embryonal carcinoma cells may also serve as an in vitro approach for classifying potential teratogens (Martin and Evans, 1975a,b; Jakobits et al., 1975).

Biotechnology-derived compounds are being developed for use in a variety of diseases. These products have created new issues in safety and toxicity assessment. Separate guidelines have, in fact, been issued by the FDA to address particular safety testing requirements for biologics. These guidelines include: Points to Consider in the Production and Testing of Interferon Intended for Investigation and Use in Humans; Points to Consider in the Manufacture of Monoclonal Antibody Products for Human Use; Points to Consider in the Characterization of Cell Lines Used to Produce Biological Products; Points to Consider in the Production of New Drugs and Biologicals Produced by Recombinant DNA Technology; and Antiviral Drugs for Non-Life-Threatening Diseases: Draft Points to Consider for Safety Evaluation Prior to Phase I Studies.

There are now numerous biotechnically produced reagents that are being used clinically and there is promise of a dramatic increase of these products in the future. The appropriate evaluation of these materials will require the toxicologist to apply more sophisticated biological and biochemical methods in the areas of the clinical laboratory, data retrieval systems, and statistical evaluation (Traina, 1983). Biotechnology, therefore, is a new and challenging field for the modern toxicologist.

Acknowledgments

The authors are indebted to Sara and A. J. Lewis for their constant ''totsicological'' evaluation, to Dr. Raymond Cox for his review and comments, and to Ms. Ruth Weberg and Ms. Patricia Jenkins for their expert secretarial assistance in preparation of this manuscript.

References

Cooper, J. A., Seraci, R., and Cole, P. (1979) Describing the validity of carcinogen screening tests. *Br. J. Cancer* **39**, 87–89.

de Serres, F. J. and Matsushima, T. (1984) Meeting report—environmental mutagenesis and carcinogenesis: Test method development, validation and utilization. *Mutation Res.* **130**, 353–359.

Dews, P. B. and Wenger, G. R. (1979) Testing for behavioral effects of agents. *Neurobehav. Toxicol.* **1** (suppl. 1) 119–127.

Dunkel, V. C., Pienta, R. J., Sinah, A., and Traul, K. A. (1981) Comparative neoplastic transformation responses of Balb/3T3 cells, Syrian hamster embryo cells and Rayscher murine leukemia virus-infected Fischer 344 rat embryo cells to chemical compounds. *J. Natl. Can. Inst.* **67**, 1303–1312.

Enslein, K. (1984) Estimation of toxicological endpoints by structure-activity relationships. *Pharmacol. Rev.* **36**, 131S–135S.

Farber, E. (1982) Chemical carcinogenesis. A biological perspective. *Am. J. Pathol.* **106**, 271–296.

Federal Register (1978) Good laboratory practice for nonclinical laboratory studies. **43**, 29–36.

Hassell, J. R. and Horigan, E. A. (1982) A model developmental system for measuring teratogenic potential of compounds. *Teratogen Carcinogen Mutagen.* **2**, 353–359.

Health and Welfare, Canada. (1981) *Preclinical Toxicologic Guidelines.*

Heinze, J. E. and Poulsen, N. .K. (1983) The optimal design of batteries of short-term tests for detecting carcinogens. *Mutation Res.* **117**, 259–269.

Hollstein, M., McCann, J., Angelosanto, F., and Nichols, W. (1979) Short-term tests for carcinogens and mutagens. *Mutation Res.* **65**, 133–226.

Jackson, E. M. (1983) Editorial. Are food, drugs and cosmetics different or the same? *J. Toxicol. Cut. Ocular Toxicol.* **2**, 79–80.

Jakobovits, A., Banda, M. J., and Martin, G. R. (1985) Embryonal Carcinoma-Derived Growth Factors: Specific Growth-Promoting and Differentiation-Inhibiting Activities, in *Cancer Cells* 3. *Growth Factors and Transformation* Cold Spring Harbor Laboratories, New York.

Kolbye, A. C. (1980) Impact of Short-Term Screening Tests on Regulatory Action, in *Applied Methods in Oncology* vol. 3 *The Predictive Value of Short-Term Screening Tests in Carcinogenicity Evaluation* (Williams, G., Kroes, R., Waaijers, H. W., and van de Poll, K. W., eds.) Elsevier North-Holland Biomedical, New York.

Kroes, R. and Feron, V. J. (1984) General toxicity testing: Sense and non-sense, science and policy. *Fund. Appl. Toxicol.* **4**, S298–S308.

Loose, L. D. (1985) Immunotoxicology—1984. in *The Year In Immunology 1984–1985.* (Cruse, J. M. and Lewis, R. E., Jr., eds.) Karger, Basel.

Martin, G. R. and Evans, M. J. (1975a) Multiple differentiation of clonal teratocarcinoma stem cells following embryoid body formation in vitro. *Cell* **6**, 467–475.

Martin, G. R. and Evans, M. J. (1976b) Differentiation of clonal lines of teratocarcinoma cells: Formation of embryoid bodies in vitro. *Proc. Natl. Acad. Sci.* **72**, 1441–1445.

Miller, E. C. and Miller, J. A. (1981) Mechanisms of chemical carcinogenesis. *Cancer* **47**, 1055–1064.

Nardonne, R. M. and Bradlaw, J. A. (1983) Toxicity testing with in vitro systems. I. Occular tissue culture. *J. Toxicol. Cut. Ocular Toxicol.* **2**, 81–98.

National Toxicology Program (1984) *Report of the ad hoc panel on chemical carcinogenesis testing and evaluation.* Board of Scientific Counselors National Toxicology Program, U.S. Department of Health and Human Services Public Health Service.

Russell, W. M. S. and Burch, R. L. (1959) *The Principles of Humane Experimental Techniques*. Methuen, London.

Rowan, A. N. (1985) Perspectives on alternatives to current animal testing techniques in preclinical toxicology. *Ann. Rev. Pharmacol. Toxicol.* **25**, 225–247.

Squire, R. A. (1984) Carcinogenicity testing and safety assessment. *Fund. Appl. Toxicol.* **4**, S326–S334.

Tilson, H. A., Mitchell, C. L., and Cabe, P. A. (1979) Screening for neurobehavioral toxicity. The need for examples of validation of testing procedures. *Neurobehav. Toxicol.* **1** (suppl. 1), 137–148.

Traina, V. M. (1983) The role of toxicology in drug research and development. *Med. Res. Rev.* **3**, 43–72.

Williams, G. M. and Weisburger, J. H. (1983) New approaches to carcinogen bioassay in safety evaluation and regulation of chemicals. *1st Int. Conf.*, Boston, Massachusetts.

Appendix

Tissues for Histopathology Examination

ORGAN	ACUTE	SUBCHRONIC	CHRONIC
Adrenals	+	+	+
Aorta		+	+
Bone Marrow (Sternum/Femur)	+	d	d
Brain	+	+	+
Cecum		c	c
Colon (Lg. Int.)	+	c	c
Duodenum	+	c	c
Esophagus		+	+
Eye(s)		a	+
Femur			
Gallbladder (if present)		+	+
Heart	+	+	+
Ileum		c	c
Jejunum		c	c
Kidney(s)	+	+	+
Lacrimal Gland			
Liver	+	+	+
Lung(s)	+	+	+
Lymph Nodes (Representative)	+	+	+
Mammary Gland		a	+
Nerve (Peripheral)			
Sciatic		+	+
Ovaries	+	+	+
Pancreas	+	+	+
Parathyroid(s)	+	+	+
Pituitary		+	+
Prostate		+	+
Rectum		c	c
Salivary Gland		+	+
Skeletal Muscle		+	+
Skin			+
Spinal Cord			
(3 levels)			
(2 levels)		a	+
Spleen	+	+	+
Sternum		+	+
Stomach	+	+	+
Testes	+	+	+
Thymus		+	+
Thyroid	+	+	+
Trachea		+	+
Urinary Bladder		+	+
Uterus	+	+	+
Gross Lesions/Tumors	+	+	+
Acces. Genital Organs		a	+
Application sites			+

[a]Tissues may be examined if indicated by signs of toxicity or target organ involvement
[b]Including articular surface
[c]Small and large intestines specified
[d]Bone marrow suggested for nonrodent; not specified for rodent

Basic Toxicology Study Designs

	ACUTE	SUBCHRONIC	CHRONIC
Strain	Commonly used laboratory strains; rat preferred.	Commonly used laboratory strains; rat preferred rodent; dog preferred nonrodent.	Strains should be well-characterized, commonly used disease resistant and free from interferring congenital defects
Age of Initiation	As soon as possible after weaning; ideally before six weeks of age but no more than eight weeks.	Rats as soon as possible after weaning; ideally before 6 weeks but no more than eight weeks of age - dogs 4-6 months, not more than 9 months.	As soon as possible after weaning and acclimation; for rats about 6 weeks.
Housing	Group or individual	Group or individual; caging should be appropriate to the species.	Group or individual house; caging should be appropriate to the species.
No./Sex/Group(a)	5/sex/group - females should be nulliparous and nonpregnant	10/sex/group - rodents; 4/sex/group - nonrodents	20/sex/group - rodents; 4/sex/group - nonrodents
No. Dose Levels/No. Control Groups(b)	3 dose levels/1 control group	3 dose levels/1 control group	3 dose levels/1 control group
Satellite Groups	1 group 5/sex, treated with high dose and observed for 14 days post-treatment.	May be used treated with high dose 10/sex and observed for 28 days post-treatment (not specified for dogs)	
Duration	14 to 20 days	28 to 90 days	3 to 24 months
Route of Administration (c)	Oral, dermal, parenteral	Oral, dermal, parenteral	Oral, dermal, parenteral
Mortality/Moribundity	2x/day	2x/day	2x/day
Clinical Observations	Daily	Daily	Daily
Physical Examinations	Weekly	Weekly	Weekly
Body Weights	Weekly	Weekly	Weekly 1-13, once every 4 weeks thereafter
Food Consumption	Weekly	Weekly	Weekly 1-13, three month invervals thereafter
Water Consumption	Weekly (d)	Weekly (d)	Not specified
Ophthalmic Examinations	Not specified	Prior to initiation and termination, preferably in all but at least in control and high doses (if changes in eyes are detected, examination should be conducted in all animals).	Not specified
Hematology No./Sex/Group Frequency	All at termination	All rodents at termination, all dogs prior to initiation then at monthly intervals.	10/sex/group - rodents and all non-rodents 3 month intervals and at termination
Clinical Chemistry No./Sex/Group Frequency	All at termination.	All rodents at termination, all dogs prior to initiation then at monthly intervals	10/sex/group - rodents and all non-rodents; intervals approximately every 6 months and at termination.
Urinalysis No./Sex/Group Frequency	(e)	(e)	10/sex/group. Rodents and in all non-rodents. 3 month and 6 month intervals and at termination.
Organ Weights No./Sex/Group Frequency	All Liver, kidneys, adrenals, testes	All Liver, kidneys, adrenals, testes; thyroid with parathyroid for dogs	10/sex/group - Rodents and in all non-rodents. Brain, liver, kidneys, gonads and adrenals; thyroid with parathyroid in nonrodents.

Histopathology Examination(g)			
Groups/Tissues	Yes, control and high dose; (see Table 5)	Yes, rodents and dogs. Control and high dose (satellite group, if used) (see Table 5)	Yes, control and high dose rodents and nonrodents (see Table 5)
Deaths and Moribund Sac's	Not specified	Not specified	Yes
Tissues	Not specified	Not specified	Same as control and high dose
Target Organs	All	All	All
Gross Lesions	All	All	All
Other Groups Examined	Satellite groups; same as control and high dose.	Yes	Not specified
Tissues		Lungs, liver, kidneys	Not specified

(a) If interim sacrifices are planned, the number should be increased by the number of animals scheduled to be sacrificed before the completion of the study.
(b) Control groups depend on chemical formulation and type of vehicle, if any.
(c) Choice of administration of test substance depends on the physical and chemical characteristics of the test substance and the form typifying exposure in humans
(d) Measurement of water consumption depends on compound route of administration, i.e., drinking water.
(e) Urinalysis is not recommended on a routine basis, but only when there is an indication based on expected or observed toxicity.
(g) Tissues for histopathology examination, based on oral route of administration; histopathology may be extended to animals of other dosage groups, if considered necessary, to further investigate changes observed in the high dose group.

Drug Delivery Systems

JOSEPH V. BONDI AND D. G. POPE

1. Introduction

The term ''drug delivery'' is truly a catch phase of the 1980s, few journals or periodicals having neglected to review the state of the art (Alper, 1984; Hildebrand, 1983; Banakar, 1984; Rogers, 1982; Check, 1984; Anderson and Kim, 1984; Zaffaroni, 1980; Inhorn, 1981; Banker and Rhodes, 1979; Senyie et al., 1985). However, although these newer concepts in drug delivery point toward a disappearance of traditional methods of drug therapy, it is likely that the bulk of pharmaceutical dosage forms, such as tablets, capsules, suspensions, and solutions for oral or parenteral use will enjoy continued wide use and acceptance. The science surrounding these dosage forms has progressed significantly in terms of the knowledge required to make better tablets, to understand the role of disintegrants, and to optimize inert ingredient ratios through sophisticated formulation design (Lachman et al., 1976; Lieberman and Lachman, 1980).

Today only a small percentage of dosage forms intended for oral use utilize special delivery principles, e.g., enteric coatings, matrix tablets, multiple coatings on pellets, floating capsules, and ion exchange resins to impart control of drug release. A much smaller percentage of dosage forms utilize membrane-controlled diffusion, diffusion through polymers, and osmotic principals to regulate drug delivery. Thus, in terms of current therapy, there is a great deal of the past, as represented by traditional dosage forms; a mix of present and future technology represented by the more novel dosage form types; and a wealth of yet to be applied literature that holds the promise of the future and a radically new approach to drug delivery (Shell, 1985; Tomlinson and Davis, 1986).

Concerns regarding bioavailability and drug side effects can be circumvented by the choice of an appropriate drug delivery system. Prodrugs to facilitate target site delivery as in the "soft-drug" concept of Bodor (1977) or to avoid first-pass metabolism or degradation in the gastrointestinal (GI) tract may radically alter the criteria used to identify product candidates at the preclinical level (Williams and Malick, this volume). Similarly, the possibility of administering a therapeutic modality in a steady-state mode from an osmotic pump or a transdermal "patch" may prevent the side effects that accrue from the administration of large drug boli in which the concentrations often far exceed those required for efficacy. Although this approach is attractive and underlines the need for the process of drug development to cross traditional temporal boundaries, the issue of possible tachyphylaxis cannot be overlooked or dismissed. One should not lose site of the fact that the need for a specialized drug delivery system implies that the disposition of the drug moiety *per se* in the body has less than desirable attributes. The intent of the "controlled delivery system" is thus to modulate body drug levels to optimize performance.

How then does one decide whether a specialized type of drug delivery system is required for a new drug entity? What technologies are available to impart controlled drug release? What in vivo data are required to aid in the design of a controlled drug delivery system? How successful have the novel delivery systems of the mid 1970s to early 1980s been to improve therapy and increase patient compliance? The intent of this chapter is to answer these questions through descriptions of the current art and an overview of the ever-expanding literature on drug delivery systems. Since many novel concepts are currently being employed in the animal health area, this field of drug delivery is also discussed, since many of these novel delivery technologies may be adapted to human drug delivery.

2. Rationale for Selecting a Drug Delivery System

The main goal of a drug delivery system is to optimize the drug input rate to achieve the optimal therapeutic response while minimizing the tendency for side effects that, by definition, are unwanted. In order to choose the most appropriate drug delivery technology for a given drug, one must be aware of system expectations. Table 1, adapted from Welling (1983), delineates the ad-

Table 1
Advantages and Disadvantages of Controlled Release

Advantages
• To achieve rapid onset and then maintain therapeutic drug levels
• To reduce dosing frequency and reduce fluctuations in plasma drug levels
• To achieve dose conservation
• To increase patient compliance
• To avoid nighttime dosing
• To obtain more uniform pharmacologic response
• To reduce GI irritation and side effects
Disadvantages
• Possibility of dose dumping
• Reduced potential for accurate dose adjustment
• Slow absorption may delay onset of activity
• Increased potential for first-pass metabolism
• Possible reduction in systemic bioavailability
• Drug release period restricted to residence time in GI tract

vantages and disadvantages for a controlled-release dosage form. These points are also applicable to controlled release drug delivery in the animal health-care market. A prime factor preventing the efficient control of certain disease states in domestic animals is the difficulty of maintaining a strict repeat-dosage schedule. Regimental dosing of an animal is time consuming, labor intensive, and hence, extremely expensive. If an animal can be treated once, and the effects of a single treatment sustained for a period sufficient to either overcome the disease state or prevent the outbreak of a disease state, significant improvement in the animal's growth rate may be achieved. A satisfactory cost–benefit ratio would ensure compliance by the farmer in the use of the controlled-release delivery system.

The various strategies applicable to a new delivery system have been discussed by Welling (1983). Through use of computer-simulated plasma level profiles, one is able to visualize how a dosage form having an immediate release fraction followed by a constant delivery rate may differ from one whose drug release follows pure zero-order kinetics. The utility of such simulations is obvious, although in practice it is rare that the design of a drug delivery system is supported with iv infusion data, demonstrating both efficacy and safety for the desired sustained plasma level profile. However, drug plasma levels do not always represent the final focus in understanding drug performance and distribution. A more desirable basis for novel dosage form design would be measurement of drug plasma levels concurrent with demonstration of the desired pharmacologic/pharmacodynamic response. With the present state of the art, it is not uncommon for the assumption to be made that if an immediately available dosage form administered three to four times a day elicits a favorable response, a controlled-release or sustained-release dosage form will not only reduce the number of daily doses, but produce the same type of pharmacologic response. This hypothesis remains to be proven absolute.

With drugs exhibiting long elimination half-lives intended for oral administration to humans, plasma levels will not fluctuate greatly, and, consequently, dosing frequency may be once or twice a day even from immediate-release preparations, such as capsules or rapidly disintegrating tablets. Also, if drug absorption occurs only from a defined area of the GI tract, a dosage form that releases its payload continuously while moving down the GI tract will obviously compromise the drug's bioavailability.

Other problems have been reported with controlled sustained-release delivery. For example, in the agricultural area, Penncap M® was introduced as a prolonged-delivery form of the pesticide, malathion (Koestler, 1980). However, because of its extended environmental half-life, toxicity to the honey bee was noted. The controlled-release pesticide adhered to the bees' feet and subsequently contaminated the hive. In the case of antibiotics and nitroglycerin therapy, the questions regarding pulsed vs continuous delivery have yet to be answered satisfactorily. In insulin therapy, it is generally accepted that glucose levels should be kept low. Somatostatin has been shown effective in reducing the body's requirements for insulin by virtue of its effect on glycogen mobilization. However, it also has an effect on growth hormone release. This may suggest that a pulsed delivery of somatostatin will elicit a more desirable pharmacologic effect than a slow prolonged release (Sherwin et al., 1983).

The rationale for selecting a drug delivery system should be: to improve drug targeting and bioavailability, to minimize drug plasma level oscillations, to reduce daily dosing frequency, and to optimize pharmacologic responses. The improvement in pharmaceutical formulation technology must be matched by new findings by the biologist and the pharmacologist to provide a better rationale for designing a drug delivery system. Furthermore, experimentation to understand the effect of drug entry rate and to actually define what the plasma level profile over a given dosing interval should be, must follow. If the medicinal chemist, molecular biologist, pharmacologist, and pharmaceutical scientist are able to collaborate to an optimal degree to gain a better understanding of the disease state, receptor site and specific molecular configurations of the modalities required to treat a disease state, little else may be required other than to get the drug into the circulatory system. At the same time, however, much can be done with the compounds presently available to improve their delivery characteristics for more immediate therapeutic use.

2.1. Terminology

As with any discipline in science, a terminology familiar to those working in the field evolves that may not be readily apparent to the novice. Since drug delivery is no exception, certain key terms are defined below:

Boundary Layer Effect—as an active agent leaves a dosage form and diffuses into a receptor phase (tissue, blood, GI fluid, saliva, and so on), a concentration gradient can form at the interface if a stagnant liquid layer forms. This layer will alter and retard the release-rate profile from that seen under in vitro conditions.

Enhancer—also referred to as an absorption or permeation enhancer. In theory, this type of agent functions in a passive manner to render a membrane (skin, GI, nasal, or rectal mucosa) more permeable to a drug. Ideally, these agents produce only transient changes in a membrane and are not irritating to it. Some examples of enhancers are: salicylic acid, diethyl-*m*-toluamide and 1-dodecylazacyclo-heptane-2-one (Azone® —Nelson Research).

Enteric Coating—a physical barrier coating around an oral dosage form that is insoluble in the acidic environment of the stomach, but dissolves readily in the more alkaline pH of the duodenum.

Hydrocolloids—high molecular weight water-soluble substances of natural synthetic or semisynthetic origin that form viscous, gel-like structures when placed in an aqueous environment.

Implant—usually refers to a device that is inserted into a tissue space as opposed to a body cavity.

Insert—a device that is placed into a body cavity such as vagina, rectum (e.g., a suppository), *cul-de-sac* of eye, buccal cavity, and so on.

Matrix—a carrier for the active principal that functions as the repository or reservoir. The physical nature of the active ingredient can be either particulate (that is, present at a concentration in excess of its solubility) or present as a concentration less than its saturation solubility.

Membrane—a physical barrier distinct in composition from that which it surrounds. Its purpose will generally be to control the egress of an active agent into an external environment. Membranes may also function as containment devices only and permit the free passage of the active principal. Additionally, terms like semipermeable, microporous, or impermeable may further describe the mechanism by which a membrane permits the transport to occur.

3. Matrix Control of Drug Release

3.1. Theory

The rate of drug release to the body system may be controlled by dispersing a drug in the matrix of a suitable carrier, the specific properties of which allow the drug to be released by either diffusion, dissolution, and/or erosion. The primary mechanism for release of the drug is usually dependent upon the type of matrix carrier used. With dissolution or erosion control, the drug-release rate is predicated upon the rate of dissolution or erosion, this being directly proportional to the surface area of the system in question. For example, for a spherical drug system or for a cylinder that has the same diameter to height ratio, a sharp decrease in the dissolution rate occurs with time. At half the time for complete dissolution, the rate has fallen by 75%. The rate of release follows a cube-root profile. Hixson and Crowell (1931a,b) recognized that the surface area of a regular particle is proportional to the two-thirds power of its volume. Hence the rate of change of the dissolving or eroding sphere can be expressed as

$$W_0^{1/3} - W_t^{1/3} = K_s t \tag{1}$$

Where W_0 is the initial weight of the dosage form, W_t is the weight of the dosage form remaining at time t, and K_s is a constant incorporating the surface volume-proportionality.

For other shaped systems, for example a long, thin cylinder, the dissolution rate decrease is essentially constant with time. As a cylinder approaches a flat disk or slab, the dissolution rate becomes constant, i.e., a zero-order rate is approached (Hopfenberg, 1976).

In drug delivery employing diffusional control, the rate-determining step is diffusion of the drug to the external surface of the matrix device. Monolithic devices containing dissolved or excess suspended drug yield release patterns that decline as the inverse of the square root of time (Higuchi, 1961; Flynn et al., 1974; Baker an Lonsdale, 1974).

When only dissolved drug is present, the square root of time function is operative during the first 60% of drug release, which is then followed by a period of exponential decay. For suspended or dispersed drug, when the total drug concentration is maintained at a value much greater than solubility in the polymeric phase, the square root of time relationship is expected to be maintained throughout the time course of drug delivery.

A comparison of square root time, first-order, and zero-order release profiles are given in Fig. 1.

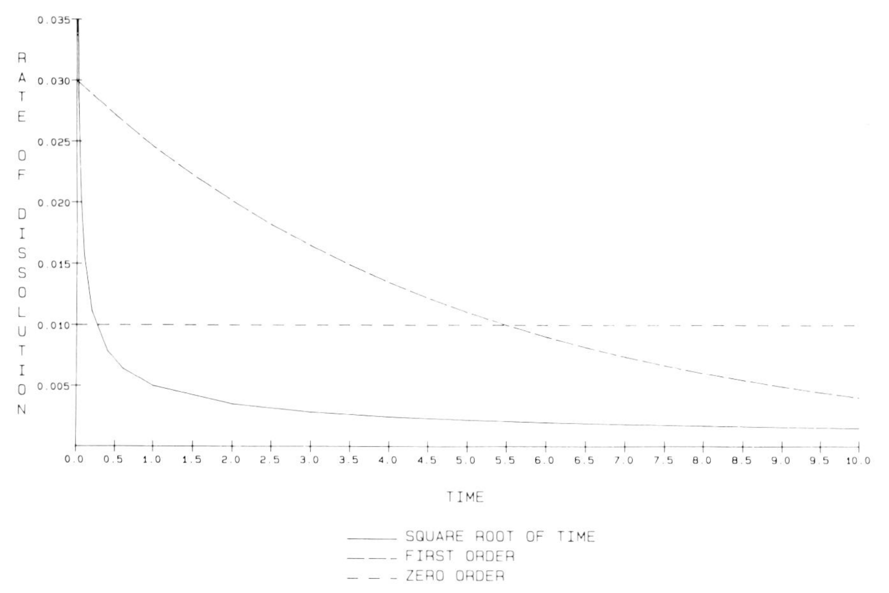

Fig. 1. Comparative rate of dissolution profiles for the most commonly characterized release mechanisms.

If the rate of transport of drug through the matrix is faster than diffusion away from the device, i.e., through the boundary diffusion layer surrounding the device, a zero-order profile will result. The boundary layer effect could be experienced in vivo even though in vitro release patterns suggest square-root time diffusional control. The reason for the difference is related to the hydrodynamic conditions at the boundary surface. For instance, with an implant, turbulance about the device in vivo is low. Hence, depending upon the drug and matrix in question, a zero-order profile may result (Roseman and Yalkowsky, 1976).

By careful control of the shape of the diffusional control devices, zero-order can be approached. By balancing increasing diffusional pathlength to increase in surface area, a zero-order profile can be achieved (Korsmeyer and Peppas, 1983). Brooke and Washkuhn (1977) and then Lipper and Higuchi (1977) proposed that the shape of a sector of a right circular cylinder would release drug at a zero-order rate. Recently Rhine et al. (1980) investigated a hemisphere-shaped device, coated with an impermeable membrane except on the central cavity. Tojo (1984) has developed the mathematics for zero-order drug release from a cylindrical matrix device with a small release hole. Because of the difficulty in manufacture of most of these specialized shapes, commercial production of these matrices has yet to become feasible.

To illustrate the current utility of matrix systems for controlled delivery, a number of both human health and animal health examples will be discussed.

3.2. Oral Matrix Release Systems

Release of the drug from the matrix by purely erosional control (Hermlin, 1959) is exhibited by such dosage forms as Tenuate Dospan® (Merrell Dow), Tepanil Ten-Tab® (Riker), and Sul-Spantab® (SKF). Erosion and diffusional control is exhibited by products such as Slow-K® (Ciba) and K-Tab® (Abbott). Strictly diffusional control is seen with products such as the Gradumet line from Abbott. Biphasic release is exhibited by systems such as Peritrate SA® (Parke-Davis), Pyribenzamine Longtab® (Ciba), and Clistin R-A® tablets (McNeil). Biphasic release, that is immediate release followed by slow release, is accomplished by containing the initial dose in either a pan coat, a press coat, or one part of a bilayered tablet.

A number of newer modifications to matrix-release control are also being investigated for their utility. For example, hydrogels (Schor and Nigalye, 1983; Schor, 1982; Kost and Langer, 1984; Kim, 1983; Good, 1978) use technology based on the use of water-soluble polymers that become gelatinous upon contact with moisture. The absorption of water unlocks the solid lattice structure and allows the drug compound to be released by diffusion. In addition, if slow erosion occurs, drug can be released by both mechanisms. An advantage to these hydrogel systems is that many of the systems have the ability to be manufactured by thermoplastic processing methods. A recent review by Graham and Wood (1982) provides an overview of the various types of polymers fitting into this category of hydrogels.

Chemical biodegradation of polymers is also a means by which entrapped active drugs can be selectively released from the matrix (Heller, 1984a,b). Another form of polymeric release is achieved using pendant chain systems. Here the drug is attached via chemical bonds to a polymer backbone. The release is controlled by the degree of hydrophilicity of the polymer backbone (Harris, 1984).

With the above-mentioned systems, delivery is only on the order of 8–12 h. This is because, in drug therapy, the time period over which controlled delivery may be maintained following oral ingestion is limited by the GI transit time. Generally in humans, residence time cannot be extended beyond 8–12 h, but this again is quite variable. Gastrointestinal transit time is dependent upon the volume of the meal, the meal's chemical composition, and other physiological factors.

The veterinary formulation pharmacist, however, has the advantage over his counterpart in human medicine, when formulating for controlled oral delivery to ruminants (cattle, sheep). With specialized design of the delivery system, the anatomy of the ruminant's GI system can be utilized to achieve sustained residence; in particular by appropriate selection of density and/or size of the delivery system (Pope, 1978, 1983; Pope et al., 1985).

One of the earliest sustained-release boluses commercially available was the cobalt bullet (Dewey et al., 1958). This formulation resulted in gradual release of cobalt by erosion in the ruminoreticulum. Release was for periods of the order of 1 yr. Accuracy of release was not essential for this delivery system since this is a trace element, hence the decreasing rate with decreasing surface area was not bothersome.

More recently, glass boluses for supplementation of sheep and cattle with selenium, cobalt, and copper over a 12-mo period have been developed in a joint effort by the University of Leeds, UK, and Chance Pilkington Glass. These cylindrical glass boluses are composed primarily of sodium phosphate with calcium and magnesium oxides and the three trace elements. Substances in the glass control the rate at which the bolus is broken down by the digestive juices in the reticulum. The trace elements are metabolized at a rate that will restore and maintain blood concentrations of minerals at the optimum level for a period of a year. The boluses, called Cosecure®, are marketed by the Wellcome Foundation.

The principle of glass dissolution in selected media could also be used for human or veterinary implant delivery of drugs. Drake and Brecklehurst (1984) have developed a water-soluble glass device with an array of cavities, therein containing the active material. As the glass dissolves, the cavity contents are released sequentially and at a predetermined rate.

Suitable alteration of the density of human health oral drug delivery systems can also prolong the delivery profile, although not as significantly as can be done in veterinary medicine. Sheth and Tossounian (1979) have described a dry homogeneous mixture of medicament with hydrophillic hydrocolloids that, in contact with gastric fluid, forms a soft, gelatinous mass on the surface of the mixture. This mixture has a density of less than one by virtue of the voids in the powder blend and is, therefore, buoyant and floats in the gastric cavity. Drug release occurs via slow erosion of the surface gel and diffusion of drug through the gelled hydrocolloid. Two products commercially available utilizing this technology for the anxiolytics Valium® and Librium® are Valrelease® and Librelease® from Roche. Empty gelatin capsules can also be overcoated with the active drug entity. The capsule shell allows the product to float on the gastric juice for an extended period (Watanabe et al., 1976). Another system based upon a flotation chamber has been described by Harrigan (1977).

Systems in which an inflatable chamber contains a liquid, e.g., ether, that gasifies at body temperature, give a postadministration inflated bouyancy chamber for stomach retention. Such systems have been described by Michaels (1974) and Michaels et al. (1975).

The advantage of the floating system approach is that the drug is continuously being metered into the upper GI tract, which is usually the principal absorption site. Disadvantages for this approach to control GI residence time are: drugs subject to degradation at low pH cannot be used and this dosage form can not function as designed in fasting individuals since stomach emptying time is on the order of 5–60 min (Welling, 1983).

High-density multiple-unit formulations have also been evaluated. Bechgaard and Ladefoged (1978) showed that an increase in density of from 1.0 to 1.6 increased the average transit time from 7 to 25 h in ileostomy patients. In ruminants, a density of greater than 1.5 g/mL results in potentially permanent ruminal residence. The ideal density is of the order of 2.5 g/mL (Riner et al., 1982).

Erosional systems of 3.5–5 d duration for oral administration to cattle include; ALBON S.R.® (sulfadimethoxine) from Roche, Hava-Span® (sulfamethazine) from Haver Lockhart, and S.E.Z.—C.R.® (sulfaethoxypyridazine) from American Cyanamid. Because the time to start delivery, the time to shut down, and the maximum and minimum plasma levels are not critical to the efficacy profile for these sulpha drugs, the erosional release exhibited by these densified boluses is acceptable.

For sustained release over periods of greater than 14 d, and especially when careful control of drug release is necessary, the compressed bolus erosion system is not a reliable mechanism because bolus erosion is dependent upon a number of factors. Riner (1982) showed that the rate of erosion is dependent upon whether the bolus is lodged within the reticulum or rumen. More abrasion of the system occurs in the smaller reticulum. Also, young animals, because of the smaller size of the ruminoreticulum, exhibit a faster erosion rate on boluses than mature animals. The more boluses that are given, the faster the erosion rate. Hardware within the ruminoreticulum will also increase the erosion rate. Of a lesser influence is the effect of the particle size, shape, and density of the foodstuffs consumed by the animal.

Even though the erosion rate may vary from animal to animal and study to study, a more constant release rate can be achieved over extended periods of time by maintaining a constant surface area for erosion with time. A number of systems have been patented that do maintain a constant erosional surface. Laby (1978, 1982) described a spring-driven device that results in the drug formulation always being present at a constant surface area orifice to the ruminoreticular environment. Another system, more simplistic in design, maintains a constant surface area by formulating a cylinder of the active ingredient and coating with a brittle resin (University of Glasgow, 1984). One end of the cylindrical core is not coated. As formulation erodes, the brittle coating is exposed without support and hence is broken off to the level of the erosional surface. In this way, the surface area and diffusional path length remain the same. This sequence of events occurs throughout the lifetime of the bolus.

3.3. Matrix Implants

The utility of matrix devices has been proposed for subdermal implants (Robertson, 1985; Folkmann and Long, 1964) or vaginal implants (Robertson, 1983), for control of conception (Nash, 1984), in the treatment of narcotic addiction (Wise, 1984), and in the treatment of ocular dysfunctions (Shell, 1985). Commercially available systems include NORPLANT® (Population Council of New York), a silastic-levonorgestrel formulation for contraception, and LACRISERT® (Merck Sharp and Dohme), a hydroxypropyl cellulose eye insert for treatment of keratitis sicca (Katz and Blackman, 1977).

Implants for animal health applications have been confined primarily to growth promotion (Tindall, 1983). Some examples are Zeranol® ("RALGRO," International Minerals & Chemical Corp.), estradiol benzoate and progesterone STEER-OID® (Bioceutic Laboratories), SYNOVEX S® (Syntex Corp.), and estradiol benzoate and testosterone SYNOVEX-H® (Syntex Corp.). Usually these are implanted in the ears of cattle during the last 60–150 d of the fattening period.

Since pellet implants do not result in a constant release of drug with time, Elanco (Greenfield, IN) developed a more sophisticated delivery system that forces a "zero-order" release from a non-zero-order release pattern. The active ingredient in this COMPUDOSE® system is estradiol-17β. This is incorporated as a matrix in silastic polymer (Hudson and Wagner, 1980). Because the rate of diffusion of estradiol-17β through the silastic matrix is dependent upon the diffusional path length, the rate will continually decrease with time in a square-root time relationship as the path length increases. However, because of the design of the COMPUDOSE® system, the path length for diffusion is limited. The limited path length for diffusion was achieved by coating only a thin layer of the drug-silastic matrix onto an inert cylindrical core. Once the thin silastic-drug outer shell becomes depleted (very little change in the diffusional path length), the device suddenly runs out of drug and hence "shuts down." Elanco has a 200- and a 400-d COMPUDOSE® system. The increased time system is achieved simply by increasing the thickness of the drug-silastic coat on the inert core.

3.4. Vaginal Inserts

The contraceptive vaginal ring may be left in place for 3 wk or longer and thus affords a relatively constant release of drug to the user. The release, generally from a silastic polymer (Long and Folkman, 1966), follows the typical matrix diffusional release profile unless special manufacturing and formulation procedures are used to force zero order. For example, a "shell" ring of active compound can be placed around an inert core so that the overall diffusional distance is kept to a minimum. This is in essence the same principle as that used in the COMPUDOSE® implant discussed previously. The theory and release profiles for vaginal inserts have been reviewed by Chien (1978) and Nash (1984).

3.5. Intrauterine Devices

The only intrauterine device (IUD) that can be classified as an erosional matrix system is the copper IUD. A pure, virgin electrolytic copper wire, sufficient to give an exposed surface area of 200 mm^2, is wrapped around a polypropylene support. The copper slowly undergoes oxidation and release of copper ions. In a device tested by Moo-Young et al. (1975), the initial rate of release was rapid. This was followed by a slow rate of decline to about 25 μg Cu/day at 750 d.

PROGESTASERT® by the ALZA Corporation is not a matrix-releasing type. It consists of a T-shaped polyethylene support whose verticle axis is filled with a sleeve containing the steroid progesterone. The sleeve consists of an inner reservoir of steroid and a silicone polymer and an outer rate-regulating layer of polyethylene-vinyl acetate. The release characteristics afforded this system are discussed in section 4 of this chapter.

3.6. Matrix Transdermal Delivery

Another area in which matrix control of drug release is utilized is in several transdermal delivery systems recently introduced onto the market. Unlike oral delivery systems, drug release from such devices must occur via diffusional processes. Only two products available in the US, Searle's Nitro-Disc® and Key's Nitro-Dur®, are examples of this technology (Dasta and Geraets, 1982).

In the Searle system, a multiphasic matrix is employed in which nitroglycerin diffuses out of a reservoir environment into an environment of silicone to the skin surface. The silicone functions as a rate-controlling membrane to yield a zero-order release of drug (Karim, 1983).

The Key system, on the other hand, employs a matrix composed of several water-soluble polymers, glycerol, and dispersed drug. Diffusion of drug follows square root of time kinetics. Key holds numerous patents for specific drug combinations, including clonidine and ephedrine. A general description of their system is contained in EP patent application 13,606 (Keith and Snipes, 1980).

Less than 15 years ago, the general notion existed that the stratum corneum was an impenetrable barrier and that systemic delivery of drugs in therapeutic quantities was unlikely. Today, the search is on to identify agents that may penetrate the skin. Delivery systems ranging from ointments to multilayered systems of considerable sophistication have been reported. Yet, practical limitations in the size of a transdermal patch that can be worn comfortably exist. Thus, despite these rapid advancements, work still continues to identify penetration enhancers (Chien, 1982; McClure and Stoughton, 1982). In the case of estradiol, ethanol has been shown to be an effective penetration enhancer (Campbell and Chandrasekaran, 1983), whereas 1-dodecylayacyclohepton-2-one (Azone®; Nelson) has been studied extensively by Stoughton (1982). A prime consideration in this area is the toxicologic and regulatory issues likely to complicate the registration of a new transdermal drug candidate that employs a penetration enhancer.

3.7. Slow-Release Pesticide Generators

The most common examples of slow-release pesticide generators employed for animal health are the dog and cat flea and tick collars and medallions, and the cattle ear tags. The same system types are employed as pest strips for elimination of flies, and so on, from the household. These slow-release pesticide generators are usually manufactured from a plasticized solid thermoplastic resin having either a solution or a microdispersion of the ectoparasiticide in the resin matrix. The ectoparasticide is slowly released to the external environment by typical matrix-release kinetics. Active ingredients include Rabon® (Miller and Morales, 1976), Dichlorvos® (Folkemer et al., 1967), Propoxur® (Grubb and Baxter, 1974; Fisch et al., 1977; Miller et al., 1977), and Temephos® (Quick, 1971).

Depending upon the parasiticide incorporated into the resin, release may be as a vapor (Dichlorvos) or as a powder (Rabon, Propoxur, Carbaryl, Temephos). The primary differences between the vapor and powder-producing collar types are:

(a) Collars made with the volatile liquid pesticides often result in dermal irritation in the neck area of the animal. An attempt to overcome this is evident in the use of a free-hanging medallion (Quick, 1971), and
(b) Collars incorporating the volatile liquid pesticides do not provide as sustained a control over ticks as do the powder-release systems.

To control horn flies, face flies, and certain ticks and to aid in the control of house flies, stable flies, and lice, cattle ear tags have become commonplace. Synthetic pyrethroids (Wright et al., 1984) and dichlorvos (Baker and Stanford, 1979) have been used successfully. To eliminate the need for double tagging and the cutting out of old ear tags, insecticidal tapes have been formulated to be attached to existing identification ear tags.

3.8. Nanoparticles

Nanoparticles (Kreuter, 1978; Sugibayashi et al., 1979) have been reported as potential delivery systems for drugs, especially for parenterally administered compounds. A nanoparticle may be defined as any solid particle that exists in the nanometer size range. Ideally, these particles are able to pass through the capillary bed without becoming entrapped. Nanoparticles may be composed of gelatin, albumin, or synthetic polymers. When proteinaceous in nature, the external surface is cross-linked and hardened to retard erosion. In most cases the mechanism for drug release is erosion of the matrix and release of drug via diffusional processes.

4. Membrane Control of Drug Release

4.1. Theory

As the name implies, drug release to an external environment is governed by a discrete membrane that is unaltered over the life of the dosage form. The advantage of such systems is that drug release is less dependent on the nature of the environment in which the dosage form resides.

In its simplest embodiment, any bilayer system in which the diffusion of drug in the layer adjacent to the environment or receptor phase is slower than the diffusion of drug through the inner or reservoir layer will in general terms fall into this category. Although it is commonly accepted that the use of this class of device is solely for the purpose of imparting a constant release profile of active agent from the device, it need not be limited to this condition.

When the thermodynamic activity of the active ingredient in the reservoir is maintained constant (most easily via maintenance of excess nondissolved drug in the reservoir), a steady-state release is achieved in accordance with Fick's law. For drug release from a planar geometry having a membrane of thickness l

$$\frac{dQ}{dt} = \frac{SKD(C_i - C_e)}{l} \tag{2}$$

where dQ/dt is the steady-state release rate, S is the surface area for drug release, K is the partition coefficient between drug in the reservoir and membrane, D is the diffusion coefficient of drug in the membrane, and $(C_i - C_e)$ is the concentration difference between the internal and external drug concentration.

The mathematics for other geometrics has been presented by Peppas (1984) and Colton (1969). In the case of a cylindrical geometry, where the entire surface is covered by a rate-controlling membrane

$$\frac{dQ}{dt} = \frac{SKD(C_i - C_e)}{ln(r_e/r_i)} \tag{3}$$

where r_e and r_i are the external and internal radii, respectively.

In the case of a sphere, the rate expression

$$\frac{dQ}{dt} = \frac{4KD(C_i - C_e)}{(r_e - r_i)/r_e r_i} \tag{4}$$

characterizes the release kinetics.

Drug release from membrane-controlled devices is further complicated by the fact that at equilibrium, because of a finite solubility of the drug in the membrane, a "burst" effect is often evident during the initial stages of release. The time course of this phenomenon is a function of the solubility and diffusion coefficient of the drug in the membrane and the membrane thickness. Alternatively, if equilibrium of drug in the membrane has not occurred at the time of drug-release rate testing, a lag time will be encountered until the steady-state release pattern is observed.

Control of drug release kinetics has also been attempted using a variety of multilayered systems with rate-controlling properties. For example, Lee et al. (1980) described a system based on poly hydroxyethyl methacrylate, whereas Good (1978) described a swelling-controlled system.

4.2. Oral Dosage Forms

Some of the older dosage forms that belong in this category of membrane control of drug release do not behave in strict accord with the theory described above. However, since a coating around the dosage form limits the availability of the drug, these dosage forms most conveniently belong here.

The use of enteric coatings, which would dissolve only in the higher pH environment of the small intestine, have been used for many years. One of the earliest published reports in 1884 (Osol and Farrar, 1947) deals with a keratin-coated product. Salol (phenyl salicylate) was discussed in 1915, but its use probably predates the 20th century (Lyman and Sprowls, 1955).

A recent product, ERYC®, from Fauldings, Australia and consigned to Parke Davis, utilizes an enteric coating around a multiparticulate dosage form of the antibiotic erythromycin. The enteric coating is applied as the last step in manufacture to protect the active ingredient, erythromycin, from enzymatic degradation in the stomach.

Another type of membrane is one that is composed of poorly soluble or water-insoluble substances. For example, a British patent issued to Lipowski (1948) described the first preparation of an encapsulated dosage form containing groups of 1–2 mm diameter pellets, having successively thicker coats of materials that would resist dissolution in gastric and/or intestinal fluids. The Blythe patent (Blythe, 1956), assigned to Smith Kline & French Laboratories, described the time release preparation for amphetamine or ephedrine derivatives. This patent is the basis of the Spansule® line of products so successful for SKF. This concept, designed to provide multiple pulses of drug release as the pellets pass through the GI tract, is manufactured by applying successive coats of drug and waxes to nonpariel sugar cores. The final product is a blend of several fractions having various coating thicknesses and, therefore, different rates of drug release. This highly successful dosage form represented by products such as COMBID® , COMPAZINE® , FEOSOL® , and CONTACT® has at least 30 sister products, approved in the US through December 1983, all bearing a similar image.

Key Pharmaceutical has utilized similar technology for their Theodur® line of slow-release theophylline, but have compressed the pellets into a tablet as the finished dosage form. One obvious advantage of this approach is cost. Tablets, by virtue of the speed of manufacture and the lack of cost associated with a capsule per se, are more economical to produce.

A second type of product is the repeat-action dosage form. As the name implies, a repeat dose is delivered to the GI tract at some time following the initial dose of drug. This is usually accomplished by enteric coating of the core tablet. Each dose within the tablet is designed for immediate release. Therefore, the advantage becomes one of a convenient once-a-day design without any effect on drug entry rate into the body. CHLORTRIMETON REPETABS® (Schering) used to treat allergies exemplifies this technique.

Membrane-control sustained release has been accomplished in the animal health field by Pfizer, with their PARATECT® , morantel tartrate, bolus for cattle. This oral system is designed to deliver the drug to the ruminoreticulum for a period of 60–90 d. The PARATECT® bolus' rate-controlling membranes, situated at both ends of the cylinder, are composed of a cellulose triacetate hydrogel contained in/on a sintered polyethylene disk for support and protection from abrasion in the rumen (Dresback, 1980a,b). The system is targeted to give an average drug delivery of approximately 90 mg morantel/d, for at least 90 d (Freedom of Information Summary, 1984). The release is not zero-order, as implied by the residue depletion data shown in Table 2.

Table 2
Residue Depletion in Cattle Liver Following Treatment with the Paratect Bolus System[a]

	Day						
	7	15	30	61	90	120	180
Average ppm	0.53	0.62	0.50	0.36	0.33	0.22	0.20

[a]Results are from a mean of three animals.

Iodine deficiency, or congenital goiter, of lambs, which causes still-births and neonatal deaths, occurs sporadically in some areas of Australia. Treatment consists of oral dosing of pregnant ewes with iodine supplements or injecting them with iodized oil during the last 2 mo of pregnancy. To improve compliance and decrease the time and labor required to maintain the dosing regimen, the Commonwealth Scientific and Industrial Research Organization (CSIRO) of Australia, in conjunction with Hortico, Ltd., has developed an oral capsule that delivers a small but constant dose of iodine to a ewe for at least 3 yr (Lehane, 1982). The capsule is retained in the ruminoreticulum following oral administration by the use of the Laby expanding wing concept (Laby, 1974). Inside the capsule, iodine solid is in equilibrium with iodine vapor. As long as there is solid iodine present, the vapor pressure remains constant.

Thus, a constant concentration gradient across a rate-controlling membrane results in a sustained zero-order delivery from the device.

4.2.1. Osmotic Control of Drug Delivery

A drug delivery device that will release its drug payload independent of the environment in which it is placed requires the device to have a built-in energy supply. Using osmotic pressure as the energy source, the ALZA Corporation has been able to achieve a generic system for drug delivery that can function almost independent of molecular size or charge (Theeuwes and Bayne, 1981). The mini-osmotic pump or ALZET is widely used in preclinical research in which precise constant infusion rates to small animals are required. One such pump has a reservoir of 170 μL and typically, this system delivers 1 μL/h. The volume flow from the pump is described by the equation

$$\frac{dv}{dt} = K \frac{S}{l} (\pi_S - \pi_l) \tag{5}$$

where K is equal to LO, the product of the mechanical permeability coefficient and the reflection coefficient, π_S is the osmotic pressure of the system, and π_l is the osmotic pressure of the environment (Theeuwes, 1980a). The concentration of drug in the pump reservoir, therefore, is used to control the drug delivery rate.

An example of the use of the mini-osmotic pump in the development of a dosage form is given by Pope et al. (1985). They evaluated the pharmacokinetics of ivermectin, a potent broad-spectrum macrocyclic lactone disaccharide parasiticide (Chabala et al., 1980; Egerton et al., 1980), which was administered orally to cattle in a specifically weighted ALZET 2ML4 mini-osmotic pump. The pump was weighted to a density of 2.7 g/cm^3 to ensure retention within the ruminoreticulum. The cage containing the mini-osmotic pump was also fitted with a radio transmitter (Kath et al., 1985) in order to monitor retention within the ruminoreticulum. The osmotic pumps in this study delivered the drug consistently over the trial time period. Steady-state plasma levels were achieved in 7–14 d, and plasma concentration depletion curves were observed starting at approximately d 35, the theoretical lifetime of the osmotic pumps. Bioavailability was estimated to be 40%, and dose rate–plasma steady-state interrelationships were shown to be linear.

For a pharmaceutical oral dosage form, the elementary osmotic pump, which is a compressed matrix of the drug and osmotic agents overcoated with an insoluble semipermeable coating through which a hole has been drilled on one surface with a laser, has been used with success. When water is imbibed into a system from the environment, displacing an equal volume of agent formulation, the rate of delivery, dm/dt, is described by:

$$\frac{dm}{dt} = \frac{KS}{l}(\pi_{\text{eff}} - \pi_{\lambda})C \tag{6}$$

where π_{eff} is the effective osmotic pressure created by the formulation in the device, C is the concentration of drug in the formulation, S is the area of the membrane, l is the membrane thickness, and K is a constant characteristic of the membrane.

This equation does not define the orifice size, although this parameter must be properly designed in order for the osmotic drug pressure ($\pi_{\text{eff}} - \pi_l$) to control drug release adequately.

Features of the osmotic pump worthy of note include: the ability to function uniformly over the range of pH conditions encountered in the GI tract, an exceptional correlation between in vitro and in vivo performance, a release rate not affected by food and the utility of the device in situations other than drug delivery, provided water is available at sufficiently high activity.

At present one marketed product, ACUTRIM® , by Ciba-Geigy, is available and other therapeutic agents are under development.

Variations of the osmotic-pressure-controlled drug delivery systems having several compartments have been discussed in the literature (Theeuwes, 1982, 1980b). One advantage of having two compartments is either the ability to eliminate a compatibility problem between the drug substance and the osmotic agent within the solid core tablet or a problem that may occur at the actual time of delivery. Further developments along these lines were described (Chien, 1982; Thombre et al., 1985; Zentner, 1985) whereby a microporous coat around the tablet was formed. In another case (Urquhart, 1977), drug and buffer were contained in a dosage form, and upon exposure to the GI environment, fluid diffusion into the dosage units caused solubilization of buffer and drug resulting in release of drug to the GI tract. To date, utilization of this method of drug delivery has been somewhat limited.

4.3. Ocular Systems

Alza introduced Ocusert®, a solid, nondissolving medication system designed to deliver the anticholinergic pilocarpine to the eye over a 7- or 14-d period for the treatment of glaucoma. By laminating a drug reservoir between two sheets of ethylene vinyl acetate, drug delivery at predetermined rates across a known surface area was achieved. Chien (1982) has provided a good overview of the details concerning the Ocusert® system.

4.4. Transdermal Systems

A system, now in common use, is the Transderm Nitro® patch from Ciba-Geigy. Much like the Ocusert®, this device also employs a drug reservoir and a membrane that permits passage of drug to the receptor site, the skin. An additional complexity to this configuration is an adhesive through which drug must pass before reaching the skin. Complexities introduced by an adhesive include: accumulation of drug in the adhesive, which can lead to an alteration of the burst effect and interaction of the drug with the adhesive, leading to loss of adhesive strength.

A variation of the use of a membrane is found in the Trans-Scope® patch, of Ciba-Geigy, for control of motion sickness. This transdermal system described by Urquhart et al. (1977) utilizes a microporous membrane inside an adhesive matrix. The microporous membrane, in addition to acting as a barrier to prevent migration of drug particles, also controls the rate at which dissolved drug can diffuse from the reservoir layer to the adhesive-delivery area.

Several patents (Bernstein et al., 1981; Hyman et al., 1972) have been issued to Herculite Protective Fabrics that describe somewhat different membrane controlled devices for drug delivery. Unlike the Ciba transdermal patch, the systems described by Kydonieus (1980) found application in the agrochemical field, as well as for pharmaceuticals.

4.5. Pesticide Generators

The Hercon dispenser (Health-Chem. Corporation) is an adhesive-backed, multiple-layered product containing a reservoir layer of dissolved insecticide (Diazinon, Chlorpyrifos, Baygon) or other ingredients (volatile pheromones) that migrate continually under a concentration gradient to the strip surface through one or more rate controlling membranes. Laboratory tests have shown that the Hercon tapes have 100% efficacy for a year or more (Chein, 1978).

5. Ion Exchange

Drug bound to an ion exchange resin of either the cationic or anionic type would be released into the milieu of the GI tract as counter ions displace the drug. This technique for oral liquid and solid dosage forms was originally developed by Strasenburgh for amphetamine and marketed as Ionamine® and Biphetamine® (Madan, 1985). Various patents (Becker and Hays, 1958; Freed and Hays, 1959; Brudney, 1959) have been issued describing this type of dosage form.

Grass and Robinson (1959) described a method for preparing slow-release particles suitable for suspension in aqueous medium. Their patent teaches a multiple-coating procedure and was used in a product no longer marketed by SKF, SUL-SPANSION®.

More recently, a patent issued to Pennwalt (Raghunatha, 1980), which describes the preparation of small, coated, ion-exchange particles, with a semipermeable coat-controlled release of the active ingredient, is thereby imparted to the product. The dosage form is referred to as Pennkinetic® and is now being utilized in both delivery of dextromethorphen as Delsym® by Pennwalt or Extend 12® by Robbins and in agriculture for methyl parathion as Penncap M® by Pennwalt.

6. Biodegradable Polymers

6.1. Chemical Biodegradation—Matrix Release

Considerable literature exists regarding novel polymers designed to hydrolyze or break down in the presence of biological fluids (Baker et al., 1984; Yolles and Sartori, 1980; Heller, 1984a,b). The impetus for this effort is clear. Control of drug delivery necessitates the design of systems that will perform in a predictable fashion regardless of environmental conditions. One approach is based on matrices and/or coatings composed of polymers that are broken down to harmless or endogenous monomers. By virtue of their chemistry, these agents are labile and subject to changes caused by moisture and local pH changes. Various polymers that can undergo hydrolytic cleavage in biological fluids include, but are not limited to, poly (lactic acid), poly (glycolic acid), poly (orthoester), poly (anhydrides), and poly (amides). Basically the intent of this strategy is to maintain an essentially constant surface area in order to provide a constant rate of drug delivery. If properly controlled, the availability of drug imbedded in such a matrix is affected by the rate at which the polymer dissolves and exposes new surface of drug.

In the development of a biodegradable drug delivery system, one must focus on the need to:

(a) maintain compatibility of the polymer with the environment in which it will function;
(b) develop chemistry so that the rate of biodegradation of the polymer can be controlled in a reproducible manner;
(c) identify monomers that when produced will be related to, or appear to the body as, endogenous material;
(d) have properties that allow for easy fabrication of dosage forms.

Fabrication of devices based on the use of such polymers will likely require special handling precautions. Also, one may find that storage of the finished dosage form may require additional concern for packaging and shelf-life limitations.

One of the earliest reports dealing with biodegradable polymer synthesis appeared in 1971 (Yolles et al., 1971). Despite the many years since this report, the first dosage form to commercially utilize this concept has yet to appear. This should not be viewed so much as a sign of failure, but rather as an indicator of the complexity involved in moving an entirely new concept forward.

6.2. Biodegradable Drug-Carrier Complexes

The use of polymeric drug systems has been reviewed (Donaruma and Vogl, 1978). The idea is straightforward; to build into a molecule a substrate configuration that will react with some endogenous enzyme (Horbett et al., 1984). The perceived utility is that one can achieve enhanced specificity of delivery of a drug and increased duration of action. To date, only in vitro or animal assessment of this approach have been carried out.

Conceivably, this drug delivery concept is most applicable to parenterally administered drugs, although its use for implants as well as orally administered drugs could find potential application.

Brownlee and Cerami (1979) demonstrated that a maltose insulin conjugate bound to concanavalin A (ConA) can be released in the presence of glucose. In their system, the glycosylated insulin is immobilized on Sepharose. More recently, Jeong et al. (1983) reported on a method whereby glycosylated insulin would be coated with a polymer membrane. The purpose of the membrane was to control the flux of insulin and glucose to provide better control of release to prevent burst effects. Overall control of release, however, was via glucose influx into the system. Kim et al. (1984) found that the saccharide group on the glycosylated insulin influences the binding affinity to the ConA.

Yet another application of this new technology is the immobilization of the enzyme glucose oxidase within a crosslinked, ionizable polymeric membrane (Horbett et al., 1983). The glucose oxidase is able to catalyze the converison of glucose to gluconic acid. The acid, by protonating the amine function in the membrane, causes the membrane structure to swell, thus resulting in faster diffusion of insulin.

Both of the insulin systems described above could function as implant devices. Considerable work remains to be done to further evaluate these approaches. The reader is referred to a compilation by Roche (1977) for additional details.

7. Cellular Methods

7.1. Artificial/Naturally Occurring Cellular Methods

Red blood cell ghosts, nylon microcapsules, and liposomes may be seen as means for packaging drugs for delivery to a living organism. Recent publications by Gregoriadis (1979a,b) present a comprehensive review of these controlled drug-delivery methods. The emphasis of this approach is toward drug targeting; both from the standpoint of bringing drugs to the target *per se,* as well as enabling a drug to permeate a target area.

The approach using cell ghosts is interesting since it has the potential for reducing the incidence of foreign body reactions sometimes associated with administration of biologically derived substances into the body. The procedure involves the lysis of the cell followed by refilling using a solution of the drug substance to be delivered. By virtue of the nature of erythrocyte biology, this carrier, when injected back into the patient, has the potential for lasting up to 120 d before complete destruction occurs (Ihler, 1979).

7.2. Liposomes

No discussion of artificial cells or drug delivery devices can be complete without some word about liposomes. Since the first identification of a lipid vesicle by Bangham et al. (1965), liposomes have been proposed as a "magic bullet" able to deliver therapeutic agents to virtually any site in the human body (Tyrrell et al., 1976). Gregoriadis (1979a,b) has presented a very thorough review of the various substances incorporated into liposomes reported in the literature and more recently has edited a three volume set, entitled "Liposomes Technology" (Gregoriadis, 1984).

What was once believed to be a curiosity has advanced to a stage of development in which vesicles responding to pH, light, and heat can be prepared. A wide variety of methods to prepare liposomes having more tightly controlled particles sizes continues to appear in the literature. During the past ten years, an innumerable array of journal articles discussing the fabrication, membrane modeling aspects, potential for drug delivery, and so on, have appeared.

The promise for liposomes to carry drugs to specific tissues has continued to be pursued with only limited success.

By far, liposomes tend to localize in liver, spleen, and lung tissue. This nonselective targeting can be used to an advantage for treating such disorders as glycogen storage disease (Colley and Ryman, 1974), heavy metal poisoning (Rahman et al., 1973), leishmaniasis (Alving et al., 1978), and carcinoma (Kimelberg, 1976). The reader is encouraged to consult this latest compilation by Gregoriadis (1984) for a comprehensive survey of the field.

Another area actively pursued in the mid 1970s was the hope that liposomes would enhance the oral absorption of macromolecular agents. Except for a few isolated reports, this later use has received little attention.

Let it be said that the age of the liposome as a practical drug delivery system has yet to come. This should not be taken to mean that it will not be one day, but rather, like the case for a biodegradable polymer noted above, it takes time.

8. Mechanical Pumps

A wide variety of mechanical pumps are in current use (Sefton et al., 1984). Some of these are simple syringe pumps and utilize a stepper motor to drive the syringe. The rate of delivery is adjusted to meet individual needs.

The CPI syringe pump (Lilly) has a microchip that allows for programming in insulin units. Auto Syringe Inc. includes an alarm to warn of motor failure or empty syringes. These devices deliver their drug payload via a subcutaneous catheter.

Alternate design features such as a peristaltic pump, Biostator® (Prestele et al., 1981), have been described. Despite the miniaturization that has been achieved, patient acceptance has been limited, primarily because of pump bulk and discomfort during wear.

Implantable pumps containing sophisticated microprocessors to monitor drug delivery, sense drug reservoir levels, and receive commands to modify dosing regimens have also been developed (Blackshear et al., 1972, 1973; Buckwald et al., 1980). These highly sophisticated devices are able to send a status report to the attending physician by merely placing a phone over the implant area. By the same token, if the dosing regimen were changed, a signal via phone could be similarly transmitted. Refilling the pump was performed simply by a subcutaneous injection into the pump.

The desirability of such sophistication cannot be denied. However, the cost of each system is likely to exceed $10,000, thereby reducing its availability in all but the most critical cases.

Sefton et al. (1984) described a simplified implantable micropump that provides basal drug delivery via diffusion through a thin hydrophilic membrane. Augmented delivery is achieved via repeated compression of a foam disk by a piston activated by a solinoid via manual application of current to the solinoid.

Work continues in this important area of therapy. With advances in electronics, blood compatible polymers, and miniaturization of instruments, microchip control of drug release has a high potential for success in the future.

9. Conclusions

The wealth of information available in the literature would tend to confirm the belief that the future for controlled release drug delivery systems is promising. The multiparticulate capsule dosage form based on the Smith Kline "Spansule®" products is synonomous with sustained-release dosage forms in both over-the-counter and prescription dosage forms.

Transdermal dosage forms are being widely prescribed and the 1985 US Market for nitroglycerin products was greater than $200 million. As new products enter the marketplace, additional usage will be enjoyed. Any means that can be employed to reduce dosing frequency will lead to increased patient compliance—one of the prime reasons for developing these types of dosage forms.

In the field of animal health, the ability to treat domestic and range animals at only widely spaced intervals and preferrably only once, cannot be overemphasized. Significant advances have been made in this field that will ultimately have an impact on human health care and the economy as a whole.

Yet despite the overwhelming desire to make every dosage form a "controlled-release" preparation, sight should not be lost of the fact that it is the final clinical response that will dictate whether a controlled-release product has therapeutic advantage over the conventional dosage form. Indeed, the possibility that control of drug-entry rate may not always be the most efficient means for drug delivery and consequent therapeutic ability is a factor that will decide the usefulness of physicochemical as opposed to pharmacological approaches to drug development.

References

Alper, J. (1984) Drug Delivery, *High Technol.* **4**(10), 89–94.

Alving, C. R., Steck, E. A., Edgar, A., Chapman, W. L., Waits, V. B., Hendricks, L. D., Schwartz, G. M., and Hanson, W. L. (1978) Therapy for leishmaniasis superior efficacies of liposome-encapsulated drugs. *Proc. Natl. Acad. Sci. USA* **75**, 2959–2963.

Anderson, J. M. and Kim, S. W., eds. (1984) *Recent Advances in Drug Delivery Systems.* Plenum, New York.

Baker, J. A. F. and Stanford, G. D. (1979) Slow Release Devices as Aids in the Control of Ticks Infesting the Ears of Cattle in the Republic of South Africa, in *Recent Advances in Acarology* vol. II (Rodiquez, J. D., ed.) Academic, N.Y.

Baker, R. W. and Lonsdale, H. K. (1974) Controlled Release: Mechanisms and Rates, in *Controlled Release of Biologically Active Agents* (Tanquary, A. C. and Lacey, R. E., eds.) Plenum, New York.

Baker, R. W., Tuttle, M. E., and Helwing, R. (1984) Novel erodible polymer for the delivery of macromolecules. *Pharm. Technol.* February, 26–30.

Banakar, U. V. (1984) Drug release mechanism of membrane-moderated drug delivery. *Pharm. Manufact.* September, 33–56.

Bangham, A. D., Standish, M. M., and Watkins, J. C. (1965) Diffusion of univalent ions across the lamellae of swollen phospholipids. *J. Mol. Bio.* **13**, 238–252.

Banker, G. S. and Rhodes, C. T. (1979) *Modern Pharmaceutics*. Marcel Dekker, New York.

Bechgaard, H. and Ladefoged, K. (1978) Distribution of pellets in the gastrointestinal tract. The influence on transit time exerted by the density or diameter of pellets. *J. Pharm. Pharmacol.* **30**, 690–692.

Becker, B. A. and Hays, E. E. (1958) Prolongation and potentiation of oral codeine analgesia in the rat. *Proc. Soc. Exp. Biol. Med.* **99**, 17–19.

Bernstein, B. S., Hyman, S., and Kapoor, R. C. (1981) US Patent 4,284,444.

Blackshear, P. J., Dorman, F. D., Blackshear, P. L., Varco, R. L., and Buchwald, H. (1972) The design and initial testing of an implantable infusion pump. *Surg. Gynecol. Obstet.* **134**, 51–55.

Blackshear, P. J., Dorman, F. D., Blackshear, P. L., Buckwald, H., and Varco, R. L. (1973) US Patent 3,731,681.

Blythe, R. H. (1956) US Patent 2,738,303.

Bodor, N. (1977) Novel Approaches for the Design of Membrane Transport Properties of Drugs, in *Design of Biopharmaceutical Properties through Prodrugs and Analogs* (Roche, E. G., ed.) American Pharmaceutical Association, Washington, DC.

Brooke, D. and Washkuhn, R. J. (1977) Zero-order drug delivery system; theory and preliminary testing. *J. Pharm. Sci.* **66**, 159–162.

Brownlee, M. and Cerami, A. (1979) A glucose-controlled insulin-delivery system: Semisynthetic insulin bound to lectin. *Science* **206**, 1190–1191.

Brudney, N. (1959) Ion-exchange resin complexes in oral therapy. *Can. Pharm. J. Sci. Sect.* **92**(5), 245–258.

Campbell, P. S. and Chandrasekaran, S. K. (1983) US Patent 4,379,454, April 12.

Chabala, J. C., Mrozik, H., Tolman, R. L., Eskola, P., Lusi, A., Peterson, L. H., Woods, M. F., Fisher, M. H., Campbell, W. C., Egerton, J. R., and Ostlind, D. A. (1980) Ivermectin: A new broad-spectrum antiparasitic agent. *J. Med. Chem.* **23**, 1134–1136.

Check, W. A. (1984) New drugs and drug delivery systems in the year 2000. *Am. Pharm.* **24**(9), 44–56.

Chien, Y. W. (1978) Method to Achieve a Sustained Drug Delivery—The Physical Approach: Implants, in *Sustained and Controlled Release Drug Delivery Systems* (Robinson, J. R., ed.) Marcel Dekker, New York.

Chien, Y. W. (1982) Ocular Controlled Release Drug Administration, in *Novel Drug Delivery Systems* (Chien, Y. W., ed.) Marcel Dekker, New York.

Colley, C. M. and Ryman, B. E. (1974) Model for lysosomal storage disease and a possible method of therapy. *Biochim. Soc. Trans.* **2**(5), 871–872.

Colton, C. K. (1969) Permeability and transport studies in batch and flow dialyzers and application to hemodialysis. Doctoral dissertation, Massachusetts Institute of Technology.

Dewey, D. W., Lee, H. J., and Marston, H. R. (1958) Provision of cobalt to ruminants by means of heavy pellets. *Nature* **18**, 1367–1371.

Donaruma, L. G. and Vogl, O., eds. (1978) *Polymeric Drugs.* Academic, New York.

Drake, C. F. and Brecklehurst, J. R. (1984) Glass encapsulated materials. US Patent 4,449,981.

Dresback, D. S. (1980a) Delivery system. US Patent 4,220,152.

Dresback, D. S. (1980b) Controlled release delivery system. US Patent 4,222,153.

Egerton, J. R., Birnbaum, J., Blair, L. S., Chabala, J. C., Conroy, J., Fisher, M. H., Mrozik, H., Ostlind, D. A., Wilkins, C. A., and Campbell, W. C. (1980) 22,23-Dihydroavermectin B1, A new broad spectrum antiparasitic agent. *Br. Vet. J.* **136**, 89–97.

Fisch, H., Angerhofer, R. A., and Nelson, J. H. (1977) Evaluation of a carbamate-impregnated flea and tick collar for dogs. *J. Am. Vet. Med. Assoc.* **171**, 269–270.

Flynn, G. L., Yalkowsky, S. H., and Roseman, T. J. (1974) Mass transport phenomena and models: Theoretical concepts. *J. Pharm. Sci.* **63**, 479–510.

Folckemer, F. B., Hanson, R. E., and Miller, A. (1967) Resin compositions comprising organophosphorus pesticides. US Patent 3,852,416.

Folkman, J. and Long, D. M. (1964) The use of silicone rubber as a carrier for prolonged drug therapy. *J. Surg. Res.* **4**, 139.

Freed, S. C. and Hays, E. E. (1959) Non-amphetamine anorectic agent. *Am. J. Med. Sci.* **238**, 55–59.

Freedom of Information Summary (1984) Paratect cartridge for use in Cattle, NADA 134-779, Pfizer, New York.

Good, W. R. (1978) Diffusion of Water Soluble Drugs From Initially Dry Hydrogels, in *Polymeric Delivery Systems* (Kostelnik, R. J., ed.) Gordon and Breach Science, New York.

Graham, N. B. and Wood, D. A. (1982) Hydrogels and biodegradable polymers for the controlled delivery of drugs. *Polymer News* **8**, 230–236.

Grass, G. M. and Robinson, M. J. (1959) US Patent 2,875,130, February 24.

Gregoriadis, G., ed. (1979a) *Drug Carriers in Biology and Medicine.* Academic, London.

Gregoriadis, G. (1979b) Liposomes, in *Drug Carriers in Biology and Medicine* (Gregoriadis, G., ed.) Academic, London.

Gregoriadis, G., ed. (1984) *Liposome Technology* vol. I–III, CRC, Boca Raton, Florida.

Grubb, L. M. and Baxter, J. K. (1974) Tick and flea collar of solid solution plasticized vinylic resin-carbamate insecticide. US Patent 3,852,416.

Harrigan, R. M. (1977) Drug delivery device for preventing contact of undissolved drug with the stomach lining. US Patent 4,055,178.

Harris, F. W. (1984) Controlled Release From Polymers Containing Pendent Bioactive Sustituents, in *Medical Applications of Controlled Release* vol. I (Langer, R. S. and Wise, D. L., eds.) CRC, Boca Raton, Florida.

Heller, J. (1984a) Bioerodible Systems, in *Medical Applications of Controlled Release* (vol. I (Langer, R. S. and Wise, D. L., eds.) CRC, Boca Raton, Florida.

Heller, J. (1984b) Controlled Drug Release From Poly(orthoesters), in *Proceedings—11th International Symposium on Controlled Release Bioactive Materials* (Meyers, W. E. and Dunn, R. L., eds.) Ft. Lauderdale, Florida, July 23–25.

Hermelin, V. M. (1959) US Patent 2,887,438.

Higuchi, T. (1961) Rate of release of medicaments from ointment bases containing drugs in suspension. *J. Pharm. Sci.* **50**, 874–875.

Hildebrand, R. (1983) Polymers release drugs continually. *High Technol.* **3**(1), 10–13.

Hixson, A. W. and Crowell, J. H. (1931a) Dependence of reaction velocity upon surface and agitation. I. Theoretical Considerations. *Indust. Engineer. Chem.* **23**, 923–931.

Hixson, A. W. and Crowell, J. H. (1931b) Dependence of reaction velocity upon surface and agitation. II. Experimental procedure in study of surface. *Indust. Engineer. Chem.* **23**, 1002–1009.

Hopfenberg, H. B. (1976) Controlled Release From Erodible Slabs, Cylinders and Spheres, in *Controlled Release Polymeric Formulations* (Paul, D. R. and Harris, F. W., eds.) ACS Symposium Series 33, American Chemical Society, Washington, DC.

Horbett, T. A., Kost, J., and Ratner, B. D. (1983) Swelling behavior of glucose sensitive membranes. *Polymer Preprints* **24**(1), 34–35.

Horbett, T. A., Ratner, B. D., Kost, J., and Singh, M. (1984) A Bioresponsive Membrane for Insulin Delivery, in *Recent Advances in Drug Delivery Systems* (Anderson, J. M. and Kim, S. W., eds.) Plenum, New York.

Hudson, J. L. and Wagner, J. F. (1980) Remarkable drug implant. US Patent 4,191,741.

Hyman, S., Bernstein, B. S., and Kapoor, R. C. (1972) US Patent 3,705,938.

Inhler, D. M. (1979) Potential Use of Erythrocytes as Carriers for Enyzmes and Drugs, in *Drug Carriers in Biology and Medicine* (Gregoriadis, G., ed.) Academic, London.

Inhorn, M. C. (1981) New dimensions in drug delivery. *Drug Topics* August 7, 38–50.

Jeong, S. Y., Sato, S., McRea, J. C., and Kim, S. W. (1983) Controlled release of bioactive glycosylated insulins. *10th International Symposium on Controlled Release of Bioactive Materials* San Francisco, California, July 24–27.

Karim, A. (1983) Transdermal absorption of nitroglycerin from microseal drug delivery (MDD) system. *Angiology* **34**(1), 11–22.

Kath, G. S., Egerton, J. R., and Geiger, R. (1985) In-dwelling rumino-reticulum bolus radiobeacon. *Am. J. Vet. Res.* **46**, 136–137.

Katz, I. M. and Blackman, W. M. (1977) A soluble sustained-release opthalmic delivery unit. *Am. J. Ophthalmol* **83**, 728.

Keith, A. and Snipes, W. (1980) European Patent Application 0013 606A2.

Kim, S. W. (1983) Hydrogels as drug delivery systems. *Pharm. Int.* April, 90–91.

Kim, S. W., Jeong, S. Y., Sato, S., McRea, J. C., and Feijen, J. (1984) Self-Regulating Insulin Delivery System—A Chemical Approach, in *Recent Advances in Drug Delivery Systems* (Anderson, J. M. and Kim, S. W., eds.) Plenum, N.Y.

Kimelberg, H. K. (1976) Differential distribution of liposome entrapped (^{3}H)-methotrexate and labeled lipids after intravenous injeciton in a primate. *Biochim. Biophys. Acta* **448**, 531–550.

Koestler, R. C. (1980) Microencapsulation by Interfacial Polymerization Techniques–Agricultural Applications, in *Controlled Release Technologies: Methods, Theory and Applications* vol. II (Kydonieus, A. F., ed.) CRC, Boca Raton, Florida.

Korsmeyer, R. W. and Peppas, N. A. (1983) Macromolecular and Modeling Aspects of Swelling Controlled Systems, in *Controlled Release Delivery Systems* (Roseman, T. J. and Mansdorf, S. Z., eds.) Marcel Dekker, New York.

Kost, J. and Langer, R. (1984) Controlled release of bioactive agents. *Trends Biochem.* **2**(2), 47–51.

Kreuter, J. (1978) Nanoparticles and nanocapsules—New dosage forms in the nanometer range. *Pharm. Acta Helv.* **53**, 33–39.

Kydonieus, A. (1980) Fundamental Concepts of Controlled Release, in *Controlled Release Technologies: Methods, Theory and Applications* vol. I (Kydonieus, A., ed.) CRC, Boca Raton, Florida.

Laby, R. H. (1974) Device for administration to ruminants. US Patent 3,844,285.

Laby, R. H. (1978) Australian Patent Application 35908/78.

Laby, R. H. (1982) Controlled release compositions for administration of therapeutic agents to ruminants. Australian Patent PCTAU80/00082.

Lachman, L., Lieberman, H. A., and Kanig, J. L., eds. (1976) *The Theory and Practice of Industrial Pharmacy* 2nd Ed. Lea & Febiger, Philadelphia, Pennsylvania.

Lee, E. S., Kim, S. W., Kim, S. H., Cardinal, J. R., and Jacobs, H. (1980) Drug release from hydrogel devices with rate controlling barriers. *J. Membr. Sci.* **7**, 293.

Lehane, L. (1982) Controlled-release capsules for improving herd health. *Rural Res.* **115**, 10–14.

Lieberman, H. A. and Lachman, L., eds. (1980) *Pharmaceutical Dosage Form: Tablets* vols. 1–3, Marcel Dekker, New York.

Lipowski, A. (1948) Australian Patent 109,438.

Lipper, R. A. and Higuchi, W. I. (1977) Analyses of theoretical behavior of a proposed zero-order drug delivery system. *J. Pharm. Sci.* **66**, 163–164.

Long, D. M., Jr., and Folkman, J. (1966) Polysiloxane carrier for controlled release of drugs and other agents. US Patent 3,279,996.

Madan, P. L. (1985) Sustained-release drug delivery systems. I. An overview. *Pharm. Manu.* February, 23–27.

McClure, W. O. and Stoughton, R. B. (1982) Enhancement of transdermal drug administration-Azone, presented at *Industrial Pharmaceutical R&D Symposium on Transdermal Controlled Release Medication* Rutgers University, College of Pharmacy, Piscataway, New Jersey, January 14–15.

Michaels, A. S. (1974) Drug delivery device with self actuated mechanism for retaining device in selected area. US Patent 3,786,813.

Michaels, A. S., Bashwa, J. D., Zaffaroni, A. (1975) Integrated device for administering beneficial drug at programmed rate. US Patent 3,901,232.

Miller, A. and Morales, J. G. (1976) Non-volatile slow release pesticidal generators. US Patent 3,944,662.

Miller, J. E., Baker, N. F., and Colburn, E. L. (1977) Insecticidal activity of propoxur- and carbaryl-impregnated flea collars against *Ctenocephalides felis. Am. J. Vet Res.* **38** 923–925.

Moo-Young, A. J., Tatum, H. J., Wan, L. S., and Lane, M. E. (1975) Copper Levels in Certain Tissues of Rhesus Monkeys and of Women Bearing Copper IUD's, in *Anal. Intrauterine Contracept., Proceedings of the International Conference* 3rd (1974), (Hefnawi, F. and Segal, S. J., eds.) Elsevier, New York.

Nash, H. A. (1984) Controlled Release Systems for Contraception, in *Medical Application of Controlled Release* vol. II (Langer, R. S. and Wise, D. L., eds.) CRC, Boca Raton, Florida.

Osol, A. and Farrar, G. E., eds. (1947) *The Dispensatory of the United States of America* Lippincott, Philadelphia, Pennsylvania, p. 1178.

Peppas, N. A. (1984) Mathematical Modeling of Diffusion Processes in Drug Delivery Systems, in *Controlled Drug Bioavailability* vol. I *Drug Product Design and Performance* (Smolen, V. F. and Ball, L. A., eds.) John Wiley, New York.

Pope, D. G. (1978) Animal Health Specialized Delivery Systems, in *Animal Health Products Design and Evaluation* (Monkhouse, D. C., ed.) American Pharmaceutical Association, Washington, DC.

Pope, D. G. (1983) Specialized Dose Dispensary Equipment, in *Formulation of Veterinary Dosage Forms* (Blodinger, J., ed.) Marcel Dekker, New York.

Pope, D. G., Wilkinson, P. K., Egerton, J. R., and Conroy, J. (1985) Oral controlled release delivery of invermectin in cattle via an osmotic pump. *J. Pharm. Sci.* **74**, 1108–1110.

Prestele, K., Franetzki, M., and Kreese, H. (1980) Development of program controlled portable insulin delivery devices. *Diabetes Care* **3**, 362–370.

Quick, H. L. (1971) Clinical evaluation of a free-hanging 20% dichlorvos (DDVP) disc for control of flea infestation of dogs. *Vet. Med.* **66**, 773–774.

Raghunatha, Y. (1980) US Patent 9,221,778.

Rahman, Y. E., Rosenthal, M. W., and Cerny, E. A. (1973) Intracellular plutonium removal by liposome-encapsulated chelating agent. *Science* **180**, 300–302.

Rhine, W., Sukhatme, V., Hsieh, D. S. T., and Langer, R. S. (1980) A New Approach to Achieve Zero-Order Release Kinetics From Diffusion-Controlled Polymer Matrix Systems, in *Controlled Release of Bioactive Materials* (Baker, R., ed.) Academic, New York.

Riner, J. L. (1982) Sustained release ruminant boluses and factors determining their release rates. Doctoral dissertation, Oklahoma State University.

Riner, J. L., Byford, R. L., Stratton, L. G., and Hair, J. A. (1982) Influence of density and location on degradation of sustained-release boluses given to cattle. *Am. J. Vet. Res.* **43**, 2028–2030.

Robertson, D. N. (1983) Release rates of levonorgestrel from silastic capsules, homogeneous rods and covered rods in humans. *Contraception* **27**, 483.

Robertson, D. N. (1985) Contraception with long-acting subdermal implants. A five-year clinical trial with silastic covered rod implants containing levonorgestrel. *Contraception* **31**, 351.

Roche, E. B., ed. (1977) Biopharmaceutical properties of drugs in design of biopharmaceutical properties through prodrugs and analogs. *American Pharmaceutical Association*, Washington, DC.

Rogers, J. A. (1982) Recent developments in drug delivery. *Can. J. Hosp. Pharm.* **25**(6), 170–196.

Roseman, T. J. and Yalkowsky, S. H. (1976) Importance of Solute Partitioning on the Kinetics of Drug Release From Matrix Systems, in *Controlled Release Polymeric Formulations* (Paul, D. R. and Haris, F. W., eds.) ACS Symposium Series 33, American Chemistry Society, Washington, DC.

Schor, J. M. (1982) US Patent 4,357,469.

Schor, J. M. and Nigalye, A. (1983) US Patent 4,389,383.

Sefton, M. V., Allen, D. G., Horvath, V., and Zingg, W. (1984) Insulin Delivery at Variable Rates From a Controlled Release Micropump, in *Recent Advances in Drug Delivery System* (Anderson, J. M. and Kim, S. W., eds.) Plenum, New York.

Senyei, A. E., Driscoll, C. F., and Widder, K. J. (1985) Biophysical drug targeting—magnetically responsive albumin microspheres. *Meth. Enzymol.* **112**, 56–67.

Shell, J. W. (1985) Opthalmic drug delivery systems. *Drug Dev. Res.* **6**, 345–261.

Sherwin, R. S., Schulman, G. A., Hendler, R., Walesky, M., Belous, A., and Tamborlane, W. (1983) Effect of growth hormone on oral glucose tolerance and circulating metabolic fuels in man. *Diabetologia* **24**(3) 155–161.

Sheth, P. R. and Tossounian, J. L. (1978) Hoffman-La Roche US Patent 4,126,672.

Stoughton, R. B. (1982) Enhanced percutaneous penetration with 1-dodecylazacycloheptan-2-one(Azone). *Arch. Dermatol.* **118**, 474–480.

Sugibayashi, K., Morimoto, Y., Nadai, T., Kato, Y., Hasegawa, A., and Arita, T. (1979) Drug-carrier property of albumin microspheres in chemotherapy. II. Preparation and tissue distribution in mice of microsphere-entrapped 5-fluorouracil. *Chem. Pharm. Bull.* **27**, 204–209.

Theeuwes, F. (1980a) US Patent 4,235,236, November 25.

Theeuwes, F. (1980b) Delivery of Drugs by Osmosis, in *Controlled Release Technologies: Methods, Theory and Applications* vol. II (Kydonieus, A., ed.) CRC, Boca Raton, Florida.

Theeuwes, F. (1982) US Patent 4,309,996.

Theeuwes, F. and Bayne, W. (1981) Controlled-Release Dosage Form Design, in *Controlled Release Pharmaceuticals* (Urquhart, J., ed.) American Pharmaceutical Association, Washington, DC.

Thombre, A. G., Zentner, Z. M., and Himmelstein, K. S. (1985) Mechanism of Transport Kinetics in a Controlled Porosity Osmotic Pump, in *Proceedings of the 12th International Symposium on Controlled Release of Bioactive Materials*. Geneva, Switzerland.

Tindall, B. (1983) Implants and growth promotion. *Anim. Nutrit. Health* Sept.–Oct., 14–20.

Tojo, K. (1984) Prolonged drug release from a cylindrical matrix device with a small release hole. *Chem. Eng. Commun.* **30**, 311–322.

Tomlinson, E. and Davis, S. S. (1986) *Site-Specific Drug Delivery* Wiley, New York.

Tyrrell, D. A., Heath, T. D., Colley, C. M. and Ryman, B. E. (1976) New aspects of liposomes. *Biochim. Biophys. Acta* **457**, 259–302.

University of Glasgow (1984) Device for introducing nutrients and/or therapeutic materials into ruminant animals. *European Patent* 97–507a.

Urquhart, J., Chandrasekaran, S. K., and Shaw, J. E. (1977) US Patent 4,031,894, June 18.

Watanabe, S., Kayamo, M., Ishino, Y., and Miyao, K. (1976) Solid therapeutic preparation remaining in stomach. US Patent 3,976,764.

Welling, P. G. (1983) Oral controlled drug administration, *Drug Devel. Indust. Pharm.* **9**(7), 1185–1225.

Wise, D. L. (1984) Controlled Release for Use in Treatment of Narcotic Addiction, in *Medical Applications of Controlled Release* vol. II (Laryer, R. S. and Wise, D. L., eds.) CRC, Boca Raton, Florida.

Wright, C. L., Titchener, R. N., and Hughes, J. (1984) Insecticidal ear tags and sprays for the control of fleas on cattle. *Vet. Record* **115**, 60–63.

Yolles, S., Eldridge, J. E., and Woodland, J. H. R. (1971) Sustained delivery of drugs from polymer/drug mixtures. *Polymer News* **1**, 9.

Yolles, S. and Sartori, M. F. (1980) Erodible Matrices, in *Controlled Release Technologies: Methods, Theory, and Applications* vol. II (Kydonieus, A. F., ed.) CRC, Boca Raton, Florida.

Zaffaroni, A. (1980) The innovators-delivering drugs. *Chemtech*. February, 82–88.

Zentner, G. (1985) Osmotic flow through controlled porosity films: An approach to delivery of water soluble compounds. *Second International Symposium on Recent Advances in Drug Delivery Systems*. Salt Lake City, Utah, February 27–March 1.

Clinical Evaluation of Drug Candidates

Scott A. Reines and Dick Fong

1. Introduction

The clinical evaluation of a new chemical entity as a therapeutic agent, and its registration for use as a new drug, are very complex procedures. More than 10 years of research and approximately $65 million may be required from the time of original identification or synthesis of a compound to its commercial introduction as a marketed drug. The majority of the time and money are usually spent on clinical development.

The thalidomide tragedy in Europe (Smithells, 1962) prompted drug regulatory agencies to play an increasing role in the development of new drugs. In the United States in 1962, the Kefauver Harris amendments to the Food, Drug and Cosmetic Act of 1938 (public law, 1962) gave the Food and Drug Administration (FDA) the authority to ensure that all new drugs be proven safe and effective prior to release for general use.

The purpose of this chapter is to provide an update and overview of the clinical research process and the regulatory atmosphere in which it takes place. For further details the reader is referred to books that are devoted entirely to this subject (Matoren, 1984; Spilker, 1984; Friedman et al., 1984).

2. Requirements and Timing for Proceeding to Human Trials

Since 1963, a sponsor wishing to undertake clinical testing of a new drug candidate in the United States is required to submit to the FDA an IND (Notice of Claimed Investigational Exemption for a New Drug). Within 30 days the FDA is obligated to act upon this notice, before human testing can begin. The IND describes in detail the chemical, pharmacological, pharmaceutical, and toxicological properties of the compound as formulated for human administration (New Drug and Antibiotic Regulations, 1985).

The FDA has issued specific guidelines defining the animal toxicity studies needed to support human use of drugs in major therapeutic areas (Sher et al., 1980). Mutagenicity studies, such as the Ames test, are conducted in bacterial strains as preliminary screens to those chemicals that might have carcinogenic or mutagenic potential. Animal toxicity studies must then define the effects of acute (several days to 2 wk), subacute (2 wk to 6 mo), and chronic (12 mo or longer) drug administration in mammals. The entire preclinical safety package requires several years of animal testing, particularly for the assessment of risk of carcinogenicity. Such extensive animal testing is not warranted until there is some evidence from early clinical studies that drug development should be continued. Single-dose studies in humans are often initiated after preclinical testing has proceeded to the completion of subacute toxicity (e. g., 5–14 wk) in two mammalian species, although guidelines vary among different pharmaceutical companies and national drug regulatory authorities. Mammalian reproduction studies to evaluate the effect of a drug on fertility, and to estimate its teratogenic potential, are generally conducted at a later stage of drug development, thereby limiting initial clinical testing to males and nonfertile females.

3. Definition and Scope of the Phases of Drug Development

Clinical testing of drug candidates in humans is conventionally divided into four or five progressive categories or phases, denoting the stage of development of the compound. The boundaries between phases are not sharply delineated, and there is generally some overlap between the completion of one phase and the initiation of the next. The following are working definitions for each of the categories.

3.1. Phase I

This phase begins with the first administration of the drug to humans, and includes very early dose-ranging studies to determining tolerability in patients. Healthy male subjects are usually the first recipients of a new drug. Occasionally, as in the case of a substance with anticipated human toxicity that is to be used in patients with serious or terminal illnesses, appropriate patients may be the first to receive the drug.

Phase I is generally considered to have been completed after the principal side effects have been elucidated in volunteers, and the maximally tolerated dose has been estimated. If no side effects are observed, the dose at which therapeutic effects are anticipated must be greatly exceeded. Some 20–100 subjects and patients are required for these determinations. Studies of drug pharmacokinetics and metabolism, which are generally completed after clinical testing has moved beyond Phase I, are nevertheless considered to be Phase I studies (General Considerations for the Clinical Evaluation of Drugs, 1978).

3.2. Phase II

Early Phase II trials generally consist of open-label, single- and multiple-dose studies in patients. These studies may record changes relative to baseline or historical control (i. e., previous clinical observation) to allow for a preliminary evaluation of efficacy. Later, Phase II trials are usually placebo- or active-drug-controlled and designed to obtain more convincing evidence of efficacy. Critical goals of the Phase II studies are the definition of a therapeutic dosage range and an appropriate dosage regimen (frequency and route of administration) to be used in future large-scale trials. Pharmacokinetic data are valuable in determining an appropriate dosage regimen. Difficulties are often encountered, however, in the attempt to establish a therapeutic dosage range, particularly in conditions that tend to remit spontaneously or for which there is a time lag between drug administration and therapeutic effects. Many psychiatric illnesses, including depression and schizophrenia, fall into these categories.

3.3. Phase III

Drug candidates that enter Phase III have usually been administered to several hundred patients and normal subjects. Preliminary evidence of efficacy, a profile of commonly occurring side effects, and the therapeutic dosage range should be tentatively established. Essentially nothing will be known about uncommon side effects (less than 2% incidence), which may ultimately be crucial to the success or failure of the drug. Phase III trials are normally controlled, multiclinic studies, enrolling a total of several hundred to more than a thousand patients, and designed to establish the efficacy of the drug and to define its adverse effect profile as precisely as possible. During this phase, some evidence of uncommon side effects and adverse laboratory changes may be obtained. Specific patient populations, such as subtypes of depression or schizophrenia, may be defined as responsive or unresponsive to therapy with the drug.

The Phase III program should be designed to produce sufficient efficacy and safety data for registration of the drug. This will be accomplished in the United States by submission of a New Drug Application (NDA), as described further below.

3.4. Phase IV

Conducted after marketing of the drug has begun, Phase IV trials have been broadly defined in two ways:

1. Postmarketing surveillance studies to elucidate uncommon side effects, and
2. All post-marketing studies, including clinical trials in patient populations not fully explored during Phase III (General Considerations for the Clinical Evaluation of Drugs, 1978).

For the sake of clarity, the latter studies are better defined as Phase V. Phase IV will then include only postmarketing surveillance (PMS) studies, in which the overall aim is "the assessment of the relative risks and relative benefits of drug therapy" (Inman, 1981). In practice these studies can hope to provide reassurance that serious side effects are relatively infrequent. To achieve this goal, long-term surveillance in the form of controlled or, more usually, uncontrolled data collection (i. e., monitoring of clinical experience) is undertaken. Such activity may require the participation of hundreds of physicians, each treating 10 or more patients, to determine the incidence of relatively rare side effects. For psychotropic, analgesic, and most other drug categories, efficacy data obtained during Phase IV will be of little scientific value because of the lack of controls in the data collection. If reports of failure of response are rare, however, the data may be taken as supportive of the efficacy of the drug.

3.5. Phase V

Phase V studies are defined here as those initiated after the marketing of a drug that are designed to evaluate comparative safety and efficacy relative to other marketed products. For example, a new analgesic might be compared to standard therapy in patients with pain syndromes.

Other studies, particularly those to explore the use of the drug in new patient populations and for new indications, might be initiated after the drug is marketed. For example, a β-receptor blocking agent with proven efficacy in patients with angina might be studied in Phase V for use in migraine headache. Such studies would be conducted under the original IND as additional Phase III trials.

4. Clinical Assessment of Safety

The adverse consequences or risks associated with the use of a particular drug may be divided into two categories.

1. Relatively common (>5% incidence), reversible unwanted effects, generally attributable to the pharmacological properties of the drug, that produce discomfort, but no serious consequences. The extent to which such effects occur determines the *tolerability* of the drug.
2. Serious and unexpected adverse effects, usually unrelated to the principle pharmacological properties of the drug, that occur infrequently (<1% incidence). Examples include aplastic anemia (chloramphenicol), hemolytic anemia (methyldopa), hepatic injury (ticrynafen), anaphylaxis (zomepirac), or neurological side effects (zimeldine). These adverse reactions are often idiosyncratic in nature, unpredictable from animal models, and non-dose-related.

During premarketing evaluation (Phases I–III), most drugs will be administered to 1–2 thousand patients. For drugs intended as chronic therapy, 50–100 patients will be treated for at least 6–12 mo. Among 16 NDAs submitted to the US Food and Drug Administration between 1975 and 1981 in four major drug categories [antimicrobials, nonsteroidal antiinflammatory drugs (NSAID), cardiovascular drugs, and psychopharmacological drugs], at least 580 patients and as many as 2294 were treated with the new drugs during Phase III clinical trials (Idanpaan-Heikkila, 1983). The average was lower for cardiovascular drugs (approximately 1100 patients) and higher for the NSAIDs (approximately 1600).

The number of patients exposed to a new drug during typical premarketing studies in the US is adequate to evaluate its acute tolerability. This is not a straightforward task, however, since the incidence and severity of side effects will depend upon the population studied, the duration of treatment, the design of the clinical trial, and the methodology used to collect adverse reaction data. For example, the incidence of dry mouth caused by an antidepressant drug may differ between inpatients and outpatients, and may be greater if specifically elicited by questionnaire than if recorded only after a spontaneous report by the patient. Furthermore, since dry mouth is commonly reported as a symptom of depression, the extent to which its occurrence can be attributed to an antidepressant medication can only be determined in adequately controlled trials. Clinical pharmacology studies may help to define the potential of a drug to produce certain adverse effects, including habituation, but the ultimate answers can come only through the treatment of sufficient numbers of patients.

The detection of uncommon but serious side effects presents a different set of problems. Such events may be detected as readily in uncontrolled clinical experience as in controlled trials, provided that sufficient numbers of patients are treated. When the event is rare, however, or if its rate of spontaneous occurrence in the population under consideration is only slightly increased by drug administration, hundreds of thousands of patients may need to receive treatment before its significance is fully recognized. Phase III, and even large-scale Phase IV, trials may not be adequate. Zomax (zomepirac sodium) was administered to an estimated 15 million persons before it was withdrawn by the manufacturer because of rare but severe anaphylactic reactions (Hollie, 1983). Eleven hundred such reactions (0.007% incidence) resulting in five deaths (0.00003% of all users) were responsible for the demise of an otherwise successful and beneficial drug.

Although it has been suggested that drug companies should "work to identify the characteristics of people likely to experience negative side effects" (Bezold, 1981), this remains much more of an ideal than a practical goal. Idiosyncratic hepatotoxicity, as observed, for example, with halothane, affects an extremely small percentage of exposed individuals, is not predicted by laboratory testing, and is not associated with known risk factors (Tillman and Koltz, 1981). The hepatotoxicity associated with the antidepressant zimeldine, a drug marketed briefly in Europe, occurred in an apparently random fashion in a small percentage of patients (Sommerville et al., 1982).

5. Assessment of Efficacy

In most countries, including the US, the efficacy of a new drug candidate in the population intended for treatment must be demonstrated to the satisfaction of the prevailing regulatory agency. Superiority over existing agents, which may be defined as an increased benefit-to-risk ratio, ostensibly is *not* required (Legrain, 1983). Nevertheless, most new drugs are designed as improvements over existing therapy, in terms of their therapeutic effects and/or safety profiles. The acceptance of a new drug over established and usually less expensive agents will depend on physicians' and patients' perceptions of its advantages over available therapy.

Assessments of efficacy should therefore be designed to detect clinically relevant differences not only between new drug and placebo, but between new drug and standard agents. In practice, it may be impossible to conduct studies that are large enough to demonstrate differences between two effective therapies. A common error in studies that compare new drugs to standards is the conclusion that both agents are "equally effective" or "therapeutically equivalent with respect to efficacy." Even large Phase III multiclinic trials often lack the statistical power to support such assertions. The observation that the two drugs are equivalent may simply result from the inability to observe meaningful differences as a result of inadequate sample size (Anello, 1974). Primarily for this reason, regulatory agencies are leaning toward placebo-controlled trials as the pivotal studies on which approval of a drug is based.

The demonstration of benefit relative to existing therapy will therefore depend, in most instances, on the better safety and tolerability of the new agent without loss of efficacy. Exceptions do occur, sometimes with enormous impact (e. g., cimetidine in ulcer disease, timolol in glaucoma). Clinical trials should incorporate standard agents as positive controls in Phase III or Phase V, with the hope rather than the expectation that superior efficacy can be demonstrated in terms of at least one parameter (speed of onset, percentage of patients responding, quality of response). Crossover studies incorporating a preference question may be of value in appropriate clinical situations, such as pain reduction, headache prophylaxis, or seizure control.

Whereas assessment of safety is relatively similar across therapeutic categories, the assessment of efficacy is much more dependent upon the illness to be treated. Endpoints may be relatively straightforward, as in the evaluation of an antihypertensive drug, or difficult to define, as in the treatment of anxiety or Alzheimer's dementia. Likewise, clinical assessments will be as varied as the illnesses for which treatments are sought. Blood pressure readings and other well-defined clinical observations at appropriate time intervals may provide evidence of efficacy for an antihypertensive agent, although for a psychotropic drug, batteries of standardized rating scales may be required. In the latter circumstance, rating scales must be validated as reasonable measures of change in appropriate patients prior to initiation of the drug trial.

Although specific assessments will vary, the following considerations may be applied to all measurements of efficacy (Schor, 1974).

1. Is the measurement clinically relevant? For example, tests of psychomotor function, although certainly important, are not enough to establish the efficacy of an antidementia agent if activities of daily living are unchanged.
2. Are the subjects appropriate in terms of the goals of the experiment? A comparison of the efficacy of a new antidepressant relative to imipramine could not be conducted fairly in patients refractory to tricyclic antidepressants, yet this population is frequently used.
3. Do the rating methods allow for meaningful comparisons? The Hamilton Depression Scale incorporates sleep disturbance as a measure of depression, thus biasing the outcome in favor of sedative antidepressants.

6. Clinical Trial Methodology

6.1. Study Design

A well designed clinical study will evaluate the safety and efficacy of a new drug in the context of its intended use. Factors to be considered in the selection of a study design include the developmental stage of the test compound, the natural history of the disease to be studied, and currently available treatments.

Most clinical studies can be categorized as either:

- Open-label or blinded
- Uncontrolled or controlled
- Parallel or crossover

Additionally, controlled studies may be:

- Single-blind or double-blind
- Placebo and/or active controlled

In open-label studies, both the patient and the investigator know the identity of the test agent. During the early stages of the development of a drug, such studies are useful to evaluate whether the compound is reasonably well tolerated and potentially effective in a limited patient population. Although rarely definitive, conclusions from open-label studies are useful in the choices of proper study designs, including dosage regimen, for the subsequent controlled trials. Changes relative to baseline or the expected course of the disease may provide some evidence of efficacy, for instance, in severely ill patients in whom little or no placebo response is anticipated. Although usually uncontrolled (i. e., no comparative treatment group is included), open-label studies occasionally employ a concurrent control group.

Open-label designs may also be useful in pharmacokinetic and metabolism studies, or in mass-balance studies in which few clinical assessments will be made.

Studies from which definitive data are sought are generally "blinded," referring to the fact that the identity of the treatment drug is unknown to some or all of the participants. The blinding technique is used to reduce the subjective bias that often occurs, in spite of the best attempts at objectivity, in open-label studies.

In a single-blind study, the investigator knows which patients are receiving the test drug and which are receiving the control agent. The patient, however, does not have access to this information. The single-blind design allows the clinician to titrate dosage with as much information as possible, to pay particular attention to patients receiving placebo, and to form early impressions of the safety and efficacy of the test agent. The principal shortcoming of this design is its inability to exclude investigator bias.

Double-blind studies are designed so that neither the investigator nor the patient knows who is receiving study drug and who is receiving control agent(s). The test drug and control agent(s) must be identical in appearance and taste. This design eliminates potential bias by the investigator and the patient, and is essential in most drug trials that are intended to generate substantial evidence of efficacy.

Controlled studies may employ placebo and/or active reference agents. As previously noted, if the latter are used the study will have low statistical power and will probably find "no difference" between the efficacy of new drug and reference agent, although a clinically meaningful difference may exist. In conditions such as anxiety and depression, active-controlled studies are generally inadequate to establish efficacy.

Crossover studies are those in which each patient receives two or more different treatments (e. g., control agent and new drug) during the trial. Clinical pharmacology studies frequently employ crossover designs, which are well suited to acute studies of pharmacological or pharmacokinetic parameters. In clinical trials, crossover designs are useful only in conditions that change little over time, such as intractable epilepsy or perhaps recurrent migraine.

In simple two-way crossover studies, after a baseline period half the patient population receives one of the possible treatments (e. g., test drug or placebo), whereas the other half receives the alternative. After a specified interval the two groups switch treatments. A washout period should be interspersed between the two treatments to minimize rebound or carry-over effects and to reestablish the baseline.

Crossover studies allow each patient to receive the test drug in addition to the control treatment(s), and thus each patient serves as his or her own control. Such studies gain statistical power, but are vulnerable to potentially confounding effects, such as dropouts and variability of the illness over time. As a result, one may observe differences based on the sequence in which test agents were administered. In a crossover study of the β-blocker timolol maleate in migraine headache prophylaxis, although timolol was clearly superior to placebo, patients who received placebo followed by timolol responded better to each treatment than did those who received timolol followed by placebo (Stellar et al., 1984). Longer duration of each treatment period might have been helpful in that particular case.

Parallel designs are those in which subjects receive only one treatment (e. g., test medication or placebo) throughout the course of the study. Typically, after a baseline washout period, one group receives test drug and the other group(s) receive control agent(s). Both placebo and active drug controls may be used in three-way comparison studies. Parallel studies are less vulnerable to the disrupting effects of dropouts or variation of the illness over time, and require less time per patients for completion. They require larger numbers of patients than crossovers, however, and may be more adversely affected by nonhomogeneity of treatment groups. Acute, nonrecurrent conditions such as posttraumatic muscle spasm require parallel rather than crossover designs.

6.2. Randomization

It is essential to attempt to enroll comparable treatment groups through random assignment of patients to each group. In crossover studies, randomization determines the order in which treatments are administered, whereas in parallel studies it establishes the sole treatment for each patient. Randomization should be performed *after* any baseline period to ensure that dropouts during baseline do not skew the initial composition of the groups. In many studies, simple randomization of patients into the two (or more) treatment groups will suffice. Some situations will require separate randomization of patients who are first stratified according to one or more key parameters (e. g., age or sex), in order to reduce the disparity between groups at entry.

6.3. Study Duration and Schedule of Observations

The disease state to be studied and the extent of prior experience with the new drug usually dictate the frequency of observations for efficacy and safety, as well as the duration of the study. For instance, to assess the efficacy of a new muscle relaxant, one may wish to evaluate the patient at baseline and again after 2, 4, 7, and 10 d of treatment. By this time, further evaluations are unlikely to be fruitful. In contrast, treatment with an antidepressant should continue for at least 4–6 wk to establish the full efficacy of the agent and to obtain some information on the persistence of the response. Anticonvulsant studies may require 6–12 mo before changes in the frequency of seizures can be measured adequately, particularly in patients who have infrequent seizures.

Factors to be considered in scheduling safety observations are the stage of evaluation and total experience with the test drug, severity of illness, and likelihood of occurrence of adverse events.

6.4. Patient Selection

To obtain an appropriate and reasonably homogeneous treatment population, patient selection must be based upon specific diagnostic and eligibility criteria. Those subjects who are studied must be as representative as possible of the patient population who will receive the new drug in actual clinical practice. However, in a clinical trial, patient safety and the need to demonstrate drug efficacy may take precedence over other considerations. For example, antidepressant studies generally include only patients with major affective disorder of moderate or greater severity, since such patients are sufficiently ill to improve substantially and are less responsive to placebo than those with mild depression. As further entry requirements, patients should achieve minimum depression scores on appropriate rating scales. Those who are elderly or who have medical illnesses may be excluded for reasons of safety. Ironically, in actual practice most patients who receive antidepressant drugs have mild forms of depression and many have other illnesses.

In addition to specific diagnostic criteria, general eligibility factors that influence patient selection include gender, age, risk of pregnancy, secondary diagnosis, and patient reliability.

Exclusion as well as inclusion criteria must be well defined and adequate to protect persons at risk from exposure to the drug and to prevent those with confounding conditions from entering the study.

6.5. Number of Subjects (Sample Size)

The number of subjects required for any clinical trial is dependent on the probability of obtaining statistically significant differences from baseline (uncontrolled studies) or between groups (controlled studies) in efficacy and safety parameters. In estimating sample size, the biostatistician should consider the number of patients expected to remain in the study for sufficient time to provide meaningful data. The larger the variability in patient response and the smaller the expected differences between test drug and control, the greater will be the number of patients required. If the sample size required to draw significant conclusions is large, a multiclinical trial will be necessary. This is commonly the case for analgesic, antiarrhythmic, antihypertensive, anticonvulsant, and psychotropic agents.

Clinical trials with insufficient power to assess drug effects may lead to unwarranted termination of a potentially useful compound. More commonly, misleading results will lead to selection of an ineffective compound for further development.

6.6. Case Report Forms

Case report forms (CRF) or data collection forms are the most important documents generated in a clinical trial. CRFs are a medium, in the case of a pharmaceutical company-sponsored trial, to transfer data from the study site to the sponsor for data processing. In principle they should record all information essential to the evaluation of the drug, and facilitate the data entry into the computer. CRFs also serve as a check of adherence to the protocol. Well-designed forms will help the investigator to execute the study according to protocol and simplify the verification of the data.

Some pharmaceutical companies provide CRFs as workbooks so that the investigator may record all relevant clinical evaluations as they are performed, and then transcribe the data to final forms. These are maintained at the study site, and copies are submitted to the sponsor on a regular basis.

Because certain types of data, particularly with regard to drug safety, are common to all studies, many companies have designed standard forms for uniform data collection. Modular CRFs, defined as units of related data elements such as vital signs that can be assembled into finished forms, have also been proposed (Cato and Cook, 1984). Many companies are experimenting with electronic data processing, including the use of optical scanners to transfer data from CRFs directly into computer files. Microcomputers at study sites are becoming more common for input of raw data, which may then be transmitted electronically or as printed copy.

6.7. Conduct and Monitoring of Clinical Trials

In anticipation of the initiation of a clinical trial, qualified investigators must be identified. Professional representatives or regionally based clinical research associates may be useful in recommending prospective investigators. A prestudy visit by the clinical monitor is necessary to determine the investigator's qualifications, the availability of a suitable patient population for the disease intended to be studied, and the adequacy of the physical facility. A research coordinator, usually a nurse, is invaluable in managing the trial from an administrative viewpoint, and should be included in the site visit. Financial arrangements should be discussed. The investigator and his or her staff are generally compensated for their time, effort, and expense in rigorously following the protocol, obtaining required tests, and completing case record forms and reports (Shapiro and Charrow, 1985).

Once a commitment to the program has been made by sponsor and investigator, the study protocol and the patient consent form are submitted by the investigator to a human research committee (HRC) or an institutional review board (IRB). Federal regulations state that all human studies require the approval or a local HRC or IRB (Code of Federal Regulations, 1981). The committee must review protocol and patient consent forms initially and every 12 mo thereafter during continuation of the clinical trial.

The sponsor usually supplies the investigators with workbooks, case report forms, and individually packaged test medications. Proposed federal regulations suggest that the study be monitored on a regular basis, generally every 4–8 wk (Code of Federal Regulations, 1977), either by the clinical monitor or a suitable representative. During the regular visits, the monitor will review the progress of the trial, focusing on compliance with the protocol, rate of patient enrollment, dropouts, adverse reactions, and clinical issues that invariably arise. Independent of monitoring visits, serious adverse events must be reported to the FDA within 15 working d, regardless of their presumed relationship to test drug. Throughout the study, the monitor or another technically qualified person should compare the clinical data on the CRFs to the patient's source records for accuracy of laboratory data and clinical observations.

At the conclusion of the study, all unused investigational drug supplies must be returned to the sponsor. All original records must be retained by the investigator. This is particularly critical since FDA audits may occur well after a study has been completed.

7. Ethical and Legal Considerations

Clinical drug development is broadly guided by the principles elucidated in the Declaration of Helsinki in 1964 (World Medical Assembly, 18th, 1964). Specific guidelines for the protection of human subjects have also been issued in the US (Federal Register, 1981).

As noted previously, in most countries authorization must be obtained from the drug regulatory authority prior to human testing of a new chemical entity. To obtain such authorization (IND) in the US, extensive preclinical data, a protocol that provides for adequate safeguards during human testing, and qualifications of the investigator to conduct clinical research must be presented. Fulfilling these requirements is merely the first step, however, in addressing the moral and legal aspects of the proposed research. Although too subtle and numerous to be considered in any detail at present, two of these issues deserve special mention. Informed consent and the use of placebos are among the most critical considerations in drug development (Hodges, 1974).

Obtaining informed consent is fundamental to the proper conduct of any type of clinical research. Information that must be imparted to potential subjects includes the aims, methods, anticipated benefits and potential hazards of the study, and acknowledgment of the right of the subject to not participate or to withdraw from the study without prejudice. Documentation of informed consent may be of great importance in legal proceedings that arise from clinical experimentation. Although this issue may appear to be straightforward, problems arise when patients have impaired judgment or cognitive deficits, sometimes caused by the condition to be studied. This is particularly relevant in psychiatric conditions, such as schizophrenia or dementia, in which the ability of the patient to make an informed decision is questionable. Even in a population of fully informed normal volunteer medical students participating in the first human studies of a new agent, we recently observed during the course of the study that most of the subjects assumed the drug had been previously administered to others. In spite of the explicit statements that had been reviewed in the consent form, they were surprised when we reinformed them after the second treatment period that they were the first people to receive the drug.

The use of placebos is scientifically the most rigorous method for separating the specific pharmacological effects of a drug from nonspecific effects, such as natural history of the disease treated, response to hospitalization or other environmental factors, or psychological effects of receiving treatment. Placebo effects have been observed in the treatment of almost every illness, and are mediated by both psychological and physiological factors (Haegerstram et al., 1982). Such effects may be beneficial or adverse, and may thus contribute to both sides of the benefit vs risk equation.

Scientific merit notwithstanding, ethical considerations often preclude the use of placebos. Patients with illnesses with significant morbidity and mortality, which are adequately treated by standard agents, cannot justifiably receive placebo as their only treatment. For example, the conclusions from the Framingham study (McGee and Gordon, 19786) persuaded most clinicians that the use of placebos in patients with moderate to severe hypertension is unethical.

Illnesses that tend to be self-limited, or for which no effective treatments are available, and in which the potential risk of injury to the patient is low, may justifiably be treated with placebos. Anxiety is an example of the former, whereas Alzheimer's disease represents the latter. Some debate exists over the use of placebos in patients with depression, with most opinion in the US in favor of their use in definitive studies that exclude suicidal patients. Even with such safeguards, this practice encounters considerable opposition in Europe. The conduct of a placebo-controlled inpatient study poses additional dilemmas because of the issue of reimbursement for hospital stay and reluctance of staff to administer inactive agents. A 4-wk depression study would typically cost $12,000 *per patient* in bed costs alone. Few companies are able to incur such costs routinely. Veterans Administration and state-funded hospitals, in addition to NIH-funded research units, are alternatives in the US to the use of private hospitals for such studies. Creative study designs that minimize hospitalization time and allow for prompt discontinuation of nonresponders must be employed.

8. New Drug Registration

8.1. Preparation of the New Drug Application in the US

Assuming that the outcome of all aspects of the safety and efficacy evaluations are satisfactory, the drug sponsor will submit a New Drug Application (NDA) to the FDA after Phase III trials are completed. Approval of the document authorizes the release of the drug for general use in the US.

Although specific requirements differ, a well-prepared NDA targeted for submission in the US will contain the essential information required by other major countries. The US NDA is a comprehensive presentation of the data derived from preclinical and clinical studies with the test drug. These data are to be organized and summarized according to a specific format, which includes an overall summary of the entire application and separate, detailed technical sections, each containing individual summaries and analyses of specific information in the following areas: clinical, preclinial (animal), chemistry, statistics, and biopharmaceutics (as well as microbiology for antiinfective drugs). The chemistry section may be submitted for review in advance of the main application. Subsequent reports updating safety data on the drug must also be filed 4 mo following the initial submission of the application, following receipt of an "approvable" letter, and, additionally, whenever requested by the FDA. The latter requirement reflects the fact that the FDA is responsible for continuing surveillance of marketed drugs and that 2% of new chemical entities introduced between 1964 and 1983 had to be withdrawn from the market for safety questions (Bakke et al., 1984).

A detailed description of the requirements for the content and format of an NDA is provided in the New Drug and Antibiotic Regulations issued in February, 1985 (Federal Register, 1985). The remainder of this section will focus on the formidable task of preparing the clinical section of the document, which may span several hundred volumes of text in its entirety. Successful completion of the clinical portion of the NDA requires a group of people who have experience in clinical pharmacology, medicine, biostatistics, data processing, and regulatory affairs.

The preparation actually begins as soon as human data are obtained. As the case report forms are received from each investigator, the monitor will review them for serious clinical adverse reactions and completeness; the encoders will then input the appropriate data into the computer. A clinical summary of each study, which accounts for every test subject, is generated. All adverse experience and laboratory safety data must be computerized for later combined presentation. Efficacy data, at least for the larger clinical trials, must also be accessible for later reanalysis and combination. Because of the magnitude of the task, it is essential that the database be generated well in advance of the target date for NDA submission. In this way data from the Phase III studies can be processed as they are received.

After completion of all individual and multiclinic study summaries, an overall summary of the Phase I, II, and III studies must be prepared. The clinicians can then document the therapeutic benefits and risks of the drug in the population for which it will be indicated and provide a thorough discussion of adverse experiences.

When the clinical portion of the NDA is in its final stages, the physician, preclinical scientists, regulatory personnel, legal staff, and marketing team will prepare the proposed new drug labeling based on the preclinical and clinical data. This material, as modified to incorporate any revisions resulting from negotiations with the FDA, comprises the ''package circular'' contained in the *Physician's Desk Reference* and in every drug package sent to the pharmacist or physician. The package circular includes the following major sections: description; clinical pharmacology; indications and usage; warnings; precautions; adverse reactions; drug abuse and dependence; management of overdosage; dosage and administration; contraindications; and how the drug is supplied.

8.2. NDA Review in the US

In 1970, the FDA initiated a classification system for INDs involving new chemical entities that ranks them according to the FDA's assessment of their potential therapeutic importance. If the investigational drug is considered to be a major therapeutic advance, the compound will receive an ''A'' classification. A ''B'' classification represents a modest or moderate therapeutic advance, and ''C'' refers to little or no therapeutic gain. The company may request that the classification be upgraded as new data become available. Priority of FDA review is dependent upon this classification (Finkel, 1984).

Many aspects of the FDA review process are actually ongoing from the time of IND submission through final NDA approval or nonapproval. The FDA provides feedback regarding design of clinical protocols as they are submitted, and notes any safety concerns. Divisions of the FDA encourage sponsors to meet with them regarding the design of Phase III and other clinical studies, based on preclinical and Phase I and II experience with the drug. The meetings may involve not only the sponsor and FDA reviewers, but also a member of the appropriate Advisory Committee. There are 15 such committees that advise the Bureau of Drugs, consisting primarily of experts in medicine, pharmacology, and statistics.

Thus by the time the NDA is submitted, certain FDA members will already be familiar with the drug. After formal receipt of the NDA, the FDA is obligated to decide within 60 d whether the application is suitable for review. If so, the FDA is then obligated to complete the bulk of its review within 180 d of the submission date. However, the 180-d limit is extended each time a substantial amendment to the NDA is requested by the FDA or voluntarily submitted by the sponsor. Following its initial (180-d) review, the FDA may ask the appropriate Advisory Committee to help determine whether the NDA contains sufficient evidence to prove, based on well-controlled clinical studies, that the investigational drug is safe and effective for its intended use. The sponsor participates in this process by presenting its case to the Advisory Committee, which then makes a recommendation by vote on the approvability of the drug (Finkel, 1984). All final decisions rest with the FDA, which considers, but is not bound by, the recommendations of the Committee.

Table 1 demonstrates the approval time of representative NDAs submitted by one firm to the FDA between 1973 and 1983 (Berglin, 1985). It also illustrates the magnitude of these documents, some of which required a truck for shipment to the FDA.

TABLE 1
Comparative Data for Selected Major NDAs

Drug agent	No. of volumes	Approximate no. of pages	Classification	Approval time
Antidepressant	427	168,413	C	NDA withdrawn
Antihypertensive	338	132,000	Not done	Under review
Antiinflammatory	307	125,000	Not done	27 mo
Analgesic	160 (5 microfiche)	58,600	B	22 mo
Antibiotic	139	52,800	Not done	14 mo
Ophthalmic	99	36,000	A	6 mo
Antihypertensive	90 (2 microfiche)	30,800	B	20 mo

A more general survey of the total clinical development time, including NDA review, was conducted for 16 new drugs for which NDAs were submitted by various companies between 1975 and 1981 (Idanpaan-Heikkila, 1983). As indicated in Table 2, the mean time interval between the first human studies of these new chemical entities (NCEs) and their final approval for marketing in the US was about 8.2 yr, which included an average of 29.1 mo for NDA review. The 1985 revised regulations offer the hope of reducing the review time, perhaps by as much as 6 mo. The physical size of NDAs will also be reduced, since sponsors are now required to submit actual CRFs only for patients who had serious adverse experiences, including any deaths, or discontinued treatment for adverse experi-

TABLE 2
Clinical Development of a Selected Group of Drugs[a]

Type of drug (NDA)	Months for phase I–II studies	Months from IND receipt to NDA receipt	Months from NDA receipt to NDA approval/non-approval	Months from IND receipt to NDA approval/non-approval
Antimicrobials				
Mean	17.3	31.6	12.8	44.4
Range	(4.6–35.3)	(23.0–40.7)	(12.0–13.6)	(35.0–53.2)
Nonsteroidal anti-inflammatory drugs				
Mean	52.7	66.4	21.2	87.6
Range	(43.3–67.9)	(53.0–84.7)	(15.4–26.9)	(75.3–100.1)
Cardiovascular drugs				
Mean	50.7	82.3	20.3	102.6
Range	(16.6–74.0)	(39.0–150.8)	(14.2–26.4)	(53.1–170.3)
Psychopharmacological drugs				
Mean	64.1	95.1	62.3	157.4
Range	(41.2–87.3)	(54.6–132.6)	(23.8–85.1)	(139.7–196.3) 196.3)
Total mean	46.2	68.9	29.1	98.0

[a]Adapted from Indanpaan-Heikkila (1983).

ences during the clinical studies. Case report forms from pivotal or other studies may later be requested by the FDA.

Sheck et al. (1984) have studied the success rate of NCEs developed as new drug candidates in the US. Figure 1 depicts the cumulative success rate as a function of the time between IND filing and NDA approval for a large number of NDAs submitted between 1963 and 1974.

During the past 20 years in the US, only about 10% of new chemical entities that *were studied in humans* have resulted in new drugs. For the successful drugs, the total development time, including FDA review and approval, ranged from 2 to 17 years.

Among the 90% of unsuccessful compounds were many clinically effective agents that nevertheless failed as drugs because

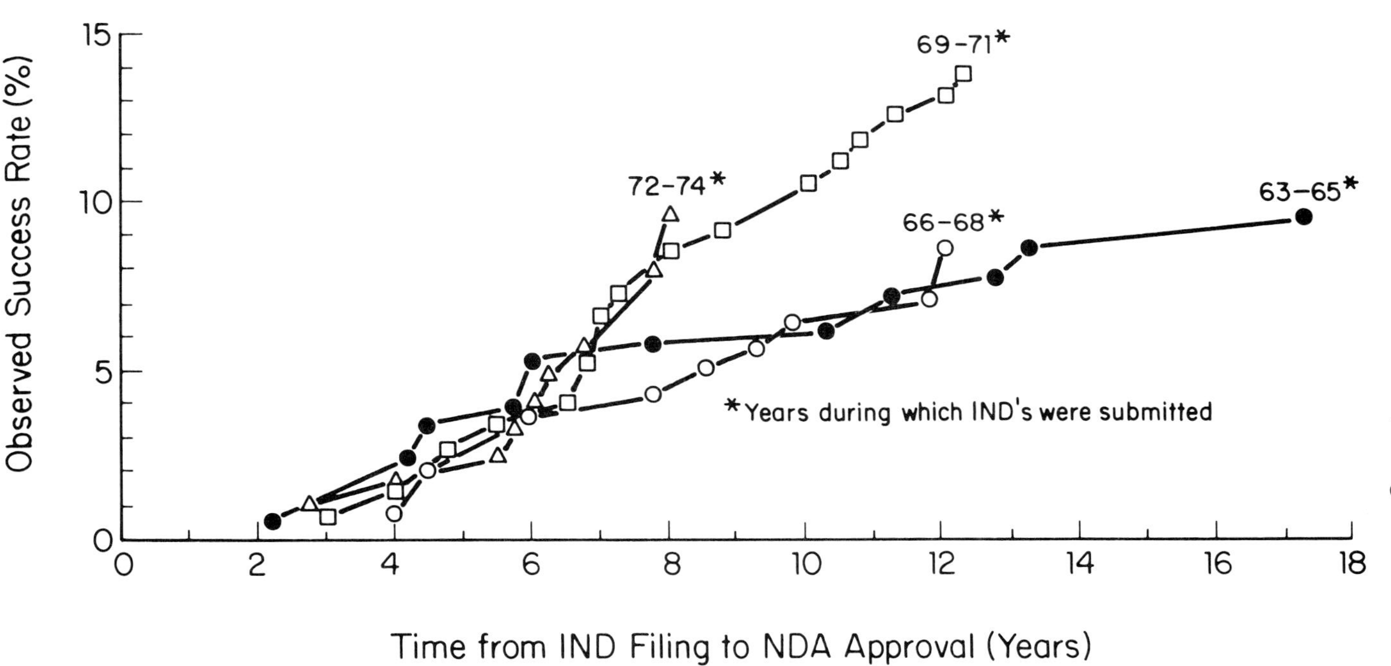

Fig. 1. Observed success rates for four cohorts of self-originated US NCE-INDs.

of undesirable effects in the patients who received them. It is likely that many other compounds were rejected because their anticipated therapeutic effects were absent or not adequately characterized, and that fewer compounds failed because they did not exhibit the pharmacological properties predicted from preclinical testing.

8.3. International Drug Registration

The required content and format of new drug applications varies throughout Western Europe, Canada, and other developed countries. There is much less variability within the developed countries, however, than there is between the developed and non-developed countries. Many companies attempt to prepare a ''European NDA'' that can serve with minor modifications for registration throughout Western Europe and Canada.

In an attempt to expedite and shorten drug-approval times, while retaining emphasis on the documentation of safety and efficacy, there is a tendency toward increased acceptance of international (i. e., foreign) data by the Western European regulatory agencies. This does not, however, apply to Japan, which has restrictive laws excluding most types of data generated outside of Japan. Among the Western European countries, some (e. g., France) tend to rely on the testimony of experts, whereas others pattern themselves after the FDA and require more objective evidence of safety and efficacy. Table 3 lists the drug regulatory agency responsible for the evaluation of new drug candidates in various major countries outside of the US.

References

Anello, C. (1974) Considerations of Significance—Clinical and Statistical, in *Principles and Techniques of Human Research and Therapeutics* vol. IV (McMahon, F. G., ed.) Futura, Mt. Kisco, New York.

Bakke, O. M., Wardell, W. M., and Lasagna, L. (1984) Drug discontinuations in the United Kingdom and the United States, 1964 to 1983; Issue of safety. *Clin. Pharmacol. Ther.* **35(5)**, 559–567.

Berglin, P. (1985) Personal communication.

Bezold, C. (1981) *The Future of Pharmaceuticals* Wiley, New York.

Cato, A. E. and Cook, L. (1984) The Protocol and Case Report Form, in *The Clinical Research Process in the Pharmaceutical Industry* (Matorem, C. M., ed.) Marcel Dekker, New York.

TABLE 3
List of Regulatory Authorities Responsible for Registration of New Drugs

Country	Authority responsible for registration
Australia	Commonwealth Department of Health (Sydney)
Canada	Bureau of Pharmaceutical Prescription Drugs, Health Protection Branch (Ottawa)
France	Ministere des Affaires sociales et de La Solidarite nationale, Direction de La Pharmacie et du medicament (Paris)
Italy	Ministero della Sanita Servizio Farmaceutico (Rome)
Japan	Pharmaceutical Affairs Bureau Ministry of Health and Welfare (Tokyo)
Netherlands	College ter beoordeling van geneesmiddelen Ministerie van Volksgezondheid (Rijswijk)
Scandinavia	Nordiska Lakemedelsnamndan Nordic Council on Medicine (Sweden)
United Kingdom	Department of Health and Social Security Medicines Division (London)
West Germany	Institut fur Arzneimittel des Bundesgesundheit-samtes (Berlin)

Code of Federal Regulations (1977) 21 (52) Proposed in *Federal Register,* pp. 49612–49630.

Code of Federal Regulations (1981) 21 (56).

Federal Register (1981) 46 (17). p. 8950.

Federal Register (1985) 50 (36), New Drug and Antibiotic Regulations.

Finkel, M. J. (1984) Role of the FDA in the Clinical Research Process, in *The Clinical Research Process in the Pharmaceutical Industry* (Matoren, G. M., ed.) Marcel Dekker, New York.

Friedman, L. M., Feinberg, C. D., and DeMets, D. L. (1984) *Fundamentals of Clinical Trials,* John Wright PSG, Littleton, Massachusetts.

General Considerations for the Clinical Evaluation of Drugs (1978) US Dept. of Health, Education and Welfare, Public Health Service, Food and Drug Administration.

Haegerstram, G., Huitfeldt, B. S., Nilsson, B., Sjovall, J., Syvalahti, E., and Whalen, A. (1982) Placebo in Clinical Drug Trials—a Multidisciplinary Review, *Meth. Find. Exp. Clin. Pharmacol* **4(4)**, 261–278.

Hodges, R. (1974) Ethical Considerations in Clinical Research, *Techniques of Human Research and Therapeutics* vol. 1 (McMahon, F. G., ed.) Futura Publishing, Mt. Kisco, New York.

Hollie, P. G. (1983) Johnson & Johnson: New Woe, *NY Times.*

Idanpaan-Heikkila, J. (1983) *A Review of Safety Information Obtained from Phases I-II and Phase III Clinical Investigations of Sixteen Selected Drugs.* US Dept. of Health, Education and Welfare, Public Health Service, Food and Drug Administration.

Inman, W. H. H. (1981) Post-Marketing Drug Surveillance, in *Risk-Benefit Analysis in Drug Research* (Cavalla, J. F., ed.) MTP, Lancaster, England.

Legrain, M. (1983) New Drug Superiority, in *Decision Making in Drug Research* (Gross, F., ed.) Raven, New York.

Matoren, G. M., ed. (1984) *The Clinical Research Process in the Pharmaceutical Industry* Marcel Dekker, New York.

McGee, D. and Gordon, T. (1976) in *The Framingham Study: an Epidemiological Investigation of Cardiovascular Disease* Publication No. (NIH) 76-1083, US Dept. of Health, Education, and Welfare, Washington, DC.

New Drug and Antibiotic Regulations (1985) in *Federal Register* 50 (36), pp. 7452–7519, Proposed rules will change the name of the IND to "Investigational New Drug Application."

Public Law (1962) No. 87-781 (76 Stat. 780).

Schor, S. (1974) Relevant Considerations to the Statistical Analysis of Clinical Data, in *Principles and Techniques of Human Research and Therapeutics* vol. IV (McMahon, F. G., ed.) Futura, Mt. Kisco, New York.

Shapiro, M. F. and Charrow, R. P. (1985) Special report: Scientific misconduct in investigational drug trials. *N. Eng. J. Med.* **312(11)**, 731–736.

Sheck, L., Cox, C., Davis, H. T., Trimble, A. G., Wardell, W. M., and Hansen, R. (1984) Success rates in the United States drug development systems. *Clin. Pharmacol. Ther.* **36(5)**, 574–583.

Sher, S. P., Bokelman, D. L., and Ditzler, W. D. (1980) Preclinical toxicity requirements for human drugs. *Drug Inform. J.* April/June, 82–97.

Smithells, R. W. (1962) Thalidomide and malformations in Liverpool. *Lancet* **i**, 1270–1273.

Spilker, B. (1984) *Guide to Clinical Studies and Developing Protocols*, Raven, New York.

Sommerville, J. M., McLaren, E. H., Campbell, L. M., and Watson, J. M. (1982) Severe headache and disturbed liver function during treatment with zimelidine. *Brit. Med. J.* **285**, 1009.

Stellar, S., Ahrens, S. P., Meibohm, A. R., and Reines, S. A. (1984) Migraine prevention with timolol. *J. Am. Med. Assoc.* **252(18)**, 2576–2580.

Tillman, C. R. and Koltz, B. E. (1981) Update on hepatotoxic drugs. *Hospital Formulary* **16**, 847–852.

World Medical Assembly (1964) 18th, Helsinki, Finland, as revised by the 29th World Medical Assembly, Tokyo, Japan, 1975.

Therapeutic Entities — From Discovery to Human Use

Cimetidine and Other Histamine H_2-Receptor Antagonists

Roger W. Brimblecombe
and C. Robin Ganellin

1. Peptic Ulcer Disease

The treatment of peptic ulcer disease (both duodenal and gastric ulcers) and other acid-related diseases has been revolutionized by the discovery of the antagonist drugs that act at the histamine H_2-receptor.

Peptic ulceration is the most common disease of the gastrointestinal tract and it is estimated that approximately 10–20% of the adult male population in Western countries will experience a peptic ulcer at some time in their lives. In 1970 for example, in the United States there were some 3.5 million peptic ulcer sufferers and 8600 deaths attributed to this disease.

Duodenal and gastric ulcers are localized erosions of the mucous membrane of the duodenum or stomach, respectively, that expose the underlying layers of the gut wall to the acid secretions of the stomach and to the proteolytic enzyme pepsin. What causes acute peptic ulcer is still not properly understood, but for many years the main medical treatment has been aimed at reducing acid production.

The stomach contains many different types of highly specialized secretory cells that are under the control of both nervous and endocrine mechanisms. Among these, the parietal cells secrete hydrocholoric acid. Prior to and during a meal, the volume of acid, pepsin, and mucus secretion increases to as much as 10 times the basal secretion rate, and the pH may fall to 1–2.

2. Chemical Messengers and the Search for New Antiulcer Drugs

Secretion of gastric acid is mediated by the autonomic nervous system via the vagus nerves, which provide parasympathetic innervation to the stomach and small intestine; the neurotransmitter released by stimulation of the vagus is acetylcholine. Branches of the vagus, innervating the antral region of the stomach, stimulate the release of the peptide hormone gastrin from special gastrin-producing "G"-cells. The presence of food in the stomach further stimulates release of gastrin, which passes into the blood stream and is carried to the parietal cells where it acts to stimulate hydrochloric acid secretion. In addition to acetylcholine and gastrin, a third chemical secretagogue—histamine—is known to be involved.

The relationship between the three secretagogues—acetylcholine, gastrin, and histamine—has been a source of considerable controversy among physiologists for many years. When it was found (in the late 1940s and early 1950s) that the antihistamine drugs did not reduce acid secretion (*see* Loew, 1947), the role of histamine was seriously questioned (e.g., Johnson, 1971). The discovery of the H_2-receptor histamine antagonists, however, clearly established that it plays a vital part in acid secretion (Code, 1974).

For many years the main medical treatment of peptic ulcers relied on the use of antacids to neutralize the gastric acid. Anticholinergic drugs (to block acetylcholine transmission) can decrease gastric acid secretion, but their use in the treatment of peptic ulceration is limited by "side effects," e.g., dry mouth, blurred vision.

When drug treatment is unsuccessful, then surgery may be required. This is designed to remove part of the acid-secretory and gastrin-producing regions of the stomach (e.g., partial gastrectomy) or to selectively cut the branches of the vagal nerve (e.g., selective vagotomy) that supply the acid-secretory region.

With this background to treatment, it is not surprising that most major pharmaceutical companies set up research programs aimed at discovering antiulcer agents, and in the 1960s, with an increased understanding of the physiology of gastric acid secretion, several companies homed in more specifically, seeking to discover drugs that would block the action of one or more of the chemical messengers.

3. Pharmacological Characterization of Histamine Receptors

Histamine is a locally acting transmitter substance that acts at specific sites that are characterized pharmacologically as receptors, although there is still very little knowledge of their molecular nature. The receptors are classified pharmacologically by means of antagonists.

The antihistamines developed in the 1940s were found to be specific competitive antagonists of the action of histamine in stimulating the contraction of smooth muscle from various organs such as the guinea pig ileum, bronchus, and uterus. It was found, however, that these drugs did not block all the pharmacological actions of histamine. They only partially blocked the vasodilator effects of large doses of histamine and completely failed to inhibit histamine-stimulated gastric acid secretion. Subsequently, two particular actions of histamine were also found to be resistant to the antihistamines, i.e., stimulation of the rate of spontaneous beating of the guinea pig isolated right atrium and inhibition of evoked contractions of the rat isolated uterus.

These facts suggested the existence of more than one type of histamine receptor, and prompted Dr. (now Sir) James W. Black to initiate in 1964 a research program in the SK&F laboratories (in the UK) to discover a specific antagonist of the histamine receptors resistant to the conventional antihistamine drugs. Two years later, Ash and Schild (1966) defined the histamine H_1-receptor as that which mediates those responses to histamine that can be competitively antagonized by antihistaminic drugs such as pyrilamine. Since there was no known antagonist for the other systems, they were left unclassified as involving non-H_1 receptors.

Attempts at SK&F to obtain H_2-receptor antagonists started with histamine and chemical modifications that seemed potentially capable of providing an antagonist. Some of the approaches taken have been summarized previously (Ganellin et al., 1976; Ganellin, 1978). The idea was to seek a molecule that would compete with histamine for its receptor site; it was anticipated that such a molecule would have to be recognized by the receptor, then bind more strongly than histamine, but not trigger the usual response.

Compounds were tested for their ability to inhibit histamine-stimulated gastric acid secretion in anesthetized rats. Routinely, the stomach of an anesthetized rat [starved and pretreated with atropine (anticholinergic) and urethane as anesthetic] was perfused with glucose solution at 37°C. The perfusate was introduced via a tube placed in the esophagus and collected via a funnel placed in the nonsecretory lumen of the stomach; the perfusate was then passed through a microflow type glass electrode system and changes in gastric pH were recorded continuously on a potentiometric pen recorder.

Initially, compounds were administered as a constrant iv infusion in the middle of a series of fixed-dose histamine responses (given by rapid iv injection). Subsequently, the system was modified and a plateau of gastric acid secretion was established by continuous iv infusion of histamine (at a dose high enough to produce a near-maximal response), and potential inhibitors were then given by rapid iv injection. Since other types of inhibitors of gastric secretion could also be picked up by this test, compounds found to be active were also tested on the isolated tissue systems to provide additional criteria for specific antagonism to histamine, i.e., on the guinea pig right atrium and the rat uterus.

4. Discovery of H_2-Receptor Antagonists

4.1. Background to Cimetidine

The search for an antagonist proved to be difficult and during the first 4 years of the program some 200 close structural analogs of histamine were synthesized and tested without indications of specific antagonism being found. The "breakthrough" to an antagonist came with the discovery that the guanidine analog of histamine, N^{α}-guanylhistamine (Fig. 1) was a partial agonist that at high doses antagonized near maximal histamine-induced gastric acid secretion (Durant et al., 1975). The development of the highly specific histamine antagonists took several more years, and many different chemical structures were investigated based on various forms of structure–activity analysis (Ganellin, 1981). The essential steps were to lengthen the side chain and replace the charged guanidine group by an uncharged thiourea group, to eventually provide the drug burimamide (Fig. 1).

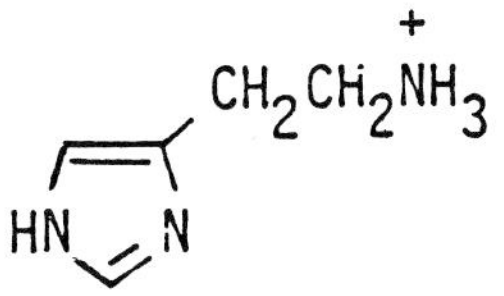

$\overset{+}{CH_2CH_2NH_3}$ (imidazole ring: HN N) — Histamine monocation

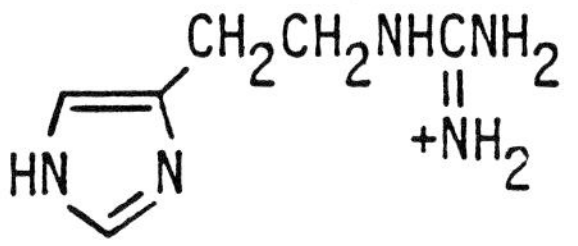

$CH_2CH_2NHC(=\overset{+}{N}H_2)NH_2$ (imidazole ring: HN N) — N^{α}-Guanylhistamine monocation, the "lead" compound (a partial agonist)

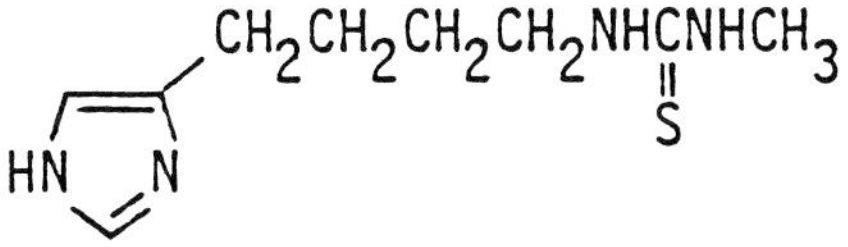

$CH_2CH_2CH_2CH_2NHC(=S)NHCH_3$ (imidazole ring: HN N) — Burimamide, the first characterized H_2-receptor histamine antagonist

Fig. 1. Chemical structures of histamine, N^{α}-guanylhistamine, and burimamide.

This compound was found to be about 100 times more active than N^{α}-guanylhistamine as an H_2-receptor antagonist and did not act as a partial agonist. The compound was also shown to be a highly specific competitive antagonist of histamine on non-H_1 tissue systems, thereby defining H_2-receptors, and characterizing burimamide as a histamine H_2-receptor antagonist (Black et al., 1972). Burimamide was also shown to be an effective inhibitor of histamine-stimulated gastric acid secretion in the anesthetized rat, cat, and conscious dog (Heidenhain pouch preparation). The human pharmacology of burimamide was studied (Wyllie et al., 1972) and it was shown to inhibit histamine-stimulated gastric acid secretion at doses similar to those found in the animal studies, thus verifying that the animal pharmacology also held for humans. The compound lacked adequate activity for exploration of its clinical potential, however, and it appeared that a more active compound was required.

In an attempt to achieve a further increase in antagonist potency, attention was focused on the imidazole ring of burimamide, and the burimamide structure was modified so as to increase the equilibrium concentration of imidazole species considered most likely to be active (Black et al., 1974). This approach was successful: Replacing a methylene group (—CH_2—) with an isosteric thioether (—S—) link in the side chain and substituting a methyl group in the ring furnished the more active compound, metiamide (Black et al., 1973; Table 1). Metiamide was found to be some 8–9 times more potent than burimamide in vitro, and 4–6 times more potent in vivo. It was shown to be highly effective clinically in reducing hypersecretion of gastric acid and proved to be of therapeutic value in duodenal ulcer disease. The occurrence of a reversible granulocytopenia in a small number of patients, however, limited the use of metiamide (Forrest et al., 1975) and led to the need to develop another compound.

Fortunately, investigation of other structures continued, particularly for alternatives that did not possess a thiourea group. One approach was to return to the use of guanidine groups and to investigate other ways of removing the positive charge, i.e., by reducing the guanidine basicity (Durant et al., 1977). The basicity of guanidines is very susceptible to substituent effects and can be markedly reduced by substituting powerful electron-withdrawing groups at the nitrogen atoms. This approach was successful and the corresponding cyanoguanidine was synthesized and found to be a potent antagonist, comparable with metiamide, and was selected for development. This compound is cimetidine (SK&F 92334; Brimblecombe et al., 1975) and contains the cyanoimino group (=N—CN) in place of the thione (=S) sulfur atom (Table 1). Cimetidine was first marketed at the end of 1976 in the United Kingdom under the trademark Tagamet® . It has since been introduced in over 100 countries and used extensively for the treatment of conditions associated with gastric hyperacidity.

Like histamine, the early compounds developed as H_2-antagonists are imidazole derivatives with structurally specific side chains, but they differ chemically from histamine in two important respects, i.e., the side chains are longer, and are not basic (the side chains are uncharged at physiological pH). These compounds also differ markedly in chemical structure from the H_1-receptor antihistamines and are less basic and much less lipophilic (they are quite polar molecules).

4.2. Development of Newer H_2-Receptor Antagonists

The highly successful development of cimetidine as a histamine H_2-receptor antagonist, useful as a drug for the treatment of conditions involving gastric hypersecretion of acid, has stimulated a search for other more potent examples from this pharmacological class of agent. Many other compounds have now been described, although, to date, only two—ranitidine (Zantac®; Table 2; Helman and Tim, 1983) and famotidine (MK 208 Gaster®; Table 3)—have made the transition to become drugs introduced into therapy. There is little doubt that other compounds will follow.

Early structure–activity studies of cimetidine and analogs at SK&F had drawn attention to the apparent special significance of the following structural features, of which cimetidine was a representative example:

- An imidazole ring or similar nitrogen heterocycle
- A flexible chain, especially $-CH_2SCH_2CH_2-$
- A planar group that is very polar and has potential for strong H-bonding, and contains the system $-NH-C-NH-$

Subsequent investigations in many other laboratories have provided examples of other structural features and considerably expanded the scope for active structures. Most of the active compounds of interest for drug development, however, still comprise an "aromatic ring" with a "flexible four-atom chain" joined to a "polar group" that characteristically displays pronounced H-bonding properties, as represented by the general formula in Fig. 2. Within this broad structural class, the compounds can be grouped into four main series according to the aromatic ring of the archetypal member of

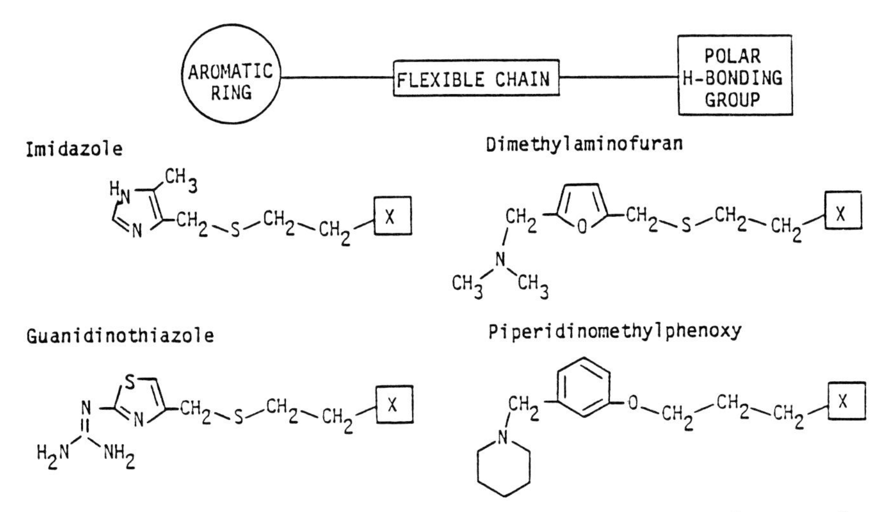

Fig. 2. General formula of histamine H_2-receptor antagonists containing a flexible connecting chain; there are four main series, according to the aromatic ring.

the series; namely, imidazoles, dimethylaminomethylfurans, guanidinothiazoles, and piperidinomethylphenoxy derivatives. These are shown with respective examples in Tables 1–4, together with references and published K_B values for antagonism in vitro of histamine-stimulated increase in the rate of beating of the right atrium from the guinea pig, and activity in vivo in rat or dog as inhibitors of histamine-stimulated gastric acid secretion.

In each series there are several examples of drugs (or potential drugs) that have been studied in human subjects or are undergoing pharmaceutical development. Reported structure–activity studies suggest it is unlikely that all of these compounds interact at H_2-receptors in a chemically similar manner, but it is not yet clear whether there really are four distinct series. An interesting further development is the discovery that the four-atom chain can be replaced by a more rigid benzenoid structure to afford active compounds, but here too there appears to be more than one chemical series.

4.2.1. Imidazole Series

All the compounds listed in Table 1 have been administered to human subjects (volunteer studies or in-patients). The archetypal structure is cimetidine with a 4(5)-methyl-substituted imidazole ring, methylthioethyl connecting chain, and methylcyanoguanidine polar group. Bristol Laboratories introduced the propargyl group to give etintidine (Cavanagh et al., 1980, 1983). Shortly after cimetidine was made, the diamino-nitroethene analog SK&F 92456 was synthesized (Durant et al., 1976), but did not appear to have sufficient improvement over cimetidine to merit development as a drug. The diamino-nitroethene group, however, was taken up by Glaxo chemists and was incorporated into the structure of ranitidine (*see* Table 3).

The range of structures for the "polar" group was extended at SK&F to ring systems that had the advantage of permitting the introduction of additional substituents, e.g., the methylenedioxybenzyl group in the 5 position of isocytosine afforded oxmetidine (Blakemore et al., 1980). This compound was taken through an active program of drug development that, unfortunately, had to be curtailed when several cases of liver sensitivity were reported after it had been administered to some 2000 patients.

In the above structures, the imidazole ring can be replaced by other heterocycles to give active compounds. A special example is the 3-methoxypyridine derivative icotidine (SK&F 93319, Table

1), which has a picolylmethyl substituent in the isocytosine ring and combines into one molecule the ability to block both H_1 and H_2 receptor-mediated actions of histamine (Blakemore et al., 1983; Harvey and Owen, 1984).

An interesting further development was the discovery that the flexible connecting four-atom chain can be replaced by a more rigid benzenoid structure to afford active diaryl compounds. One such diaryl structure is mifentidine (DA 4577 from De Angeli; Donetti et al., 1984). Here the imidazole ring is retained, the flexible chain is replaced by phenylene, and the "polar" group is a formamidine in the position *para* to the imidazole ring.

4.2.2. Furan Series

Chemists at Glaxo, UK investigated the possibility that the basic imidazole ring in cimetidine might not be essential for H_2-receptor blockade and replaced it with alternative ring systems to which a basic function was attached as a substituent, i.e., moving the nitrogen from within the ring to an exocyclic position. They found that replacement of the imidazole ring by furan substituted by a dimethylaminomethyl group afforded compounds of potency comparable with those of the imidazoles, e.g., in cyanoguanidine and thioureas (Price and Daly, 1983). Replacement of the "polar" group by diamino-nitroethene, however, afforded a substantial increase in activity compared with the imidazole derivative. This compound was selected for drug development and is ranitidine (AH 19065, Table 2; Bradshaw et al., 1979). Although ranitidine is somewhat more active than cimetidine as an inhibitor of stimulated gastric acid secretion, its time course for inhibition is very similar.

A further development at SK&F combined the dimethylaminomethylfuran ring with the picolylmethylisocytosine group that had been developed in the imidazole and pyridine series (e. g., icotidine, SK&F 93319) to obtain a potent antagonist, lupitidine (SK&F 93479; Blakemore et al., 1981), which also showed evidence of prolonged duration of action relative to cimetidine in vivo as an inhibitor of stimulated gastric acid secretion. A sustained response to lupitidine was also obtained in human subjects, but further trials were suspended in 1983 after the observation of mucosal changes in the forestomachs of rats that had been given very high doses of drug for at least 6 mo (Betton and Salmon, 1984).

Chemists at Bristol Laboratories (Algieri et al., 1982), and almost simultaneously at Merck Sharp and Dohme (Lumma et al., 1982), identified 1,2-diamino-thiadiazole sulfoxide as an alternative

TABLE 1
Imidazole Series of H_2-Antagonists

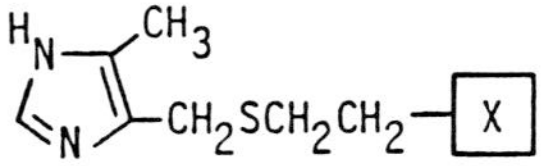

Name	X	Guinea pig atrium K_B[a], μmol/kg	Inhibition of gastric acid secretion: ID_{50},[b] μmol/kg	Route	Species	Preparation[c]	Reference
Metiamide	S	0.92	1.6	iv	Rat	GS	Black et al. (1973)
SK&F 92058	‖		3.1	iv	Dog	HP	Black et al. (1973)
	−NHCNHCH$_3$		16	po	Dog	HP	Black et al. (1973)
Cimetidine	NCN	0.79	1.37	iv	Rat	GS	Brimblecombe et al. (1975)
SK&F 92334	‖		77	iv	Rat	pl	Takeda et al. (1982)
	−NHCNHCH$_3$		58	po	Rat	pl	Takeda et al. (1982)
			1.70	iv	Dog	HP	Brimblecombe et al. (1975)
			10[d]	po	Dog	HP	Brimblecombe et al. (1975)
Etintidine	NCN	0.25	3.9	ip	Rat	gf	Cavanagh et al. (1983)
BL 5641 A	‖		0.97	iv	Dog	gf	Cavanagh et al. (1980)
	−NHCNHCH$_2$C≡CH		1.3	po	Dog	HP	Cavanagh et al. (1980)
SK&F 92456	CHNO$_2$	1.4	1.0	iv	Rat	GS	Ganellin (1981)
	‖						
	−NHCNHCH$_3$						
Oxmetidine		0.2[g]	0.92	iv	Rat	GS	Blakemore et al. (1980)
SK&F 92994		(DR = 2)	0.5[e]	iv	Dog	HP	Blakemore et al. (1980)
			5	po	Dog	HP	Blakemore et al. (1980)

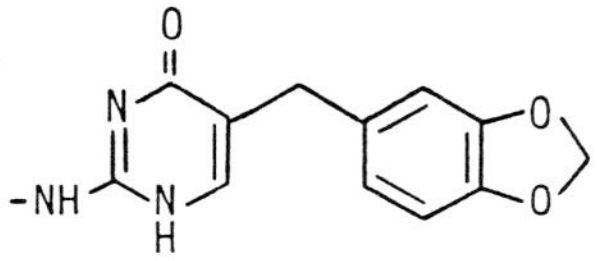

Pyridine analog						
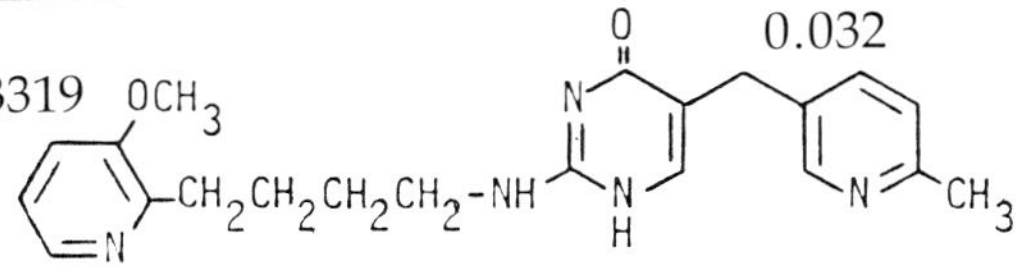 Icotidine	0.032	0.21	iv	Rat	GS	Blakemore et al. (1983)
SK&F 93319		0.5[f]	iv	Dog	HP	Blakemore et al. (1983)
		2.5[d]	po	Dog	HP	Blakemore et al. (1983)
Semirigid phenylene analog						
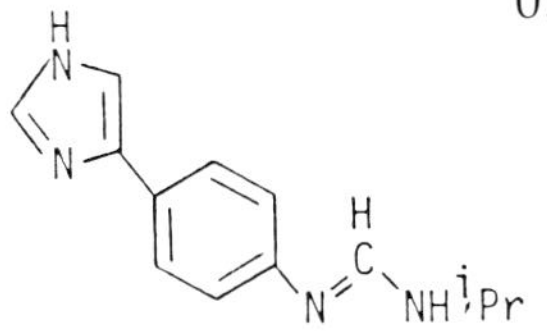 Mifentidine	0.024	0.063	iv	Rat	GS	Donetti et al. (1984)
DA 4577		0.08	iv	Dog	HP	Donetti et al. (1984)
		0.23	po	Dog	HP	Donetti et al. (1984)

[a]The K_B value is that published in the given reference, for antagonism of histamine- or dimaprit-stimulated increase in rate for the spontaneously beating guinea pig right atrium in vitro. It represents the apparent dissociation constant for the antagonist-receptor interaction, and is the concentration of antagonist that requires a doubling of the dose of agonist (histamine or dimaprit) to match the agonist response obtained in the absence of antagonist.

[b]The ID_{50} represents the dose in μmol/kg to inhibit by 50% a near-maximal level of histamine-stimulated gastric acid secretion, by the route shown, in rat or dog, as published in the given reference. Various preparations have been used and the results between preparations are not strictly comparable; in particular, the pylorus-ligated (pl) rat is not very sensitive to H_2-receptor antagonists, since the assay is for basal secretion of acid (i. e., not specifically stimulated by histamine).

[c]Abbreviations: GS, the modified Ghosh-Schild preparation of the perfused lumen of the anesthetized and atropinized rat; HP, Heidenhain Pouch conscious dog; gf, gastric fistula conscious rat or dog; pl, pylorus ligated anesthetized rat; PP, Pavlov Pouch dog.

[d]70% inhibition.

[e]60% inhibition.

[f]40% inhibition.

[g]K_B not available, but comparison at DR = 2 indicates that it has approximately 16 times the potency of cimetidine.

TABLE 2
Furan Series of H_2-Antagonists[a]

Me_2NCH_2–(furan)–$CH_2SCH_2CH_2$–X

Name	X	Guinea pig atrium K_B, μM	Inhibition of gastric acid secretion: ID_{50}, μmol/kg,	Route	Species	Preparation	Reference
Ranitidine AH 19065	$-NHC(=CHNO_2)NHCH_3$	0.063	0.37	iv	Rat	GS	Bradshaw et al. (1979)
			0.20	iv	Dog	HP	Brittain and Daly (1981)
			0.66	po	Dog	HP	Bradshaw et al. (1979)
Lupitidine SK&F 93479		0.016	0.14	iv	Rat	GS	Blakemore et al. (1981)
			0.25[b]	iv	Dog	HP	Blakemore et al. (1981)
			0.625[b]	po	Dog	HP	Blakemore et al. (1981)
BMY 25271		0.026	0.041	iv	Rat	pl	Algieri et al. (1982)
		0.006					Lumma et al. (1982)
			0.018	iv	Dog	HP	Buyniski et al. (1984)
Thiazole analog		Rat uterus K_B					
Nizatidine LY 139037		0.08	1.48	iv	Rat	pl	Lin et al. (1983)
			0.081	iv	Dog		LIn et al. (1983)
			0.18	po	Dog		Lin et al. (1983)

Nizatidine: Me_2NCH_2–(thiazole)–$CH_2SCH_2CH_2$–$NHC(=CHNO_2)NHCH_3$

[a]*See* Table 1 for abbreviations.

TABLE 3
Guanidinothiazole Series of H_2-Antagonists[a]

Name	X (general structure: $H_2N(NH_2)C{=}N$–thiazolyl–$CH_2SCH_2CH_2$–X)	Guinea pig atrium K_B, μM	Inhibition of gastric acid secretion: ID_{50} μmol/kg	Route	Species	Preparation	Reference
Tiotidine ICI 125211	NCN ‖ $-NHCNHCH_3$	0.015	0.2[b]	iv	Dog	gf	Yellin et al. (1979)
			0.69	po	Dog	HP	Cavanagh et al. (1983)
BL 6341 A	1-oxo-1,2,5-thiadiazole ring with –NH– and NH_2	0.027					Cavanagh et al. (1981)
			0.075	iv	Rat	pl	Algieri et al. (1982)
			0.040	iv	Dog	HP	Buyniski et al. (1984)
			0.35	po	Dog	HP	Cavanagh et al. (1981)
Famotidine YM 11170	NSO_2NH_2 ‖ $-CNH_2$	0.017[c]	1.3	iv	Rat	pl	Takeda et al. (1982)
			1.3	po	Rat	pl	Takeda et al. (1982)
			2.6	id	Rat	pl	Takeda et al. (1982)
			0.011	iv	Dog	HP	Takagi et al. (1982)
			0.031	po	Dog	HP	Takagi et al. (1982)
Pyrazole analog							
ICI 162846	$H_2N(CF_3CH_2NH)C{=}N$–pyrazolyl–$N(CH_2)_4$–$CONH_2$						

[a] *See* Table 1 for abbreviations.
[b] 70% inhibition.
[c] SK&F result.

TABLE 4
m-Piperidinomethylphenoxy Series of H_2-Antagonists[a]

Name	X	Guinea pig atrium K_B, μM	Inhibition of gastric acid secretion: ID_{50}, μmol/kg	Route	Species	Preparation	Reference
Lamtidine AH 22216	1-methyl-1,2,4-triazole: CH_3–N–N, –NH, N, NH_2	0.07 (DR = 2)	0.049	iv	Rat	GS	Brittain et al. (1982)
			0.055	iv	Dog	HP	Humphray et al. (1982)
			0.084	po	Dog	HP	Brittain et al. (1982)
Loxtidine	1-methyl-1,2,4-triazole: CH_3–N–N, –NH, N, CH_2OH	NC[b]	0.039	iv	Dog	HP	Stables and Humphray (1983)
AH 23844			0.075	po	Dog	HP	Stables and Humphray (1983)
TAS (Wakamoto)	1,3,4-thiadiazole: N–N, –NH, S, NH_2	0.054	3.2	id	Rat	pl	Tsuritani et al. (1984)
TZU 0460 (Teikoku)	$-NHC(=O)CH_2OC(=O)CH_3$	0.27	1.1	iv	Rat	GS	Tarutani et al. (1985a,b)
			0.75	iv	Dog	HP	Tarutani et al. (1985b)
			0.91	po	Dog	HP	Tarutani et al. (1985b)

General structure: piperidine–N–CH_2–(*m*-phenylene)–$OCH_2CH_2CH_2$–X

BMY 25260	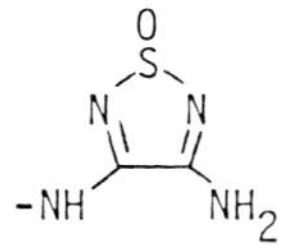	0.04					Algieri et al. (1982)
L 643441		0.02	0.036	iv	Dog	gf	Lumma et al. (1982)
			0.88	po	Dog	gf	Torchiana et al. (1983)
BMY 25368	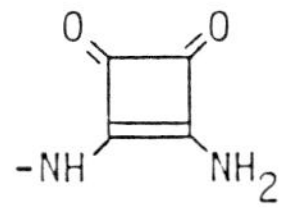	0.013	0.023	iv	Dog	HP	Buyniski et al. (1984)
Wy 45086	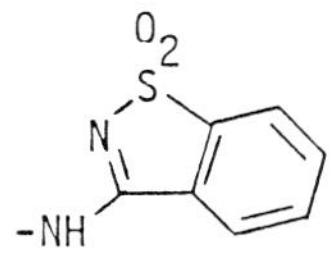	0.008	2.4		Rat	pl	Nielsen et al. (1984)
			2.4		Dog	PP	Nielsen et al. (1984)

[a] *See* Table 1 for abbreviations.
[b] Noncompetitive inhibition.

"polar" group, and the dimethylaminomethylfuran derivative (BMY 25271) appears to be under further study at Bristol (Buyniski et al., 1984).

Nizatidine (LY 139037) under development by Eli Lilly (Lin et al., 1983), is structurally a close analog of ranitidine, in which dimethylaminomethylfuran has been replaced by dimethylaminomethylthiazole.

4.2.3. Guanidinothiazole Series

The discovery of guanidinothiazole as an imidazole-replacement (Table 3) was made at ICI as a result of a screening program in which compounds were selected from the chemist's collection of samples on file (Gilman et al., 1982). Activity was found for 2-guanidino-4-methylthiazole and investigation of suitable derivatives revealed that the cimetidine analog (i.e., with the methylthioethyl-cyanoguanidine side chain) tiotidine (ICI 125211) was very potent (Yellin et al., 1979). Unfortunately, clinical studies with this compound had to be discontinued when lesions were discovered in the gastric mucosa of rats after chronic administration at high doses.

The guanidinothiazole group appears to confer high affinity at the H_2-receptor, and many active derivatives are now known. The "polar" group encompasses a wider diversity of chemical structures than obtains for the imidazole series, and examples of compounds under active development are BL 6341A (Bristol Laboratories; Algieri et al., 1982) and famotidine (YM 11170/MK208 from Yamanouchi; Takagi et al., 1982). BL 6341A also shows an extended duration of action in vivo (relative to cimetidine) as an inhibitor of stimulated gastric acid secretion, and is not easily washed out from the guinea pig atrium (Buyniski et al., 1984).

A trifluorethylguanidinopyrazole analog with a valeramide side chain (ICI 162,846) appears to be under development by ICI (Yellin and Gilman, 1982; Hardie et al., 1984; Wilson et al., 1986).

4.2.4. Phenoxy Series

A further development from the Glaxo chemists was the replacement of the furan ring by phenoxy; in effect, moving the oxygen atom from within the ring to outside of the ring (Brittain et al., 1982). Activity is also dependent upon the amine substituent, and a *m*-piperidinomethyl group was found to be particularly favorable. Use of *m*-piperidinomethylphenoxpropyl attached to the "polar" group considerably expanded the scope for active structures (Table 4), e.g., lamtidine (AH 22216; Brittain et al., 1982) and loxtidine (AH 23844; Stables and Humphray, 1983), and this finding has been applied by chemists in many other pharmaceutical companies, e.g., pifatidine (TZU 0460; Teikoku; Tarutani et al., 1985a,b), TAS (Wakamoto; Tsuritani et al., 1984), BMY 25260 (Bristol Laboratories; Algieri et al., 1982), L 643441 (Merck Sharp and Dohme; Lumma et al., 1982; Torchiana et al., 1983), BMY 25368 (Buyniski et al., 1984), and Wy 45086 (Wyeth; Nielsen et al., 1984).

It appears that *m*-piperidinomethylphenoxypropyl confers high affinity at the H_2-receptor and many active derivatives have been made. Particularly noteworthy was the finding by Teikoku researchers that the "polar" group need only contain one N atom, and they reported (Shibata, 1984) that a simple derivative such as —$NHCOCH_3$ was active; clearly, this may be related to the valeramide (ICI 162846) of Table 3.

Depending upon the "polar" group, there is also a considerable change in the kinetics of interaction, and it appears that compounds such as lamtidine, loxtidine, BMY 25260, and BMY 25368 are long lasting in vivo as inhibitors of stimulated acid secretion. They are also very difficult to displace by washing in vitro from guinea pig atria and appear to induce a nearly irreversible blockade of receptor function (Torchiana et al., 1983; Buyniski et al., 1984).

5. Toxicology of Burimamide, Metiamide, and Cimetidine and More Recently Developed H_2-Receptor Antagonists

Toxicity studies with burimamide, the results of which were unremarkable, were limited to those to support human volunteer studies (Wyllie et al., 1972). These volunteer studies confirmed in humans the predictions from animal pharmacology that burimamide was an effective antisecretory agent. Because of inadequate oral activity, development of burimamide ceased and no further toxicology work was done.

With metiamide, toxicity studies in rats and dogs of up to 1 yr duration were completed. Results of these studies were reported by Brimblecombe et al. (1973, 1974). The findings of greatest significance were pathological changes in liver and kidney in both rats and dogs given high doses, which were not seen in subsequent human studies; deaths from acute pulmonary edema and pleural effusions in a proportion (<10%) of beagle dogs given single oral doses of metiamide in excess of 50 mg/kg with again no such effect being observed in humans; a non-dose-related incidence of reversible granulocytopenia in a small percentage of dogs caused apparently by maturation arrest in the myeloid series in the bone marrow; a similar effect was seen in a few of the human subjects receiving metiamide (Forrest et al., 1975) and led to cessation of development of metiamide.

It was considered probable that the bone marrow toxicity of metiamide was related to the presence of a thioureido group in the molecule rather than to H_2-antagonism, and so the development of a nonthiourea antagonist, cimetidine (Brimblecombe et al., 1975), which was already being studied, was accelerated. A plethora of "standard" and "nonstandard" toxicity tests has subsequently been carried out with cimetidine. It has probably been studied more extensively than any other drug. Its toxicology has been summarized by Brimblecombe et al., (1985). In the early days, the objective was to complete adequate toxicology to allow clinical pharmacological and clinical studies to be instituted. The results of these toxicity studies were unremarkable and, in particular, no evidence was seen of bone marrow toxicity similar to that produced by metiamide. Belief that cimetidine lacked this toxicity was reinforced when a patient with the Zollinger-Ellison syndrome who developed agranulocytosis when being treated with metiamide showed a reversal of this effect when switched to cimetidine (Burland et al., 1975), confirming the view that the effect did not result from H_2-antagonism. Subsequently, in widespread clinical use the virtual absence of bone marrow toxicity with cimetidine has been demonstrated. By February 1981, when it was estimated that over 11 million patients had received cimetidine, there had been 192 worldwide reports of a fall in leukocyte count associated with the drug (Rowley-Jones and Flind, 1981). In only two of these cases were no other factors, such as other drugs being used concomitantly or serious illness, present.

There have been very few significant toxicological findings with cimetidine. In both rats and dogs in repeat-dose toxicity studies there has been reported a reduction in size and delay in maturation of secondary sex organs, which results from the very weak antiandrogenic activity possessed by cimetidine (Sivelle et al., 1982). Although it can be speculated that the very low incidence (0.10–0.2%) of gynecomastia seen in the widespread use of cimetidine is associated with this antiandrogenic activity, there is no real evidence for this association. Cimetidine given in high doses in reproductive toxicity studies has no effect on potency or fertility (Leslie and Walker, 1977).

Two separate 2-yr rat carcinogenicity studies have been completed with cimetidine. The results of the first study was described in detail by Leslie et al. (1981). The only finding of note was a slight increase in the incidence of benign Leydig cell hyperplasia and tumors. This was considered to be of little toxicological significance, since the strain of rat used has a high spontaneous incidence of such tumors, and this incidence appears to be increased by chemical insult with high doses of unrelated compounds.

Of particular importance in these 2-yr studies was the absence of development of tumors of the gastric mucosa. The relevance of this is that since 1979 there have been anecdotal reports of gastric cancer occurring in patients treated with cimetidine (Elder et al., 1979). These carcinomas were probably present, but undiagnosed, before cimetidine treatment commenced, but there are hypothetical possibilities of linking cimetidine with gastric cancer. First, the nitrosated derivative of cimetidine (*N*-nitrosocimetidine) bears a structural resemblance to *N*-methyl-*N*′-nitro-*N*-nitrosoguanidine (MNNG). MNNG is a gastric carcinogen in some species of laboratory animals (Sugimura et al., 1966).

The second hypothesis states that reduced gastric acidity would create an intragastric milieu in which bacteria, including nitrate-reducing organisms, could survive. The activity of these bacteria would result in increased nitrite concentrations in the gastric juice, which would in turn lead to increased levels of potentially carcinogenic *N*-nitroso compounds (Drasaar et al., 1969).

Many studies designed to prove or disprove these hypotheses have been carried out in our own laboratories and elsewhere. They are summarized by Brimblecombe et al. (1985). Perhaps the two most significant findings are, first, that *N*-nitrosocimetidine is not carcinogenic in long-term studies in animals (Habs et al., 1982; Lijinsky, 1982) and, second, there is still no generally accepted evidence that increased levels of carcinogenic *N*-nitroso compounds occur in the gastric juice of patients receiving therapeutic doses of cimetidine. There is no reason, therefore, to doubt the validity of the negative results in the 2-yr rat carcinogenicity studies. To those studies should be added a 7-yr study in dogs that were given 144 mg/kg/d of cimetidine. Biopsies of gastric mucosa taken at endoscopy during the study have shown no changes attributable to cimetidine treatment (Crean et al., 1981a,b). Similarly, careful histological examination of the gastric mucosa at the end of the study has revealed no treatment-related effects (Walker et al., 1986).

The wide variety of toxicological studies carried out with cimetidine indicated that the drug would be, by any standards, very safe and, as is indicated below, this has proven to be the case.

Toxicological studies with ranitidine (AH 19065), the second H_2-receptor antagonist to be marketed, have not been so extensively reported as for cimetidine, but available data indicate a satisfactory ratio between the human therapeutic dose and doses that produce adverse effects in subacute and chronic toxicity studies in rat and dogs. As with cimetidine, no lesions of the gastric mucosa have been reported (Poynter et al., 1982; Tamura et al., 1983a,b; Takeuchi et al., 1983; Nagata et al., 1983).

Famotidine (YM 11170/MK208, trademark Gaster®) has recently (mid-1985) been marketed in Japan. Data so far published on the toxicology of famotidine indicate a low level of toxicity, comparable to that of cimetidine and ranitidine. One finding worth noting is that there was an increased incidence of focal eosinophilic cytoplasmic granularity of the chief cells of the stomach in rats receiving large oral doses of famotidine (Bokelman, 1984). The same author reviewed the preclinical safety of the drug, but reported no other remarkable findings. Another paper reporting results of 3- and 6-mo oral toxicity studies in rats is by Suzuki et al. (1983) and Shiobara (1983); again, the results were unremarkable.

Thus, cimetidine, ranitidine, and famotidine, from data so far available, all seem to present acceptable safety profiles from the results of toxicity tests. It should be noted that all these drugs, at equipotent doses, have comparable duration of action in comparison to loxtidine (AH 23844) and lupitidine (SK&F 93479), which are discussed below.

One other compound that must be mentioned here is tiotidine (ICI 125211), development of which was terminated when it proved to be a gastric carcinogen. Findings described by Streett et al. (1984) were taken from two studies, a 6-mo oral and dietary study in male rats, and a 12-mo oral toxicity study combined with a 2-yr feeding carcinogenicity study also done in rats. Significant findings in the stomach included fundic glandular dilatation, parietal cell atrophy, metaplasmic fundic cells, fundic nuclear clusters, and dysplasia/carcinoma lesions. The incidence of the dysplasia/carcinoma lesions was 17 from 828 treated rats, compared to none in control rats. The current view appears to be that the carcinogenicity of tiotidine results from a direct action unrelated to its H_2-receptor antagonist activity.

More recently, evidence relating to the toxicology of newer, more-potent, longer-acting H_2-receptor antagonists has begun to emerge, although not much detailed information has yet been published.

Wormsley (1984) reported that life-span oral administration (greater than 2 yr) of loxtidine to rats and mice produced marked dysplasia and malignancy of the gastric glandular mucosa of a type not described previously. These tumors are believed to be fundic, with proliferating cells containing neurosecretory granules, and are probably enterochromaffin-like cell carcinoids of the oxyntic mucosa. Virtually identical changes have been found in rats following 2-yr treatment with the potent, long-acting H_2-antagonist lupitidine (SK&F 93479), at 1000 mg/kg/d (Betton and Salmon, 1984).

Interestingly, very similar gastric changes in rats have been reported following 2 yr administration of the H^+/K^+-ATPase inhibitor, omeprazole (Wormsley, 1984), which produces profound and long-lasting inhibition of acid secretion by a mechanism not associated with blockade of H_2-receptors. A provisional view, currently gaining support, is that gastric anacidity results in hypergastrinemia that has been shown to be associated with the development of mixed endocrine tumors (Larsson et al., 1978). Thus, the carcinogenicity seen following administration of lupitidine, loxtidine, and omeprazole may well not be a direct action of these drugs, but rather secondary to prolonged anacidity and thus of no relevance to the human use of such drugs in therapeutic doses.

6. Safety and Efficacy of Cimetidine in Widespread Clinical Use

The widespread use of cimetidine has provided an excellent opportunity to determine how well preclinical studies have predicted the efficacy and safety of the drug in humans. In a paper read in 1976, the year of the launch of cimetidine, Brimblecombe and Duncan (1977) concluded that "at the current state of knowledge it can be concluded that there have been no significant discrepancies between results in experimental animals and in man." Nine years later, Brimblecombe et al. (1985) wrote "over 30 million patients have so far been treated with cimetidine and the prediction from the animal studies that it would be an extremely safe therapeutic agent has been borne out in practice."

It is, however, axiomatic that no effective, pharmacologically active agent can be completely "safe," i.e., completely free from danger or risks, and a low incidence of side effects has been reported for cimetidine. These have been reviewed by, among others, Flind et al. (1980) and Rowley-Jones and Flind (1981).

As mentioned above cimetidine possesses weak antiandrogenic activity that may be responsible for the very low incidence of gynecomastia (about 1%) that has been reported. In the majority of cases, this is of nuisance value only.

Neither animal studies nor clinical trials revealed any important cimetidine-related effects on the central nervous system, but as the drug became more widely used, it became apparent that occasional episodes of mental confusion occurred, usually in the elderly or very ill.

Cimetidine shows a moderate binding by a ligand interaction (probably via the imidazole ring) to the cytochrome P450 enzyme system (Rendic et al., 1979) by which many drugs are metabolized. Thus cimetidine has the potential to prolong the half-lives of drugs that are metabolized in this way and have lower affinities toward the enzyme. The most important resulting interactions are those with oral anticoagulants, such as warfarin, and with anticonvulsants, such as phenytoin.

Rendic et al. (1982) subsequently reported that ranitidine interacts with cytochrome P450 qualitatively similarly to cimetidine, although the extent of binding is lower and a different group is coordinated to the iron of the cytochrome.

As mentioned above, animal studies indicated that cimetidine, unlike metiamide, did not produce bone marrow toxicity. Although there have been reports of bone marrow depression in patients receiving cimetidine, in all but a very small number other potential causative factors were present. The possibility exists that administration of cimetidine to a patient with an already compromised bone marrow could produce an adverse effect, but this complication is unlikely.

Apart from spontaneous side effect reporting, formal postmarketing surveillance studies have been carried out with cimetidine in both the United States and the United Kingdom. The US study (Humphries et al., 1984) involved initially nearly 10,000 patients and, in a second phase, a follow-up on over 7000 of these patients. The adverse effects reported were not different from those previously reported in clinical studies, published reports, or via the spontaneous reporting system. In the UK study, almost 10,000 patients taking cimetidine were compared with a similar number of controls (Colin-Jones et al., 1982, 1983). From analyses so far reported, no evidence of any fatal adverse effects of cimetidine emerged and the findings do not support the hypothesis that cimetidine treatment predisposes to gastric cancer.

Thus, the only side effects of significance not detected in animal studies are mental confusion, for which no analog probably exists in animals, and the propensity to produce certain drug interaction, which is also only reliably detected in man. In vitro studies to show inhibition of cytochrome P450 and in vivo studies in animals are not necessarily predictive of the human response. Work on isolated human hepatocytes currently on-going in a number of laboratories may prove helpful in this respect.

As other H_2-antagonists are used more widely it will be interesting to analyze their side-effect profiles. In this way it should be possible to ascertain which effects are associated with H_2-receptor antagonism and which are peculiar to the particular molecules.

7. Therapeutic Use of H_2-Receptor Antagonists—Lessons Learned

The adage ''no acid, no ulcer'' was still much in vogue at the time of the launch of cimetidine and governed to some extent the choice of dose. This was 300 mg four times daily (1.2 g/d) in North America and 200 mg three times daily and 400 mg at bedtime (1 g/d) in other parts of the world. This difference was based largely on a view among North American gastroenterologists that acid should be inhibited to a greater degree than their non-American colleagues deemed necessary. It should be noted that neither of these dosage regimens produce anacidity. Clinical pharmacological studies in which 24-h intragastric acidity was studied in subjects on a normal regime of meals showed that acid secretion throughout the 24-h period was reduced, but that there were still increases, albeit blunted, in response to meals. Thus, Pounder et al. (1977) studied cimetidine 200 mg four times daily and 400 mg four times daily each in three patients. Cimetidine decreased mean acidity compared with placebo in 44 of the 46 hourly sampling periods. The 0.8 g/d regimen decreased mean 24-hr intragastric H^+ activity from 42.2 to 18.9 mmol/L (–55%), whereas 1.6 g/d regimen caused a reduction from 48.7 to 15.8 mmol/L (–67%). It is of interest that doubling the dose of cimetidine from 0.8 to 1.6 g/d resulted in doubling the peak blood concentrations, but there was no difference in daytime intragastric activity in subjects in the two dose groups. Nocturnal acid secretion was inhibited more by the higher dose.

A subsequent clinical pharmacological study (Burland et al., 1980) showed that in healthy subjects, cimetidine, 400 mg twice daily, reduced intragastric acidity as effectively as 1 g given in the four times daily regimen (54.3 vs 55.8%). Mahachai et al. (1982) subsequently showed that cimetidine, 600 mg twice daily, was actually superior to the standard North American regimen of 300 mg four times daily, in suppressing intragastric H^+ activity in asymptomatic patients with a history of duodenal ulcer disease.

This led naturally to clinical trials in which twice daily dosing was compared with the then accepted four times a day regimens. In a multicenter study presented by Kerr (1981), a healing rate of duodenal ulcers of 69% was achieved following 4 wk treatment with cimetidine 1 g/d. Cimetidine, 400 mg twice daily, produced a similar healing rate of 66%.

More recently, partly in response to an original suggestion by Dragstedt and Owens (1943) that nocturnal acid secretion was the most important single factor in the pathogenesis of duodenal ulcer, single nocturnal doses of H_2-receptor antagonists were studied. Gledhill et al. (1983) studied either placebo, cimetidine 400 mg twice daily (bd), ranitidine 150 mg bd, cimetidine 800 mg at night (nocte), or ranitidine 300 mg nocte, in duodenal ulcer patients in remission for effects on 24-h intragastric acidity and nocturnal acid output. The results, shown in Table 5, led to an obvious suggestion that single nocturnal doses of H_2-antagonists should be evaluated in clinical trials in patients with duodenal ulcer. This was done with impressive results.

This brief history of the way in which the dose and dosing schedule of cimetidine have changed over the years indicates how careful study of the clinical pharmacology of a drug can lead to more rational treatment and better understanding of the pathogenesis of a disease entity, in this case duodenal ulcer. The conventional wisdom has moved from "no acid, no ulcer" to a belief that minimal inhibition of nocturnal acid secretion concomitant with adequate therapeutic response is the treatment of choice for most patients with duodenal ulcers. Whether the same is true for other acid-related diseases—gastric ulcer, gastroesophageal reflux disease, stress ulceration, and so on—is less certain and requires more work.

In any event, the discovery that existing H_2-antagonists are effective in a once-a-day dosing regimen in duodenal ulcer disease

TABLE 5
Data Showing That Inhibition of Acid Secretion by Cimetidine or Ranitidine Is Effective After Single Nocturnal Doses[a,b]

	% Reduction in mean 24-h H^+ activity	% Reduction in mean nocturnal acid output
Ranitidine 150 mg bd	63	73
Cimetidine 400 mg bd	30	37
Ranitidine 300 mg nocte	62	84
Cimetidine 800 mg nocte	40	69

[a]A number of studies (Delattre and Dickson, 1984; Capurso et al., 1984; Lacerte et al., 1984) have compared cimetidine 800 mg nocte with cimetidine 400 mg bd in the treatment of duodenal ulcer. Healing rates for the nocte regimen ranged from 84 to 96%, and for bd dosing, from 68 to 73%. These healing rates were achieved with presumably quite modest reductions in nocturnal acid output and little effect on daytime output.

[b]These results indicate that cimetidine is effective when given as a single night-time dose in healing duodenal ulcers. The same has been shown for ranitidine. Improved patient compliance will presumably result from this dosing regimen.

and the fact that newer, potent, long-acting H_2-antagonists have run into problems with toxicity has led to reconsideration of the possible role of the latter compounds.

More profound, longer-lasting inhibition of acid secretion may be beneficial in gastroesophageal reflux disease, and parenterally administered agents of high potency and long duration of action may be more effective than existing drugs on the prevention or treatment of stress ulceration or in preventing gastric reflux during anesthesia. This remains to be seen.

It is also possible that a safe, potent, long-acting compound could produce a higher percentage healing rate of duodenal ulcers in a shorter time. The benefit of this will have to be weighed against any potential risks with such an agent.

8. Therapeutic Use of H_2-Receptor Antagonists in Indications Unrelated to Hypersecretion of Gastric Acid

The discovery of H_2-receptor antagonists led not only to a new and important class of therapeutic agents, but also provided new tools to study the physiological and pathological roles of histamine.

It is now clear that H_2-receptors are present on blood vessels, in the heart, skin, central nervous system, and respiratory system, and on basophils. Over- or under-stimulation of these receptors could lead, in theory at least, to be a wide variety of pathological conditions. This matter has been reviewed in detail by Burland and Mills (1982) and will not be dealt with at any length here.

Attempts so far to use H_2-antagonists alone or in combination with H_1-antagonists in the treatment of, for example, vascular headaches and burns have been disappointing. The possibility still exists that histamine antagonists may be beneficial in inflammatory bowel disease, pruritis, urticaria, and inflammatory skin disease. Investigations are in progress.

The brain and the immune system still represent largely unexplored therapeutic areas. It remains to be seen whether histamine antagonists have any role to play in these areas.

References

Algieri, A. A., Luke, G. M., Standridge, R. T., Brown, M., Partyka, R. A., and Crenshaw, R. R. (1982) 1,2,5-Thiazole 1-oxide and 1,1-dioxide derivatives. A new class of potent histamine H_2-receptor antagonists. *J. Med. Chem.* **25**, 210–212.

Ash, A. S. F. and Schild, H. O. (1966) Receptors mediating some actions of histamine. *Br. J. Pharmac. Chemother.* **27**, 427–439.

Betton, G. R. and Salmon, G. K. (1984) Pathology of the forestomach in rats treated for 1-year with a new histamine H_2-receptor antagonist, SK&F 93479 trihydrochloride. *Scand. J. Gastroenterol.* **19** (suppl. 101), 103–108.

Black, J. W., Duncan, W. A. M., Durant, G. J., Ganellin, C. R., and Parsons, M. E. (1972) Definition and antagonism of histamine H_2-receptors. *Nature* **236**, 385–390.

Black, J. W., Duncan, W. A. M., Emmett, J. C., Ganellin, C. R., Hesselbo, T., Parsons, M. E., and Wyllie, J. H. (1973) Metiamide—an orally active histamine H_2-receptor antagonist. *Agents Actions* **3**, 133–137.

Black, J. W., Durant, G. J., Emmett, J. C., and Ganellin, C. R. (1974) Sulphur-methylene isosterism in the development of metiamide, a new histamine H_2-receptor antagonist. *Nature* **248**, 65–67.

Blakemore, R. C., Brown, T. H., Durant, G. J., Emmett, J. C., Ganellin, C. R., Parsons, M. E., and Rasmussen, A. C. (1980) SK&F 92994: A new histamine H_2-receptor antagonist. *Br. J. Pharmacol.* **70**, 105P.

Blakemore, R. C., Brown, T. H., Durant, G. J., Ganellin, C. R., Parsons, M. E., Rasmussen, A. C., and Rawlings, D. A. (1981) SK&F 93479:

A potent and long-acting histamine H_2-receptor antagonist. *Br. J. Pharmacol.* **74**, 200P.

Blakemore, R. C., Brown, T. H., Cooper, D. G., Durant, G. J., Ganellin, C. R., Ife, R. J., Parsons, M. E., Rasmussen, A. C., and Sach, G. S. (1983) SK&F 93319: A specific antagonist of histamine at H_1- and H_2-receptors. *Br. J. Pharmacol.* **80**, 437P.

Bokelman, D. L. (1984) Famotidine: A potent histamine H_2-receptor antagonist. Summary of preclinical safety. *Ital. J. Gastroenterol.* **16**, 176–177.

Bradshaw, J., Brittain, R. T., Clitherow, J. W., Daly, M. J., Jack, D., Price, B. J., and Stables, R. (1979) AH 19065: A new potent, selective histamine H_2-receptor antagonist. *Br. J. Pharmacol.* **66**, 464P.

Brimblecombe, R. W. and Duncan, W. A. M. (1977) The Relevance to Man of Preclinical Data for Cimetidine, in *Cimetidine: Proceedings of the Second International Symposium on Histamine H_2-Receptor Antagonists* (Burland, W. L. and Simkins, M. A., eds.) Excerpta Medica, Oxford/Amsterdam.

Brimblecombe, R. W., Duncan, W. A. M., and Walker, T. F. (1973) Toxicology of Metiamide, in *Proceedings of International Symposium on Histamine H_2-Receptor Antagonists* London 1973 (Wood, C. J. and Simpkins, S. A., eds.) Welwyn Garden City, Hertfordshire.

Brimblecombe, R. W., Duncan, W. A. M., and Parsons, M. E. (1974) Metiamide, a histamine H_2-receptor antagonist. Pharmacology and toxicology. *S. African Med. J.* **48**, 2253–2255.

Brimblecombe, R. W., Duncan, W. A. M., Durant, G. J., Emmett, J. C., Ganellin, C. R., and Parsons, M. E. (1975) Cimetidine—a non-thiourea H_2-receptor antagonist. *J. Int. Med. Res.* **3**, 86–92.

Brimblecombe, R. W., Leslie, G. B., and Walker, T. F. (1985) Toxicology of cimetidine. *Human Toxicol.* **4**, 13–25.

Brittain, R. T. and Daly, M. J. (1981) A review of the animal pharmacology of ranitidine—a new, selective histamine H_2-antagonist. *Scand. J. Gastroenterol.* **16**, (suppl. 69), 1–9.

Brittain, R. T., Daly, M. J., Humphray, J. M., and Stables, R. (1982) AH 22216, a new long acting histamine H_2-receptor antagonist. *Br. J. Pharmacol.* **76**, 195P.

Burland, W. L. and Mills, J. G. (1982) The Pathophysiological Role of Histamine and Potential Therapeutic Uses of H_1- and H_2-Antihistamines, in *Pharmacology of Histamine Receptors* (Ganellin, C. R. and Parsons, M. E., eds.) Wright, Bristol, London, Boston.

Burland, W. L., Sharpe, P. C., Colin-Jones, D. G., Turnball, P. R. G., and Bowskill, P. (1975) Reversal of metiamide-induced agranulocytosis during treatment with cimetidine. *Lancet* **ii**, 1085.

Burland, W. L., Brunet, P. L., Hunt, R. H., Melvin, M. A., Mills, J. G., Vincent, D., and Milton-Thompson, G. J. (1980) Comparison of the effects on 24 h intragastric acidity of SK&F 92994 and two dose regimens of cimetidine (abstract). *Hepatogastroenterol.* (suppl.) **259** XI Int. Cong. Gastroenterol., Hamburg.

Buyniski, J. P., Cavanagh, R. L., Pircio, A. W., Algieri, A. A., and Crenshaw, R. R. (1984) Structure–Activity Relationships Among Newer Histamine H_2-Receptor Antagonists, in *Highlights in Receptor Chemistry* (Melchiorre, C. and Giannella, M., eds.) Elsevier, Amsterdam, New York, Oxford.

Capurso, L., Dal Monte, P. R., Mazzeo, F., Menardo, G., Morettini, A., Saggioro, A., and Tafner, G. (1984) Comparison of cimetidine 800 mg once daily and 400 mg twice daily in acute duodenal ulceration. *Br. Med. J.* **289**, 1418–1420.

Cavanagh, R. L., Usakewicz, J. J., and Buyniski, J. P. (1980) Comparative activities of three new histamine H_2-receptor antagonists—BL-5641, ranitidine and ICI-125,211. *Fed. Proc.* **39**, 426.

Cavanagh, R. L., Usakewicz, J. J., and Buyniski, J. P. (1981) BL-6341A: A new selective histamine H_2-receptor antagonist with more prolonged gastric antisecretory activity than cimetidine and ranitidine (abstract). *Fed. Proc.* **40**, 693.

Cavanagh, R. L., Usakewicz, J. J., and Buyniski, J. P. (1983) A comparison of some of the pharmacological properties of etintidine, a new histamine H_2-receptor antagonist, with those of cimetidine, ranitidine and tiotidine. *J. Pharmacol. Exp. Ther.* **224**, 171–179.

Code, C. F. (1974) New antagonists excite an old histamine prospector. *N. Eng. J. Med.* **290**, 738–740.

Colin-Jones, D. G., Langman, M. J. S., Lawson, D. H., and Vessey, M. P. (1982) Cimetidine and gastric cancer: Preliminary report from postmarketing surveillance study. *Br. Med. J.* **285**, 1311–1313.

Colin-Jones, D. G., Langman, M. J. S., Lawson, D. H., and Vessey, M. P. (1983) Postmarketing surveillance of the safety of cimetidine: 12 month mortality report. *Br. Med. J.* **286**, 1713–1716.

Crean, G. P., Morson, B. C., Leslie, G. B., and Roe, F. J. C. (1981a) Cimetidine further evidence of noncarcinogenicity in dogs. *N. Engl. J. Med.* **304**, 672.

Crean, G. P., Leslie, G. B., Walker, T. F., Whitehead, S. M., and Roe, F. J. C. (1981b) Safety evaluation of cimetidine: 54 month interim report on long-term study in dogs. *J. Appl. Toxicol.* **1**, 159–164.

Delattre, M. and Dickson, B. (1984) Cimetidine once daily. *Lancet* **i**, 625.

Donetti, A., Cereda, E., Bellora, E., Gallazzi, A., Bazzano, c., Vanoni, P., Del Soldato, P., Micheletti, R., Pagani, F., and Giachetti, A. (1984) (Imidazolylphenyl) formamidines. A structurally novel class of potent histamine H_2-receptor antagonist. *J. Med. Chem.* **27**, 380–386.

Dragstedt, L. R. and Owens, F. M. (1943) Supradiaphragmatic section of the vagus nerves in treatment of duodenal ulcer. *Proc. Soc. Exp. Biol.* **53**, 152–154.

Drasaar, B. S., Shiner, M., and McLeod, G. M. (1969) The bacterial flora of the gastrointestinal tract in healthy and achlorhydric persons. *Gastroenterology* **56**, 71–79.

Durant, G. J., Parsons, M. E., and Black, J. W. (1975) Potential histamine H_2-receptor antagonists, 2.N^{α}-guanylhistamine. *J. Med. Chem.* **18**, 830–833.

Durant, G. J., Emmett, J. C., Ganellin, C. R., and Prain, H. D. (1976) Heterocyclic substituted-1,1-diamino-ethylene derivatives, methods for their preparation and compositions containing them. Smith Kline and French Laboratories, British Patent 1 421 792.

Durant, G. J., Emmett, J. C., Ganellin, C. R., Miles, P. D., Prain, H. D., Parsons, M. E., and White, G. R. (1977) Cyanoguanidine-thiourea equivalence in the development of the histamine H_2-receptor antagonist, cimetidine. *J. Med. Chem.* **20**, 901–906.

Durant, G. J., Brown, T. H., Emmett, J. C., Ganellin, C. R., Prain, H. D., and Young, R. C. (1982) Some Structure–Activity Considerations in H_2-Receptor Antagonists, in *The Chemical Regulation of Biological Systems* (Creighton, A. M. and Turner, S., eds.) special publication no. 42, The Royal Society of Chemistry, London.

Elder, J. B., Ganculi, P. C., and Gillespie, I. E. (1979) Cimetidine and gastric cancer. *Lancet* **i**, 1005–1006.

Flind, A. C., Rowley-Jones, D., and Backhouse, J. N. (1980) The Safety of Cimetidine: A Continuing Assessment, in *Proceedings of European Symposium* Italy, 1979, H_2-Receptor Antagonists in Peptic Ulcer Disease and Progress in Histamine Research, (Torsoli, A., Lucchelli, P. E., Brimblecombe, R. W., eds.) Excerpta Medica, Oxford, Amsterdam.

Forrest, J. A. H., Shearman, D. J. C., Spence, R., and Celestin, L. R. (1975) Neutropenia associated with metiamide. *Lancet* **i**, 392–393.

Ganellin, C. R. (1978) Chemistry and Structure–Activity Relationships of H_2-Receptor Antagonists, in *Handbook of Experimental Pharmacology* XVIII/2 (Rocha e Silva, M., ed.) Springer Verlag, Berlin, Heidelberg, New York.

Ganellin, R. (1981) Medicinal chemistry and dynamic structure–activity analysis in the discovery of drugs acting at histamine H_2-receptors. *J. Med. Chem.* **24**, 913–920.

Ganellin, C. R., Durant, G. J., and Emmett, J. C. (1976) Some chemical aspects of histamine H_2-receptor antagonists. *Fed. Proc.* **35**, 1924–1930.

Gilman, D. J., Jones, D. F., Oldham, K., Wardleworth, J. M., and Yellin, T. O. (1982) 2-Guanidinothiazoles as H_2-Receptor Antagonists, in *The Chemical Regulation of Biological Systems* (Creighton, A. M. and Turner, S., eds.) Special Publication No. 42, The Royal Society of Chemistry, London.

Gledhill, T., Howard O. M., Buck, M., Paul, A., and Hunt, R. H. (1983) Single nocturnal dose of an H_2-receptor antagonist for the treatment of duodenal ulcer. *Gut* **24**, 904–908.

Habs, M., Schmahl, D., Eisenbrand, G., and Preussmann, R. (1982) Carcinogenesis Studies with *N*-Nitrosocimetidine. 2. Oral Administration to Sprague-Dawley Rats, in *Nitrosamines and Human Cancer* (Banbury report 12) (Magee, P. N., ed.) Cold Spring Harbor Laboratories, New York.

Hardie, J., Platt, R., Hutton, J., Warren, P., Lee, S. A., Harris, G. D., and Davies, E. P. (1984) Polymorphs (Imperial Chemical Industries plc) European Patent 115,114.

Harvey, C. A. and Owen, D. A. A. (1984) Cardiovascular studies with SK&F 93319, An antagonist of histamine at both H_1- and H_2-receptors. *Br. J. Pharmacol.* **83**, 427–432.

Helman, C. A. and Tim, L. O. (1983) Pharmacology and clinical efficacy of ranitidine, a new H_2-receptor antagonist. *Pharmacother.* **3**, 185–192.

Humphray, J. M., Daly, M. J., and Stables, R. (1982) Inhibition of gastric acid secretion by AH 22216, a new long-acting histamine H_2-receptor antagonist (abstract). *Gut* **23**, A899.

Humphries, T. J., Myerson, R. M., Gifford, L. M., Aeugle, M. E., Josie, M. E., Wood, S. L., and Tannenbaum, P. J. (1984) A unique post-market outpatient surveillance program of cimetidine: Report on Phase II and Final Summary. *Am. J. Gastroent.* **79**, 593–596.

Johnson, L. R. (1971) Control of gastric secretion: No room for histamine? *Gastroenterology* **61**, 106–118.

Kerr, G. D. (1981) Cimetidine: Twice Daily Administration in Duodenal Ulcer—Results of a UK and Ireland Multicentre Study, in *Cimetidine in the 80's* (Baron, J. H., ed.) Churchill Livingstone, Edinburgh.

Lacerte, M., Rousseau, B., Parent, J. P., Pare, P., Levesque, D., and Falutz, S. (1984) Single daily dose of cimetidine for the treatment of symptomatic duodenal ulcer: Results of a comparative two-centre trial. *Curr. Ther. Res.* **35**, 777–782.

Larsson, L., Rehfeld, S., Stockbrugger, R., Blohme, G., Schoon, I. M., Lundqvist, G., Kindblom, L. G., Save-Soderberg, J., Grimelius, L., and Olbe, L. (1978) Mixed endocrine gastric tumors associated with hypergastrinemia of antral origin. *Am. J. Pathol.* **93**, 53–68.

Leslie, G. B. and Walker, T. F. (1977) A Toxicological Profile of Cimetidine, in *Cimetidine: Proceedings of the Second International Symposium on Histamine H_2-Receptor Antagonists* (Burland, W. L. and Simkins, M. A., ed.) Excerpta Medica, Oxford, Amsterdam.

Leslie, G. B., Noakes, D. N., Pollit, F. D., Roe, F. J. C., and Walker, T. F. (1981) A two-year study with cimetidine in the rat: Assessment for chronic toxicity and carcinogenicity. *Tox. Appl. Pharmacol.* **61**, 19–137.

Lijinsky, W. (1982) Carcinogenesis Studies with Nitrosocimetidine, in *Nitrosamines and Human Cancer* (Banbury Report 12) (Magee, P. N., ed.) Cold Spring Harbor Laboratories, New York.

Lin, T. M., Evans, D. C., Warrick, M. W., Pioch, R. P., and Ruffolo, R. R. (1983) Nizatidine, a new specific H_2-receptor antagonist. *Gastroenterology* **84**, 1231.

Loew, E. R. (1947) Pharmacology of antihistamine compounds. *Physiological Rev.* **27**, 542–573.

Lumma, W. C., Anderson, P. S., Baldwin, J. J., Bolhofer, W. A., Hakecker, C. N., Hirshfield, J. M., Pietruszkiewicz, A. M., Randall, W. C., Torchiana, M. L., Britcher, S. F., Clineschmidt, B. V., Denny, G. H., Hirschmann, R., Hoffman, J. M., Phillips, B. T., and Streeter, K. B. (1982) Inhibitors of gastric acid secretion: 3,4-diamino-1,2,5,-thiadiazole 1-oxides and 1,1-dioxides as urea equivalents in a series of histamine H_2-receptor antagonists. *J. Med. Chem.* **25**, 207–210.

Mahachai, V., Thomson, A. B. R., Grace, M., Cook, D., and Symes, A. (1982) Comparison of two cimetidine regimes on 24-hour intragastric acidity in patients with duodenal ulcer disease (abstract). *Gastroenterology* **82** (5), 122.

Nagata, T., Nagata, T., Hamabata, M., Sato, M., Murakiami, U., Enomoto, M., and Tamura, J. (1983) Oral toxicity studies of ranitidine hydrochloride in beagles. *J. Toxicol. Sci.* **8** (suppl. 1), 51–83.

Nielsen, S. T., Dove, P., Palumbo, G., Sandor, A., Buonato, C., Schiehser, G., Santilli, A., and Strike, D. (1984) Two H_2-receptor antagonists as inhibitors of gastric acid secretion (abstract). *Fed. Proc.* **43**, 1074.

Pounder, R. E., Williams, J. G., Hunt, R. H., Vincent, S. H., Milton-Thompson, G. J., and Misiewicz, J. J. (1977) The Effects of Oral cimetidine on Food-Stimulated Gastric Acid Secretion and 24-Hour Intragastric Acidity, in *Cimetidine: Proceedings of the Second International Symposium on Histamine H_2-Receptor Antagonists* (Burland, W. L. and Simkins, M. A., eds.) Excerpta Medica, Oxford, Amsterdam.

Poynter, D., Pick, C. R., Harcourt, R. A., Sutherland, M. F., Spurling, N. W., Ainse, G., Cook, J., and Gatehouse, D. (1982) Evaluation of Ranitidine Safety, in *The Clinical Use of Ranitidine, Second International Symposium on Ranitidine* (Misiewicz, J. J. and Wormsley, K. G., eds.) Med. Publ. Foundation.

Price, B. J. and Daly, M. J. (1983) Ranitidine and Other H_2-Receptor Antagonists—Recent Developments, in *Progress in Medical Chemistry* (Ellis, G. P. and West, G. B., eds.) vol. 20. Elsevier, New York.

Rendic, S., Sunjic, V., Toso, R., and Kajfez, F. (1979) Interaction of cimetidine with liver microsomes. *Xenobiotica* **9**, 555–564.

Rendic, S., Alebic-Kolbah, T., and Kajfez, F. (1982) Interaction of ranitidine with liver microsomes. *Xenobiotica* **12**, 9–17.

Rowley-Jones, D. and Flind, A. C. (1981) Continuing Evaluation of the Safety of Cimetidine, in *Cimetidine in the 80's*. (Baron, J. H., ed.) Churchill Livingstone, Edinburgh.

Shibata, T. (1984) Mannich-phenoxy reversed acetyloxyacetamide (TZU-0460) (abstract 29N 1-2S). Communication to 104th Ann. Meetg. Japan Pharmaceutical Soc., Sendai.

Sivelle, P. C., Underwood, A. H., and Jelly, J. A. (1982) The effects of histamine H_2-receptor antagonists on androgen action *in vivo* and dihydrotestosterone binding to the rat prostate androgen receptor *in vitro*. *Biochem. Pharmacol.* **31**, 677–684.

Stables, R. and Humphray, J. M. (1983) Antisecretory activity of loxtidine, a long acting H_2-receptor blocking drug. Abstr. 012, 12th Meeting European Histamine Research Society, Brighton.

Streett, C. S., Cimprich, R. E., and Robertson, J. L. (1984) Pathologic findings in the stomach of rats treated with the H_2-receptor antagonist in tiotidine. *Scand. J. Gastroenterol.* **19** (Suppl. 101), 109–117.

Sugimura, T., Nagao, M., and Okada, Y. (1966) Carcinogenic action of *N*-methyl-*N'*-nitro-*N*-nitrosoguanidine. *Nature* **210**, 962–963.

Suzuki, H. and Shiobara, Y. (1983) Acute toxicity of famotidine (YM-11170) in mice and rats. *Oyo Yakuri* **26**, 147–150.

Takagi, T., Takeda, M., and Maeno, H. (1982) Effect of a new potent H_2-blocker, 3-[[[2[(diaminomethylene)amino]-4-thiazolyl]methyl]-thio]-N^2-sulfamoylpropionamidine (YM-11170) on gstric secretion induced by histamine and food in conscious dogs. *Arch. Int. Pharmacodyn.* **256**, 49–58.

Takeda, M., Takagi, T., Yashima, Y., and Maeno, H. (1982) Effect of a new potent H_2-blocker, 3-[[[2[(diaminomethylene)amino]-4-thiazolyl] methyl]thio]-N^2-sulfamoylpropionamidine (YM-11170) on gastric secretion, ulcer formation and weight of male accessory sex organs in rats. *Arzneim. Forsch.* **32**, 734–737.

Takeuchi, M., Kaga, M., Kiguchi, M., Twata, M., Yamaguchi, M., and Shimpo, K. (1983) Chronic toxicity study of ranitidine hydrochloride orally administered in rats. *J. Toxicol. Sci.* **8**, (suppl. I), 25–49.

Tamura, J., Sato, N., Ezaki, H., Miyamoto, H., Oda, S., Hirsi, K., and Tokado, H. (1983a) Rabbits and subacute oral toxicity of ranitidine in rats. *J. Toxicol. Sci.* **8** (suppl. I), 1–24.

Tamura, J., Sato, N., Ezaki, H., and Yokoyama, S. (1983b) Teratological study on ranitidine hydrochloride in rabbits. *J. Toxicol. Sci.* **8** (suppl. I), 141–150.

Tarutani, M., Sakuma, H., Shiratsuchi, K., and Mieda, M. (1985a) Histamine H_2-receptor antagonistic action of *N*-{3-[3-(1-Piperidinylmethyl) phenoxy]propyl}acetoxyacetamide hydrochloride (TZU-0460). *Arzneim. Forsch.* **35**, 703–706.

Tarutani, M., Sakuma, H., Shiratsuchi, K., and Mieda, M. (1985b) Effects of *N*-{3-[3-(1-Piperidinylmethyl)phenoxy]propyl}-acetoxyacetamide hydrochloride (TZU-0460), a histamine H_2-receptor antagonist, on gastric acid secretion and ulcer formation. *Arzneim. Forsch.* **35**, 844–848.

Torchiana, M. L., Pendleton, R. G., Cook, P. G., Hanson, C. A., and Clineschmidt, B. V. (1983) Apparent irreversible H_2-receptor blocking and prolonged gastric antisecretory activities of 3-*N*[3-(1-piperidinomethyl)phenoxy]propyl]amino-4-amino-1,2,5-thiadiazole-1-oxide (L-643,441). *J. Pharmacol. Exp. Ther.* **224**, 514–519.

Tsuritani, M., Matsukawa, H., Aoki, H., and Seya, M. (1984) Inhibitory effects of 2-*N*-[3-[3-(1-piperidinomethyl)phenoxy]propyl]amino-5-amino-1,3,4-thiadiazole (TAS). Abstract 0-92, 57th Annual Meeting of Japanese Pharmacological Soc., Kyoto.

Walker, T. F., Whitehead, S. M., Leslie, G. B., Crean, G. P., and Roe, F. J. C. (1987) Safety evaluation of cimetidine: Report at the termination of a 7½ year study in dogs. *Human Toxicol.* **6**, 159–164.

Wilson, J. A., Johnston, D. A., Penston, J., and Wormsley, K. G. (1986) Inhibition of human gastric secretion by ICI 162,846—a new histamine H_2-receptor antagonist. *B r. J. Clin. Pharmacol.* **21**, 685–689.

Wormsley, K. G. (1984) Assessing the safety of drugs for the long-term treatment of peptic ulcers. *Gut* **25**, 1416–1423.

Wyllie, J. H., Hesselbo, T., and Black, J. W. (1972) Effects in man of histamine H_2-receptor blockade by burimamide. *Lancet* **ii**, 1117–1120.

Yellin, T. O. and Gilman, D. J. (1982) Guanidine derivatives (ICI Americas, European Patent 60,094.

Yellin, T. O., Buck, S. H., Gilman, D. J., Jones, D. F., and Wardleworth, J. M. (1979) ICI 125,211: A new gastric antisecretory agent acting on histamine H_2-receptors. *Life Sci.* **25**, 2001–2009.

Atypical Psychotropic Agents

Trazodone and Buspirone

MICHAEL S. EISON, DUNCAN P. TAYLOR,
AND LESLIE A. RIBLET

1. Introduction: Identifying Atypical Agents

The discovery and development of truly atypical psychotropic agents are by far more challenging endeavors than synthesizing an additional member of a prototypical class of compounds and demonstrating that it retains a desired pharmacological profile. The identification of novel molecules with potential for clinical application is often hampered by the preconceived notions of structure–activity requirements, desired neurochemical effects, and empirically validated behavioral predictors of clinical success that comfort those seeking to develop second generations of an existing class of drug. Despite dramatic progress in the neurosciences, we still do not understand the biological bases of psychopathology enough to design drugs that will selectively reverse those aberrant states of neural activity that are associated with many disorders. Rather, in order to rationally conceptualize future avenues of pharmacotherapy, we must rely upon imperfect, evolving knowledge about disease states and inferences we may make based upon a similarly incomplete appreciation of the mechanism of action of drugs known to be effective against clinical targets.

When structurally novel molecules are synthesized, perhaps the products of medicinal chemistry theories of what is required to achieve the therapeutic action sought, animal and biochemical models help identify those that are safe and believed to be effective in humans. However, the criteria against which we gage the probability of success in humans are often derived from assumptions based upon the activity of known, standard agents. Thus the preclinical models relied upon to predict the utility of new and potentially atypical psychotropic agents may be uniquely sensitive to the pharmacological actions of the prototypical classes of molecules that, by virtue of their clinical success, validated the use of particular screening tests. The development of atypical psychotropics is a high-risk venture because the empirical relationships between preclinical pharmacology and clinical efficacy have not yet been established for unique structural series. Should a new class of molecule prove to be inactive or only weakly active in the screening tests routinely used to characterize the actions of classical agents, then either the compound has little potential for success or the tests are not appropriate to the type of compound under investigation.

The clinic is the final arbiter of this dilemma; only after clinical trials have supported or disproven preclinical predictions of therapeutic activity can the distinction between inactive compound and inappropriate test be made. However, broad clinical screening of compounds would be both impractical and inappropriate. The combination of rational design, targeted biochemical and behavioral screening, and appropriately planned pharmacological studies that has contributed to the development of atypical antidepressants and anxiolytics is described in the case histories to follow.

2. Antidepressants

2.1. An Atypical Antidepressant: Trazodone

Trazodone (Desyrel®) was the first nontricyclic antidepressant to be registered for use in the United States. The complete story of trazodone's discovery has been reviewed previously (Silvestrini et al., 1981). Clinical trials have consistently shown trazodone to be an effective antidepressant when compared to placebo and at least as effective as the tricyclic antidepressants (TCA) in treating a variety of depressive disorders (for a review, *see* Bryant and Ereshefsky, 1982). Moreover, clinical reports suggested it might possess a rapid onset of action (3–7 d; Kellems et al., 1979) in a wide range of subtypes of depression (for a review, *see* Georgotas et al., 1982). In contrast to the TCAs, trazodone represents a major therapeutic step forward in its elimination of anticholinergic side effects, reduction of cardiac toxicity, and apparent safety in overdose (Gershon and Newton, 1980; Himmelhoch, 1981; Bryant and Ereshefsky, 1982; Henry and Ali, 1983). The side effect most commonly encountered with trazodone treatment is drowsiness. In this regard, it is roughly as sedative as amitriptyline and more so than imipramine (Newton, 1981). U'Prichard and Snyder (1977) have suggested that for the TCAs there exists a direct correlation between potency in in vitro radioligand binding at α_1-adrenergic sites and sedative potential in clinical use. Comprehensive in vitro and in vivo animal testing of trazodone reveals potent interactions at α_1-adrenergic receptors. For instance, the IC_{50} value for trazodone at this site in in vitro radioligand binding assays is 73 n*M* (Riblet and Taylor, 1981). It therefore appeared that the elimination or reduction of the sedative component of trazodone's profile might yield an agent with superior efficacy, utility, and patient compliance.

2.2. Discovery and Development of Nefazodone

Work with TCAs had demonstrated that chronic administration to rats resulted in decreased in vitro binding of radiolabeled ligands for β-adrenergic sites (Bannerjee et al., 1977) and type 2 serotonin ($5\text{-}HT_2$) sites (Peroutka and Snyder, 1980). The "down-regulation" of neurotransmitter receptors by chronic antidepressant administration has been linked to their therapeutic efficacy in clinical trials because of the time course involved in both phenomena. Similar chronic experiments with trazodone revealed that it could depress β-adrenergic binding (Clements-Jewery, 1978), and that this decrease occurred in the presence of reduced $5\text{-}HT_2$ binding (Riblet and Taylor, 1981). By adjusting the dose, route, and duration of trazodone administration, it was possible to produce a decrease in $5\text{-}HT_2$ binding without the decrease in β-adrenergic binding (Enna and Kendall, 1981; Riblet and Taylor, 1981). In addition to decreasing β-adrenergic and $5\text{-}HT_2$ binding after chronic administration, trazodone inhibited acute in vitro $5\text{-}HT_2$, but not β-adrenergic binding (Riblet and Taylor, 1981).

To employ a screening system that looks at changes in the characteristics of these binding sites after chronic administration of all newly synthesized candidate compounds would be excessively time-, labor-, and drug supply-intense. What was desired was a relatively rapid, efficient method that would use minimal quantities of compound, such as an in vitro receptor binding assay. Although with typical agents it would prove a poor policy to infer antidepressant efficacy from potency in the inhibition of in vitro 5-HT_2 binding (*see*, for example, Peroutka and Snyder, 1979), credibility for this approach was derived from investigations of trazodone, which is inactive in most traditional models of depression (Silvestrini et al., 1981). Specifically, within a congeneric series derived from trazodone, the inhibition of in vitro 5-HT_2 binding might be expected to indicate pharmacologic similarities to the parent compound with a high degree of correlation. Therefore, it seemed reasonable to employ this screen as a predictor of efficacy in our search for antidepressant analogs of trazodone.

The triazolopyridinone nucleus found in trazodone offered an excellent starting place for chemical modification in a search for new agents. We postulated that modifying this heterocyclic moiety would influence the degree of binding to 5-HT_2 and α-adrenergic sites. We recently discussed the relationship of structural changes in this series to effects on biological activity in preclinical screens for efficacy and safety (Taylor et al., 1986). These few structural modifications coupled with other observations suggested that the α-adrenergic blocking component of trazodone-like molecules could be dissociated from its serotonergic actions. Initial efforts along this line were directed at chemical elaboration of the triazolyl nucleus of etoperidone. A key compound produced from this class was the α-phenoxy derivative of etoperidone, MJ 13754, later given the nonproprietary name nefazodone (Temple and Lobeck, 1982; *see* Fig. 1). Its ability to reverse reserpine ptosis was detected early in its development, and it was subsequently found to be as potent as

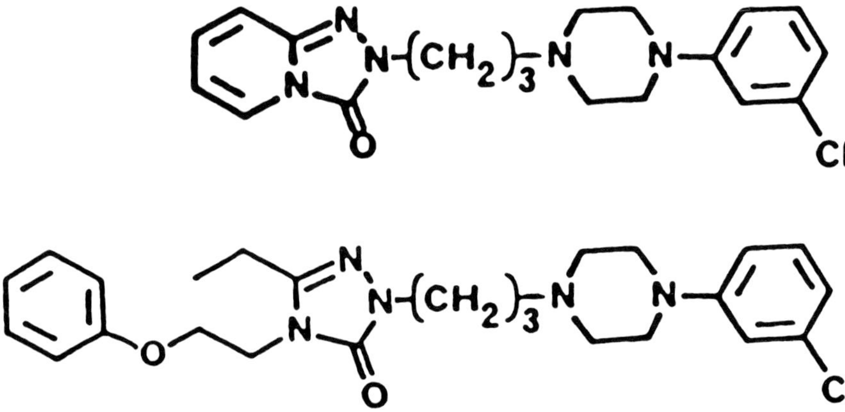

Fig. 1. Structures of trazodone (top) and nefazodone (bottom).

trazodone in 5-HT_2 binding in vitro (Taylor et al., 1982a; *see* Tables 1 and 2).

Table 1
Effects of Triazolyl Antidepressant Agents on Selected In Vivo Tests[a]

	ED_{50}, mg/kg, po	
Test system	Trazodone	Nefazodone
Prevention of reserpine-induced ptosis	>160	19.2
Inhibition of conditioned avoidance responding	45	76
Inhibition of norepinephrine-induced lethality	38	260
Inhibition of yohimbine-induced lethality	>160	154
Inhibition of histamine-induced lethality	>400	>400

[a]Data from Taylor et al. (1986).

The finding that nefazodone was equipotent with trazodone in its ability to inhibit in vitro 5-HT_2 binding was not in itself confirmation that an improved agent was in hand. However, nefazodone had significantly less potency than trazodone in the inhibition of in vitro α_1-adrenergic binding. In vivo testing confirmed a significant attenuation of α_1-adrenergic blocking activity in the prevention of norepinephrine-induced lethality. Further testing for reduction of spontaneous motor activity and potentiation of CNS depressant-induced loss of the righting reflex in rodents provided firm grounds for the belief that nefazodone would prove less sedating in clinical trials (Taylor et al., 1986; *see* Table 3). A pharmacologic profile of this type was considered desirable.

Experiments in which nefazodone was chronically administered to rats were conducted in order to investigate whether changes in receptor binding would occur. When a dose of 100 mg/kg was given either intraperitoneally (ip) for 11 d or orally (po) for 28 d, a decrease in in vitro 5-HT_2 binding in the cerebral cortex was observed (Taylor et al., 1986). This decrease was shown by ligand saturation experiments to be caused by a decrease in the maximum number of binding sites, whereas the affinity of the site for ^3H-spiperone was

Table 2
Effects of Triazolyl Antidepressant Agents on Selected In Vitro Receptor Binding Assays[a]

Receptor binding site, ^{3}H-ligand employed	IC_{50}, n*M*	
	Trazodone	Nefazodone
Type 2 serotonin (Spiperone-cortex)	42	43
α_1-Adrenergic (WB-4101)	73	212
α_2-Adrenergic (Clonidine)	860	5,500
β-Adrenergic (Dihydroalprenolol)	54,000	>1,000
H_1-Histamine (Pyrilamine)	640	805
D_2-Dopamine (Spiperone-striatum)	4,700	53,000
Benzodiazepine (Diazepam)	>100,000	>100,000
$GABA_A$ (Muscimol)	>100,000	>100,000
Muscarinic cholinergic (Quinuclidinyl benzilate)	75,000	17,000
μ-Opiate (Naloxone)	15,000	32,000
Calcium Channel (Nitrendipine)	96,000	13,500

[a]Data from Taylor et al. (1986).

unchanged. No effect was seen on β-adrenergic binding, although a slightly different dosing regimen (50 mg/kg, ip, 21 d) has been reported to produce decreases in both $5\text{-}HT_2$ and β-adrenergic binding (Scott and Crews, 1984, 1986). These indications of antidepressant potential were corroborated in a behavioral study in which nefazodone was found to reverse "learned helplessness" with an ED_{50} of 50 mg/kg (Johnson et al., 1982). Recent open-label studies in depressed individuals have shown improvement in depressive symptomatology (unpublished observations). A final determination of efficacy awaits the results of properly controlled double-blind trials now in progress.

In vitro radioligand binding studies were also used to identify and evaluate nefazodone's potential to produce side effects other than sedation. As expected from previous studies with trazodone (Hyslop and Taylor, 1980; Taylor et al., 1980), other triazolo compounds, including nefazodone, displayed very little in vitro affinity for muscarinic cholinergic binding sites. Moreover, this relative lack of affinity is reflected in a complete absence of protection of animals from the lethal effects of physostigmine (Taylor et al., 1986). It is

Table 3
Profiles of Antidepressant Agents in Tests of Sedation[a]

Agent	ED_{50}, mg/kg, po, mouse: Inhibition of spontaneous motor activity	Potentiation of loss of righting reflex with threshold depressants: Hexobarbital	Ethanol
Imipramine	2.7	3.8	6.5
Amitriptyline	38	4.6	6.1
Trazodone	38	8.9	5.4
Nefazodone	95	10.9	20.1

[a]Data from Taylor et al. (1986).

therefore expected that clinical trials will indicate that nefazodone lacks the anticholinergic side effects associated with TCA therapy.

The TCAs have very high affinity for the H_1 histamine binding site in vitro (Richelson, 1979), whereas nefazodone does not. Nefazodone is only 0.0012 as potent as the TCA doxepin. This low affinity in vitro is reflected in vivo, where doses of nefazodone as high as 400 mg/kg, po, do not protect laboratory animals from histamine-induced lethality (Taylor et al., 1986). Thus, nefazodone's potential for the induction of histamine-related side effects typical of TCAs, such as sedation and appetite enhancement with weight gain, appears to be minimal.

Trazodone is free of monoamine oxidase (MAO) inhibitory activity in vitro and in vivo. Similarly, nefazodone lacks affinity for type A or type B MAO in vitro (*see* Table 4). This is also reflected in vivo by its inability to potentiate the behavioral effects of tryptamine. These data suggest that clinical use of nefazodone will not be associated with the potential for life-threatening dietary interactions (Hyslop et al., submitted). Nefazodone displays no potency (IC_{50} values greater than 10,000 n*M*) in a variety of other in vitro receptor binding assays (dopamine, γ-aminobutyric acid, benzodiazepine, β-adrenergic, μ-opiate, glycine, and glutamate; Taylor et al., 1986). This further supports a prediction of minimal side effects.

In view of the highly selective inhibition of serotonin uptake by trazodone (Riblet et al., 1979), the ability of nefazodone to inhibit biogenic amine uptake was evaluated. An index of selectivity

Table 4
Profiles of Antidepressant Agents in Tests Predicting Mechanism of Action

Agent	Inhibition of in vitro activity, IC_{50}, n*M*					
	Monoamine oxidase activity[a]		^{3}H-Amine uptake[b]		^{3}H-Ligand binding[c]	
	Type A	Type B	Norepinephrine	Serotonin	Desipramine	Imipramine
Clomipramine			7,000	20	33	1
Trazodone	>100,000	>100,000	61,000	910	100,000	103
Amitriptyline	>100,000	>100,000	10,000	460	46	13
Imipramine	>100,000	>100,000	4,700	240	19	11
Nefazodone	>100,000	>100,000	8,700	3,200	3,400	980
Desipramine			2,700	1,400	2	114

[a]Data from Hyslop et al., submitted.
[b]Data from Taylor et al. (1986).
[c]Data from Yocca et al. (1985).

for the ability to inhibit the uptake of serotonin compared to that of norepinephrine was generated, which ranks clomipramine and trazodone as being fairly selective for serotonin (Taylor et al., 1986). Amitriptyline and imipramine are known to interact equally at both uptake sites (Riblet et al., 1979), and desipramine displays a preference for norepinephrine uptake sites. When such an index is calculated for nefazodone, it suggests a much greater preference for norepinephrine uptake sites than is exhibited by trazodone. That nefazodone's uptake inhibition profile might be less selective than trazodone's was suggested behaviorally by its ability to prevent reserpine-induced ptosis. Similarly, it has recently been shown that nefazodone is less selective than trazodone when in vitro binding of ^{3}H-imipramine and ^{3}H-desipramine is used to assess selectivity for inhibition of uptake at serotonin and norepinephrine transport sites, respectively (Yocca et al., 1985).

In addition to their potency in the in vitro inhibition of 5-HT_2 binding, both trazodone and nefazodone display affinity for 5-HT_1 binding sites of rat hippocampal membranes with IC_{50} values of 580 and 480 n*M*, respectively. It has been observed that serotonin receptor antagonists possess higher affinity for ^{3}H-tetrahydrotrazodone binding sites than do serotonin receptor agonists. Trazodone and nefazodone block the binding of ^{3}H-tetrahydrotrazodone in vitro (IC_{50} values of 50 and 39 n*M*, respectively; Kendall et al., 1983). Physiological antagonism has been observed in intact preparations of guinea pig tracheae where serotonergically-induced contractions are blocked with K_B values of 9.8 and 6.6 n*M* for trazodone and nefazodone, respectively (Hyslop et al., 1984; unpublished observations). Therefore, both trazodone and nefazodone display serotonin antagonist properties.

These studies suggest that nefazodone should be a clinically effective antidepressant without anticholinergic, antihistamine, or sedative side effects. Although this work has ruled out some mechanisms for the antidepressant efficacy of both trazodone and nefazodone, such as the inhibition of monoamine oxidase, the exact pharmacologic basis of their therapeutic action remains to be elucidated.

3. Anxiolytics

3.1. Discovery and Development of Buspirone

Buspirone is a nonbenzodiazepine anxiolytic with clinical potency comparable to diazepam (Goldberg and Finnerty, 1979; Rickels et al., 1982) and clorazepate (Goldberg and Finnerty, 1982). As clinical efficacy for this novel anxiolytic has been established, a retrospective look at the processes that led to its discovery and development will illustrate the evolution of an atypical psychotropic agent.

Buspirone (BuSpar®), known chemically as 8-(4-[4-(2-pyrimidinyl)-1-piperazinyl[butyl)-8-azaspiro-[4,5]-decane-7,9-dione hydrochloride, was originally designed to be a molecule of good fit to brain dopamine receptors (Eison, 1984). The first series of screening tests performed with buspirone generated a profile of activity that was consistent with the expectations for a major tranquilizer (antipsychotic) that were popularly accepted by the pharmaceutical industry at the time; that is, buspirone selectively inhibited the conditioned avoidance response in a barrier jump test, suppressed free-operant behavior in a continuous (Sidman, 1956) avoidance schedule, protected against the lethal effects of amphetamine-aggregation stress, and antagonized the emetic effects of apomorphine (Allen et al., 1974). Since blockade of dopamine-mediated stereotyped behavior is also considered a predictor of antipsychotic potential, buspirone's ability to partially antagonize the stereotyped effects of amphetamine in dogs, albeit an action of short duration (Sathananthan et al., 1975), encouraged a clinical investigation of its effects in schizophrenic patients. With a mean maximum dose of 1470 mg (ranging from 600 to 2400 mg), 60% of the patients exhibited a deterioration of their symptoms, 20% showed minimal improvement with no change in clinical status, and complete remission of symptoms was observed in only 20% of the patients tested (Sathananthan et al., 1975).

Although buspirone did not prove to be an effective antipsychotic in the clinic, subsequent preclinical investigations continued to indicate that it interacted with the dopamine system. Buspirone was found to block apomorphine-induced stereotypy in rats and increase dopamine turnover as reflected in dopamine metabolite levels (Stanley et al., 1979). Indications of clinical efficacy against anxiety neurosis (Goldberg and Finnerty, 1979) suggested that buspirone was an atypical anxiolytic, nonbenzodiazepine in structure and pharmacology. A series of studies comparing buspirone to the prototypical anxiolytic, diazepam, next ensued. It was learned that, like diazepam, buspirone reduced aggressivity in rhesus monkeys; however, whereas diazepam was active only at doses that also caused ataxia, buspirone tamed the aggressive monkey at doses free from motor impairments (Tompkins et al., 1980). Since an ability to attenuate punishment-induced response suppression is a reliable property of benzodiazepine anxiolytic drugs, buspirone was next investigated in conflict tests. It was found that like diazepam, buspirone attenuated conflict behavior in rats and monkeys responding for food (Hartmann and Geller, 1981; Geller and Hartmann, 1982). Buspirone was also found active in a conflict test utilizing drinking behavior, and in reducing shock-elicited aggression in mice (Riblet et al., 1982).

The atypical nature of buspirone was emphasized by its lack of the ancillary pharmacologic properties characteristic of the benzodiazepine anxiolytics. Unlike diazepam, buspirone did not exhibit anticonvulsant properties, failing to protect against seizures induced by pentylenetetrazol, bicuculline, picrotoxin, strychnine, and maximal electroshock (Riblet et al., 1982). Furthermore, buspirone only minimally potentiated the sedative effects of other CNS depressants when compared to diazepam or clorazepate. In addition, although the dose of diazepam required to cause a lethal outcome in half the animals tested was significantly lowered by coadministration of ethanol, the lethality of buspirone was essentially unchanged (Riblet et al., 1982). Abuse potential and dependence liability are also associated with many benzodiazepine anxiolytics. Buspirone proved to be atypical in that it was not active in tests predictive of these properties. For example, buspirone did not substitute for cocaine in rhesus monkeys trained to self-administer this drug of abuse (Balster and Woolverton, 1982). Furthermore, it is unlikely that a drug that does not produce interoceptive cues that are readily recognized will be abused; buspirone does not serve as a discriminable cue in drug discrimination tests, nor will it substitute for oxazepam or pentobarbital in animals well trained to recognize these anxiolytics (Hendry et al., 1983). These preclinical predictors, which suggested that buspirone would lack abuse potential, appear to be validated by clinical studies that indicate recreational drug abusers do not find buspirone a desirable drug to take (Cole et al., 1982). When drugs known to cause dependence in man, such as barbiturates, opiates, or benzodiazepines, are chronically administered to rats and then abruptly withdrawn, an abstinence syndrome is observed that includes reductions in daily body weight change and frank weight loss during the withdrawal period (Yanaura et al., 1975; Rosenberg and Chiu, 1982). The weight loss associated with withdrawal from chronic diazepam induced by cessation of treatment or administration of the benzodiazepine antagonist Ro 15-1788 (McNicholas and Martin, 1982) is not observed in rats chronically treated with buspirone (Riblet et al., 1982; Taylor et al., 1984; Eison, 1986).

The atypical nature of buspirone is further emphasized by its possession of several pharmacological properties that appear to be unique among clinically proven anxiolytics. For example, although buspirone does not induce a neuroleptic-like catalepsy in rats, it potently reverses a previously established catalepsy (Riblet et al., 1982; McMillen and Mattiace, 1983), recent evidence suggests that buspirone is also capable of preventing chronic neuroleptic-induced dopamine receptor supersensitivity (McMillen, 1983). The anti-neuroleptic effects of buspirone do not, however, predict its effects on cholinergically induced catalepsy. Buspirone has been reported to enhance the catalepsy induced by the cholinergic agonists oxotremorine and arecoline, whereas diazepam does not (Eison, 1984). Unlike the anticonvulsant benzodiazepine anxiolytics, buspirone reduces the convulsant threshold to picrotoxin, bicuculline, and strychnine (Eison and Eison, 1984).

3.2. Novel Mechanisms of Action

It is now well accepted that benzodiazepine-induced alterations in γ-aminobutyric acid (GABA) binding at the benzodiazepine receptor complex modulate chloride anion flux and, thereby, neuronal excitability (Williams, 1983). As may be expected, buspirone's atypical, nonbenzodiazepine pharmacology is associated with similarly nonbenzodiazepine actions upon the brain. Specifically, buspirone lacks affinity for benzodiazepine binding sites (Riblet et al., 1982, 1984) and for GABA or picrotoxin binding sites (Riblet et al., 1982; Tunnicliff and Welborn, 1984). Furthermore, buspirone's anticonflict effects in rats are not blocked by the benzodiazapine antagonists Ro 15-1788 or CGS 8216 (Oakley and Jones, 1983; Weissman et al., 1984).

Buspirone profoundly affects brain monoamine systems. It increases the activity of brain dopamine neurons (Riblet et al., 1982; McMillen et al., 1983), increases levels of dopamine metabolites (Kolasa et al., 1982; Cimino et al., 1983), and increases striatal tyrosine hydroxylase activity (McMillen and McDonald, 1983). Buspirone binds to dopamine receptors (Riblet et al., 1982) and appears to act as a selective presynaptic dopamine antagonist in that it does not block apomorphine-induced rotation in rats unilaterally lesioned with the neurotoxin 6-hydroxydopamine (McMillen et al., 1983; Eison and Eison, 1984). Although benzodiazapines typically reduce the activity of noradrenergic neurons in the locus coeruleus (Redmond and Huang, 1979; Grant et al., 1980), buspirone increases their firing (Sanghera et al., 1983; Sanghera and German, 1983). Buspirone also interacts potently with brain serotonin neurons, inhibiting the activity of cells in the dorsal raphe nucleus following either systemic or iontophoretic administration (VanderMaelen and Wilderman, 1984). The benzodiazepines have been reported to inhibit the firing of serotonergic dorsal raphe cells in unanesthetized animals (Trulson et al., 1982) at doses higher than those required of buspirone to inhibit serotonergic activity in anesthetized preparations. Buspirone has been reported to lower brain levels of serotonin and its principle metabolites (Hjorth and Carlsson, 1982) and to bind to serotonin receptors (Glaser and Traber, 1983; Riblet et al., 1984). The cholinergic system is also influenced by buspirone; it reduces striatal acetylcholine levels in rat brain (Kolasa et al., 1982).

The complex neurochemical effects of this atypical anxiolytic have stimulated reevaluation of our assumptions of what the biological basis of anxiety really is (Gershon and Eison, 1983). Although the importance of the dopaminergic component to buspirone's clinical effects is not known, several hypotheses regarding the role of dopamine in the mediation of anxiety have been proposed (Taylor et al., 1982b, 1983). Our concept of how buspirone works to relieve anxiety without the pharmacological properties and liabilities of the benzodiazepines continues to evolve as new data become available. The possibility that buspirone acts as a "midbrain modulator," simultaneously altering multiple neurochemical systems in a manner consistent with anxiolysis in the absence of other benzodiazepine actions, has been proposed (Eison and Eison, 1984).

3.3. Second-Generation Compounds: Gepirone

One technique used for identifying those components of buspirone's mechanism of action that are critical to its anxiolytic properties, has been to synthesize and investigate a series of chemical analogs that retain those pharmacologic actions of buspirone predictive of antianxiety effects, while significantly altering its neurophysiological profile. Gepirone, also known as BMY 13805, is the product of a synthetic effort to retain buspirone's anticonflict- and catalepsy-reversing properties, while eliminating its direct interactions with the dopamine system (Eison et al., 1985; *see* Fig. 2). Gepirone is equipotent with buspirone in conflict tests in rats, inhibits shock-elicited aggression in mice, and like buspirone, reverses neuroleptic-induced catalepsy (Eison et al., 1982, 1985). However, unlike buspirone, it does not antagonize apomorphine-induced stereotyped behavior (Eison et al., 1982) or bind to dopamine receptors (Eison et al., 1985). Although the above observations suggest that buspirone's impact on the dopamine system may not be important to its anxiolytic effects, there is evidence to suggest that gepirone indirectly influences dopaminergic activity. Gepirone does increase striatal levels of dopamine metabolites, though it does so less potently than buspirone (McMillen and Mattiace, 1983). It does not, however, alter tyrosine hydroxylase activity nor block the effects of apomorphine upon this enzyme (McMillen and Mattiace, 1983).

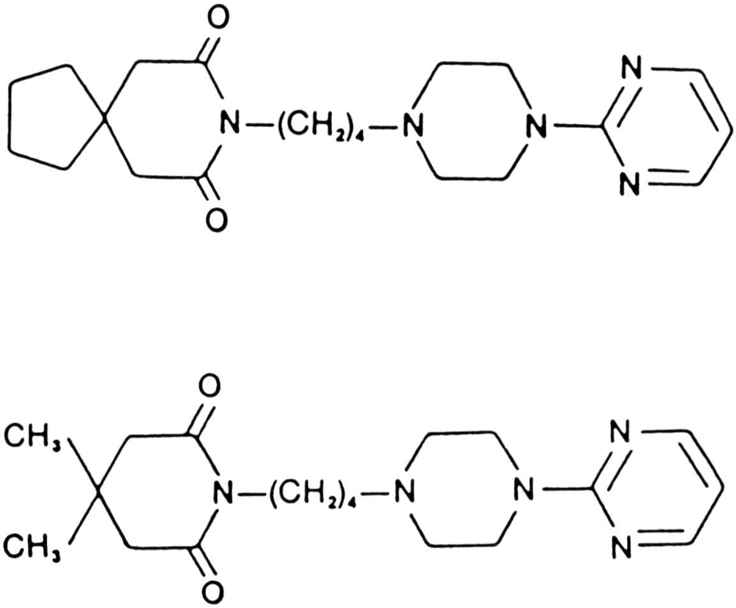

Fig. 2. Structures of buspirone (top) and gepirone (bottom).

Like buspirone, gepirone interacts with the serotonin system. Gepirone induces a behavioral syndrome similar to that observed following administration of serotonin agonists, and induces agonist-like rotation in rats with unilateral serotonin lesions (Eison et al., 1983). Chronic oral administration of gepirone down-regulates serotonin receptors without changing dopamine receptors (Eison et al., 1984b), whereas continuous administration achieved through osmotic minipump implants reduces both the number of serotonin receptors and the animals' behavioral responsiveness to serotonergic agonists (Eison and Yocca, 1985). Like buspirone, gepirone also inhibits the activity of serotonergic cells in the dorsal raphe nuclei (Eison et al., 1984a). Buspirone and gepirone also have similar effects upon noradrenergic activity in the locus coeruleus (Sanghera and German, 1983), and both lack anticonvulsant and muscle relaxant properties (Eison et al., 1985). Like buspirone, gepirone does not bind to benzodiazepine receptors or potentiate the actions of central nervous system depressants, such as alcohol or phenobarbital (Eison et al., 1982).

Gepirone is currently undergoing clinical trials in patients suffering from generalized anxiety disorder. Should gepirone prove to be effective in relieving anxiety, then the analog approach to developing second generation atypical psychotropics that differ from the first of a new class of psychotherapeutic agents will be validated.

4. Conclusions

The first generation of antidepressants were discovered serendipitously. The clinical profile of the second-generation antidepressant trazodone suggested that nontricyclic agents could be developed that provided a significant step forward in therapy by the elimination of anticholinergic and cardiotoxic side effects. A deliberate effort to synthesize trazodone analogs and employ receptor binding techniques to predict the potential for efficacy (acutely or chronically) and side effects resulted in the discovery of nefazodone. Assuming that preliminary trends toward efficacy are satisfactorily demonstrated in appropriate clinical trials, the identification of nefazodone will represent the successful application of rational chemical design and target-oriented biological testing with emphasis on the techniques of receptor binding.

The evolution of buspirone as an anxiolytic has challenged prevailing concepts of both the biochemical and behavioral requisites for effective agents in this psychotropic class. Through a combination of appropriate pharmacologic testing and directed chemical synthesis, the anxioselective analog gepirone was developed. We await the results of clinical trials (Cott et al., 1985) to confirm that the freedom to challenge conventional thinking concerning psychotropic drug action can produce atypical new drugs like gepirone.

References

Allen, L. E., Ferguson, H. C., and Cox, R. H., Jr. (1974) Pharmacologic effects of MJ 9022-1, a potential tranquilizing agent. *Arzn. Forschung* **24**, 917–922.

Balster, R. L. and Woolverton, W. L. (1982) Intravenous buspirone self-administration in rhesus monkey. *J. Clin. Psychiat.* **43**, 34–37.

Bannerjee, S. P., Kung, L. S., Riggi, S. J., and Chanda, S. K. (1977) Development of beta-adrenergic receptor subsensitivity by anti-depressants. *Nature* **268**, 455–456.

Bryant, S. G. and Ereshefsky, L. (1982) Antidepressant properties of trazodone. *Clin. Pharm.* **1**, 406–417.

Cimino, M., Ponzio, F., Achilli, G., Vantini, G., Perego, C., Algeri, S., and Garattini, S. (1983) Dopaminergic effects of buspirone, a novel anxiolytic agent. *Biochem. Pharmacol.* **32**, 1069–1074.

Clements-Jewery, S. (1978) The development of cortical β-adrenoceptor subsensitivity in the rat by chronic treatment with trazodone, doxepin, and mianserine. *Neuropharmacology* **17**, 779–781.

Cole, J. P., Orzack, M. H., Beake, B., Bird, M., and Bar-Tal, Y. (1982) Assessment of the abuse liability of buspirone in recreational sedative users. *J. Clin. Psychiat.* **43**, 69–74.

Cott, J. M., Kurtz, N. M., and Robinson, D. S. (1985) Gepirone (BMY 13805): A new analog of buspirone with anxiolytic potential. *IVth Wld. Congr. Biol. Psychiat. Abstr.* 213.

Eison, M. S. (1984) Use of animal models: Toward anxioselective drugs. *Psychopathol.* **17**, 37–44.

Eison, M. S. (1986) Lack of withdrawal signs of dependence following cessation of treatment or Ro.15-1788 administration to rats chronically treated with buspirone. *Neuropsychobiology,* in press.

Eison, M. S. and Eison, A. S. (1984) Buspirone as a midbrain modulator: Anxiolysis unrelated to traditional benzodiazepine mechanisms. *Drug Dev. Res.* **4**, 109–119.

Eison, A. S. and Yocca, F. D. (1985) Reduction in cortical 5-HT_2 receptor sensitivity after continuous gepirone treatment. *Eur. J. Pharmacol.* **111**, 389–392.

Eison, M. S., Taylor, D. P., Riblet, L. A., New, J. S., Temple, D. L., Jr., and Yevich, J. P. (1982) MJ 13805-1: A potential nonbenzodiazepine anxiolytic. *Soc. Neurosci. Abstr.* **8**, 470.

Eison, A. S., Eison, M. S., Riblet, L. A., and Temple, D. L., Jr. (1983) Indications of serotonergic involvement in the actions of a potential nonbenzodiazepine anxiolytic: MJ 13805. *Soc. Neurosci. Abstr.* **9**, 436.

Eison, M. S., Eison, A. S., Taylor, D. P., VanderMaelen, C. P., Riblet, L. A., and Temple, D. L. Jr. (1984a) Preclinical indications of antidepressant potential in a serotonergic anxiolytic candidate, MJ 13805. *IUPHAR (Lond.) Abstr.*, 2018P.

Eison, M. S., Taylor, D. P., Eison, A. S., VanderMaelen, C. P., Riblet, L. A., and Temple, D. L. Jr. (1984b) Pharmacologic effects of chronic administration of the nonbenzodiazepine antidepressant-anxiolytic candidate, BMY 13805. *Soc. Neurosci. Abstr.* **10**, 259.

Eison, M. S., Yevich, J. P., and Farney, R. F. (1985) Gepirone hydrochloride. *Drugs of the Future* **10**, 456–457.

Enna, S. J. and Kendall, D. A. (1981) Interaction of antidepressants with brain neurotransmitter receptors. *J. Clin. Psychiat.* **1**, 12S–16S.

Geller, I. and Hartmann, R. J. (1982) Effects of buspirone on operant behavior of laboratory rats and cynomolgus monkeys. *J. Clin. Psychiat.* **43**, 25–32.

Georgotas, A., Forsell, T. L., Mann, J. J., Kim, M., and Gershon, S. (1982) Trazodone hydrochloride: A wide spectrum antidepressant with a unique pharmacological profile. *Pharmacotherapy* **2**, 255–265.

Gershon, S. and Eison, A. S. (1983) Anxiolytic profiles. *J. Clin. Psychiat.* **44**, 45–56.

Gershon, S. and Newton, R. (1980) Lack of anticholinergic side effects with a new antidepressant—trazodone. *J. Clin. Psychiat.* **41**, 100–104.

Glaser, T. and Traber, J. (1983) Buspirone: Action on serotonin receptors in calf hippocampus. *Eur. J. Pharmacol.* **88**, 137–138.

Goldberg, H. L. and Finnerty, R. J. (1979) The comparative efficacy of buspirone and diazepam in the treatment of anxiety. *Am. J. Psychiat.* **136**, 1184–1187.

Goldberg, H. L. and Finnerty, R. J. (1982) Comparison of buspirone in 2 separate studies. *J. Clin. Psychiat.* **43**, 87–91.

Grant, S. J., Huang, Y. H., and Redmond, D. E. Jr. (1980) Benzodiazepines attenuate single unit activity in the locus coeruleus. *Life Sci.* **27**, 2231–2236.

Hartmann, R. J. and Geller, I. (1981) Effects of buspirone on conflict behavior of laboratory rats and monkeys. *Proc. West. Pharmacol. Soc.* **24**, 179–181.

Hendry, J. S., Balster, R. L., and Rosecrans, J. A. (1983) Discriminative stimulus properties of buspirone compared to central nervous system depressants in rats. *Pharmacol. Biochem. Behav.* **19**, 97–101.

Henry, J. A. and Ali, C. J. (1983) Trazodone overdosage: Experience from a poisons information service. *Human Toxicol.* **2**, 353–356.

Himmelhoch, J. M. (1981) Cardiovascular effects of trazodone in humans. *J. Clin. Psychopharmacol.* **1**, 76S–81S.

Hjorth, S. and Carlsson, A. (1982) Buspirone: Effects of central monoaminergic transmission. Possible relevance to animal experimental and clinical findings. *Eur. J. Pharmacol.* **83**, 299–303.

Hyslop, D. K. and Taylor, D. P. (1980) The interaction of trazodone with rat brain muscarinic cholinoceptors. *Br. J. Pharmacol.* **71**, 359–361.

Hyslop, D. K., Becker, J. A., Eison, M. S., and Taylor, D. P. (1984) Pharmacologic studies of tetrahydrotrazodone. *Fed. Proc.* **43**, 941.

Johnson, J., Sherman, A., Petty, F., Taylor, D., and Henn, F. (1982) Receptor changes in learned helplessness. *Soc. Neurosci. Abstr.* **8**, 392.

Kellams, J. J., Klapper, M. H., and Small, J. G. (1979) Trazodone, a new antidepressant: Efficacy and safety in endogenous depression. *J. Clin. Psychiat.* **40**, 390–395.

Kendall, D. A., Taylor, D. P., and Enna, S. J. (1983) [^{3}H]Tetrahydrotrazodone binding. Association with serotonin binding sites. *Mol. Pharmacol.* **23**, 594–599.

Kolasa, K., Fusi, R., Garattini, S., Consolo, S., and Ladinsky, H. (1982) Neurochemical effects of buspirone, a novel psychotropic drug, on the central cholinergic system. *J. Pharm. Pharmacol.* **34**, 314–317.

McMillen, B. A. (1983) Comparative effects of sub-chronic buspirone or neuroleptics on rat brain dopamine functions. *Soc. Neurosci. Abst.* **9**, 157.

McMillen, B. A., Matthews, R. T., Sanghera, M. K., Shepard, P. D., and German, D. C. (1983) Dopamine receptor antagonism by the novel antianxiety drug, buspirone. *J. Neurosci.* **3**, 733–738.

McMillen, B. A. and Mattiace, L. A. (1983) Comparative neuropharmacology of buspirone and MJ-13805, a potential antianxiety drug. *J. Neural. Trans.* **57**, 255–265.

McMillen, B. A. and McDonald, C. C. (1983) Selective effects of buspirone and molindone on dopamine metabolism and function in the striatum and frontal cortex of the rat. *Neuropharmacology* **22**, 273–278.

McNicholas, L. F. and Martin, W. R. (1982) The effects of a benzodiazepine antagonist, RO-15,1788 in diazepam dependent rats. *Life Sci.* **31**, 731–737.

Newton, R. (1981) The side effect profile of trazodone in comparison to an active control and placebo. *J. Clin. Psychopharmacol.* **1**, 89S–93S.

Oakley, N. R. and Jones, B. J. (1983) Buspirone enhances [^{3}H]flunitrazepam binding in vivo. *Eur. J. Pharmacol.* **87**, 499–500.

Peroutka, S. J. and Snyder, S. H. (1979) Multiple serotonin receptors: Differential binding of ^{3}H-5-hydroxytryptamine, ^{3}H-lysergic acid diethylamide and ^{3}H spiroperidol. *Molec. Pharmacol.* **16**, 687–699.

Peroutka, S. J. and Snyder, S. H. (1980) Chronic antidepressant treatment lowers spiroperidol-labeled serotonin receptor binding. *Science* **210**, 88–90.

Redmond, D. E. and Huang, Y. H. (1979) New evidence for a locus coeruleus norepinephrine connection with anxiety. *Life Sci.* **25**, 2149–2162.

Riblet, L. A. and Taylor, D. P. (1981) Pharmacology and neurochemistry of trazodone. *J. Clin. Psychopharmacol.* **1**, 17S–22S.

Riblet, L. A., Gatewood, C. F., and Mayol, R. F. (1979) Comparative effects of trazodone and tricyclic antidepressants on uptake of selected neurotransmitters by isolated rat brain synaptosomes. *Psychopharmacology* **63**, 99–101.

Riblet, L. A., Taylor, D. P., Eison, M. S., and Stanton, H. C. (1982) Pharmacology and neurochemistry of buspirone. *J. Clin. Psychiat.* **43**, 11–16.

Riblet, L. A., Eison, A. S., Eison, M. S., Taylor, D. P., Temple, D. L., Jr., and VanderMaelen, C. P. (1984) Neuropharmacology of buspirone. *Psychopathology* **17**, 69–78.

Richelson, E. (1979) Tricyclic antidepressants and neurotransmitter receptors. *Psychiatr. Ann.* **9**, 16–25.

Rickels, K., Weisman, N., Norstad, N., Singer, M., Stoltz, P., Brown, A., and Danton, J. (1982) Buspirone and diazepam in anxiety: A controlled study. *J. Clin. Psychiat.* **43**, 81–86.

Rosenberg, H. D. and Chiu, T. H. (1982) An antagonist-induced benzodiazepine abstinence syndrome. *Eur. J. Pharmacol.* **81**, 153–157.

Sanghera, M. K. and German, D. C. (1983) The effects of benzodiazepine and non-benzodiazepine anxiolytics on locus coeruleus unit activity. *J. Neural. Trans.* **57**, 267–279.

Sanghera, M. K., McMillen, B. A., and German, D. C. (1983) Buspirone, a non-benzodiazepine anxiolytic, increases locus coeruleus noradrenergic neuronal activity. *Eur. J. Pharmacol.* **86**, 106–110.

Sathananthan, G. L., Sanghvi, I., Phillips, N., and Gershon, S. (1975) MJ 9022: Correlation between neuroleptic potential and stereotypy. *Curr. Ther. Res.* **18**, 701–705.

Scott, J. A. and Crews, F. T. (1984) Studies on the mechanism of serotonin$_2$ receptor down-regulation by antidepressants. *Soc. Neurosci. Abstr.* **10**, 892.

Scott, J. A. and Crews, F. T. (1986) Down-regulation of serotonin$_2$, but not beta-adrenergic receptors during chronic amitriptyline treatment is independent of serotonin and beta-adrenergic receptor stimulation. *Neuropharmacology*, in press.

Sidman, M. (1956) Drug-behavior interaction. *Ann. N.Y. Acad. Sci.* **65**, 282–302.

Silvestrini, B., Lisciani, R., and De Gregorio, M. (1981) Trazodone, in *Pharmacological and Biochemical Properties of Drug Substances* vol. 3 (Goldberg, M. E., ed.) *American Pharmaceutical Association, Academy of Pharmaceutical Sciences,* Washington, DC.

Stanley, M., Russo, A., and Gershon, S. (1979) The effect of MJ 9022-1 on striatal DOPAC and apomorphine-induced stereotyped behavior in the rat. *Res. Commun. Psychol. Psychiat. Behav.* **4**, 127–134.

Taylor, D. P., Hyslop, D. K., and Riblet, L. A. (1980) Trazodone, a new nontricyclic antidepressant with anticholinergic activity. *Biochem. Pharmacol.* **29**, 2149–2150.

Taylor, D. P., Eison, M. S., Riblet, L. A., Temple, D. L., Jr., Yevich, J. P., and Smith, D. W. (1982a) Selective 5-HT_2 receptor blockade: Pharmacologic studies of MJ 13754, a nontricyclic antidepressant candidate. *Soc. Neurosci. Abstr.* **8**, 465.

Taylor, D. P., Riblet, L. A., Stanton, H. C., Eison, A. S., Eison, M. S., and Temple, D. L., Jr. (1982b) Dopamine and antianxiety activity. *Pharm. Biochem. Behav.* **17**, 25–35.

Taylor, D. P., Riblet, L. A., and Stanton, H. C. (1983) Dopamine and Anxiolytics, in *Anxiolytics: Neurochemical, Behavioral, and Clinical Perspectives* (Malick, J. B., Enna, S. J., and Yamamura, H. I., eds.) Raven, New York.

Taylor, D. P., Allen, L. E., Becker, J. A., Crane, M., Hyslop, D. K., and Riblet, L. A. (1984) Changing concepts of the biochemical action of the anxioselective drug, buspirone. *Drug Dev. Res.* **4**, 95–108.

Taylor, D. P., Smith, D. W., Hyslop, D. K., Riblet, L. A., and Temple, D. L. Jr. (1986) Receptor Binding and Atypical Antidepressant Drug Discovery, in *Receptor Binding in Drug Research* (O'Brien, R. A., ed.) Marcel Dekker, New York.

Temple, D. L. and Lobeck, W. G. (1982) US Pat. No. 4,388,317.

Tompkins, E. C., Clemento, A. J., Taylor, D. P., and Perhach, J. L. Jr. (1980) Inhibition of aggressive behavior in rhesus monkeys by buspirone. *Res. Commun. Psychol. Psychiat. Behav.* **5**, 337–352.

Trulson, M. E., Preussler, D. W., Howell, G. A., and Frederickson, C. J. (1982) Raphe unit activity in freely moving cats: Effects of benzodiazepines. *Neuropharmacol.* **21**, 1045–1050.

Tunnicliff, G. and Welborn, K. L. (1984) The action of structural analogs of γ-aminobutyric acid on binding sites in mouse brain. *Drug Dev. Res.* **4**, 51–59.

U'Prichard, D. and Snyder, S. H. (1977) Therapeutic and Side Effects of Psychotropic Drugs: The Relevance of Receptor Binding Methodology, in *Animal Models in Psychiatry and Neurology* (Hanin, I. and Usdin, E., eds.) Pergamon, Oxford.

VanderMaelen, C. P. and Wilderman, R. C. (1984) Iontophoretic and systemic administration of the non-benzodiazepine anxiolytic drug buspirone causes inhibition of serotonergic dorsal raphe neurons in rats. *Fedn. Proc.* **43**, 947.

Weissman, B. A., Barrett, J. E., Brady, L. S., Witkin, J. M., Mendelson, W. B., Paul, S. M., and Skolnick, P. (1984) Behavioral and neurochemical studies on the anticonflict actions of buspirone. *Drug Dev. Res.* **4**, 83–93.

Williams, M. (1983) Anxioselective anxiolytics. *J. Med. Chem.* **26**, 619–628.

Yanaura, S. E., Tagashira, E., and Suzuki, T. (1975) Physical dependence on morphine, phenobarbital, and diazepam in rats by drug-admixed food ingestions. *Jap. J. Pharmacol.* **25**, 453–463.

Yocca, F. D., Hyslop, D. K., and Taylor, D. P. (1985) Nefazodone: A potential broad spectrum antidepressant. *Trans. Am. Soc. Neurochem.* **16**, 115.

Calcium Channel Antagonists

ROBERT J. GOULD

1. Introduction

1.1. Calcium and Cellular Function

In 1883, Sidney Ringer described the importance of calcium in cardiac contraction (Ringer, 1883). Since that seminal observation, the importance of calcium as a transducer for coupling of biological signals has become evident. The signal transduction role of calcium is made possible by a large, inward-directed gradient of ionized calcium across the plasma membrane. Cytosolic concentrations of calcium in a resting, nonstimulated cell are some 10,000-fold lower than external concentrations (approximately 10^{-7} *M* vs 10^{-3} *M*) (Fozzard et al., 1985; Tsien et al., 1984). This extreme gradient is maintained by an intrinsic low permeability of the plasma membrane to calcium, active exchange mechanisms and pumps that remove calcium, sequestration of calcium by intracellular organelles such as mitochondria and the endoplasmic reticulum, and buffering by cytosolic calcium-binding proteins. Elevation of cytosolic calcium contractions to 1–10 μM initiates physiologic responses appropriate to the cell type. In muscle, interaction with troponin C or other calcium-binding regulatory proteins initiates contraction. In neuonal, exocrine, and endocrine tissue, secretion ensues. The diversity of responses initiated, the energy expenditure to maintain low cytosolic levels, and the multiplicity of ways to deplete cytosolic calcium all highlight the biologic importance of this messenger system.

How, then, is cytosolic calcium elevated? Calcium can be released from intracellular stores such as mitochondria or the endomitochondrial, intracellular stores is currently thought to be mediated by inositol 1,4,5-trisphosphate (for review, *see* Berridge and Irvine, 1984). Inositol 1,4,5-trisphosphate appears to be a mechanism of receptor-activated calcium mobilization in tissues as diverse as liver and cultured smooth muscle cells (Burgess et al., 1984; Yamamoto and Van Breeman, 1985). Calcium may also be mobilized from extracellular sources via receptor-operated channels and/or voltage-operated channels. The stimulus for opening of the voltage-operated channels is membrane depolarization via, for example, elevated extracellular K^+, whereas that for receptor-operated channels is the binding of agonist to its receptor (Bolton, 1979). It is not known at this time if the inositol phosphates play a role in these receptor-operated channels. Many tissues appear to have a complex, interacting mixture of these two functionally defined channels (Karaki and Weiss, 1984).

1.2. Therapeutic Utility of Blocking Calcium Entry

Cellular events that depend on elevated cytosolic calcium may then be regulated by modulating Ca^{2+} entry via receptor- or voltage-operated channels, by modulating Ca^{2+} interaction with calcium binding proteins, by altering Ca^{2+} sequestration into intracellular organelles, or by controlling Ca^{2+} extrusion from the cell. Thus far, the most therapeutic utility has been seen by blocking Ca^{2+} entry via voltage-operated channels. The compounds that effect this blockade, variously known as Ca^{2+} antagonists, Ca^{2+} blockers, Ca^{2+} entry blockers, or Ca^{2+} channel antagonists, may be or have been effective in treating cardiac arrhythmias, hypertrophic cardiomyopathy, myocardial failure, angina, hypertension, peripheral vascular disease, migraine, subarachnoid hemorrhage, dysmenorrhea, premature labor, myometrial hyperactivity, achalasia, "nutcracker" esophagus, irritable bowel disease, and asthma (Henry, 1980; Stone et al., 1980; Singh et al., 1983; Landmark et al., 1982; Theroux et al., 1983; Spivack et al., 1983; Kambara et al., 1981; Emanuel, 1979; Rodeheffer et al., 1983; Gelmers, 1983; Allen et al., 1983; Sandahl et al., 1979; Ulmsten et al., 1980; Forman et al., 1981; Bartoletti and Labo, 1981; Richter, 1984; Narducci et al., 1985; Cerrina et al., 1981). In general, these compounds are effective in disorders of cardiac muscle contraction and rhythm, vascular smooth muscle contraction, gastrointestinal smooth muscle contraction, and respiratory smooth muscle contraction: all tissues dependent on entrance of calcium from extracellular pools.

It is the purpose of this chapter to review (1) the history of the discovery of these drugs, (2) the use of them as tools in defining the physiological consequences of calcium channel blockade, and (3) how their potent and selective characteristics have enabled some biochemical characterization of the channel and the description of potentially new calcium channel antagonists.

2. Discovery and Physiological Effects of Calcium Channel Blockers

2.1. Historical Perspective

It is fitting perhaps that just as the importance of calcium in contractility was first appreciated in the heart by Ringer, the ability of compounds to act as calcium channel antagonists was first defined in the heart by Fleckenstein (1983). In the early 1960s, Fleckenstein discovered that epinephrine could reverse the decreased contractility of frog ventricles effected by a decrease in the calcium concentration of the bathing solution to 0.5 m*M* (*see* Fig. 14 in Fleckenstein, 1983). In a further investigation into how catecholamines ''intensify the Ca-dependent utilization of high-energy phosphates,'' it was noted that high concentrations of β-adrenoceptor blockers could mimic calcium withdrawal in isolated tissue or intact hearts in four ways: (1) the force of contraction was diminished; (2) high-energy phosphate utilization was reduced; (3) oxygen consumption was reduced; and (4) the effects were reversed by increasing the calcium concentration in the bathing medium (Fleckenstein, 1983). It was further demonstrated that β-adrenoceptor blockade by several beta-blockers occurred at doses some tenfold lower than the doses that inhibited the calcium-dependent contraction. Thus, through the tedium of dose–response relationships, they suggested that negative inotropism could be achieved by calcium antagonism. As discussed in Fleckenstein's historical perspective on the evolution of the concept of pharmacological calcium antagonism, ''it seemed reasonable to search for drugs that exert Ca antagonism in a specific form'' (p. 35, Fleckenstein, 1983). In the early 1960s, data on the negative inotropic effect of two compounds, prenylamine and verapamil (Fig. 1), were published (Lindner, 1960; Haas and Hartfelder, 1962). The mechanism of these two drugs was unknown, however (Fleckenstein, 1971). In 1969, Fleck-

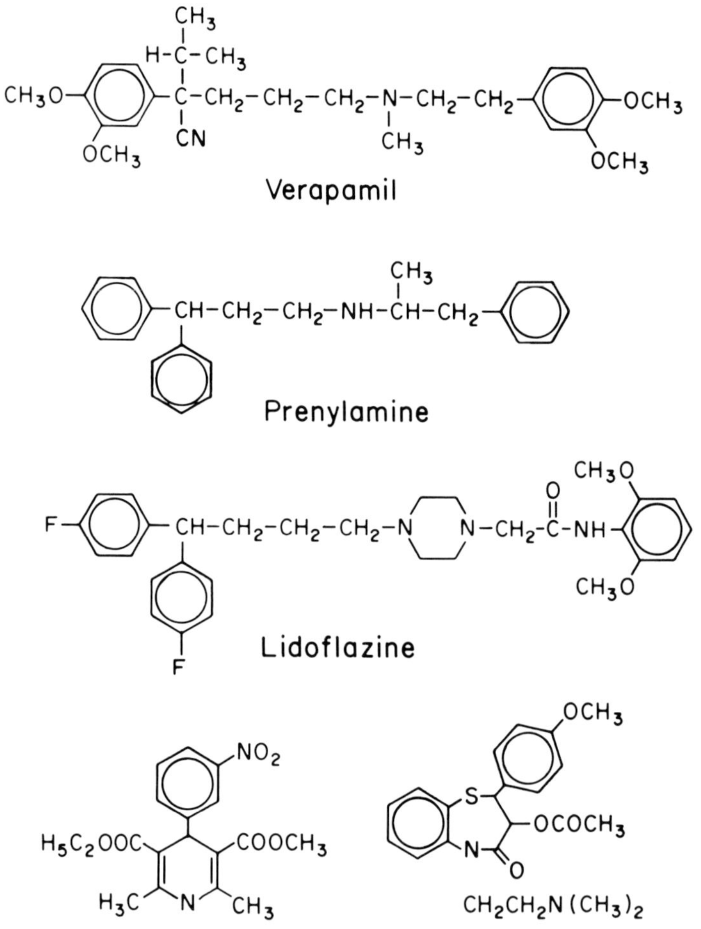

Verapamil

Prenylamine

Lidoflazine

Nitrendipine

Diltiazem

Fig. 1. Chemical structures of representative calcium channel antagonists.

enstein and co-workers reported in abstract form that verapamil (Iproveratril), prenylamine, and a methoxy derivative of verapamil, gallopamil (D600), were competitive antagonists of calcium in papillary muscle and uterus (Fleckenstein et al., 1969; Fleckenstein and Grun, 1969). In relatively short order, Fleckenstein recalls, they obtained samples of a drug, Bay a 1040, which had a pharmacological profile similar to verapamil. Bay a 1040 belongs to a different chemical class, the 1,4-dihydropyridines, an example of which, nitrendipine, is shown in Fig. 1. At the request of Professor Kroneberg of Bayer, Fleckenstein's group established that the 1,4-dihydropyridines were much more potent calcium channel antagonists than any of the previously studied compounds (Fleckenstein, 1971, 1984).

During this time, it also became possible to measure the action potential in cardiac tissue and to resolve it into at least two components, one an excitatory sodium current (i_{Na}) and the second an inactivating inward current (i_{si}) carried by Ca^{2+} and requiring extracellular Ca^{2+} (Reuter, 1967, 1968; Vitek and Trautwein, 1971). As anticipated for a calcium channel antagonist, verapamil, gallopamil, and nifedipine blocked contractility concomitant with blocking i_{si}, but had little or no effect on i_{Na} (Kohlhardt et al., 1972; Kass and Tsien, 1975; Fleckenstein, 1971). By contrast, prenylamine also affected the upstroke velocity (i_{Na}), and thus it was concluded to be less specific than the other Ca^{2+} channel antagonists (Fleckenstein, 1971). These data suggested a membrane site of action. Using skinned muscle preparations, it has been verified that the Ca^{2+} channel antagonists act at the membrane level (Metzger et al., 1982; Cauvin et al., 1983). This has been substantiated by patch-clamp analysis of dialyzed whole heart cells that demonstrates blockade of calcium channels in the plasma membrane by calcium channel antagonists (Lee and Tsien, 1983).

The definition of these compounds as calcium channel antagonists as opposed to nonspecific calcium antagonists is based on five criteria:

1. The compounds could prevent damage to the heart induced in vivo by high doses of isoproterenol even though they are not β-adrenoceptor antagonists. Moreover, isoproterenol could restore contractility inhibited by the Ca^{2+} channel antagonist (Fleckensten, 1981).
2. Increased Ca^{2+} can overcome the inhibition of contraction in smooth or cardiac muscle.
3. Calcium-induced contractions in potassium-depolarized smooth or cardiac muscle can be blocked.
4. The Ca^{2+}-dependent component of the action potential is reduced, whereas the Na^{+}-dependent component or upstroke velocity is unaffected.

5. The site of action of these compounds is at the plasma membrane, as shown by skinned muscle preparations or patch-clamp analysis of single channels.

Historically, the first four criteria were used to define the compounds. More recently, only the last four have been used, and in practice only the middle three.

At the time of Fleckenstein's conceptual breakthrough, only three chemical classes of calcium channel antagonists were known. Verapamil and gallopamil are phenylalkylamines. Prenylamine is a diphenylalkylamine and nifedipine and nitrendipine are 1,4-dihydropyridines (Fig. 1). A potent coronary vasodilator, diltiazem, was reported to have a direct effect on the coronary vascular bed producing dramatic vasodilation (Sato et al., 1971). It was not acting through antagonizing adenosine or β-adrenoceptor agonists, and its mechanism was not known (Sato et al., 1971). It was soon established that diltiazem, a 1,5-benzothiazepine, blocked calcium-induced contraction in potassium-depolarized smooth muscle, that increased calcium could overcome the negative inotropic effects of diltiazem, and that it diminished the action potential plateau caused by the influx of Ca^{2+} (Nagao et al., 1972, 1977; Nakajima et al., 1975; Saikawa et al., 1977). It thus filled three of the criteria set forth above for classification as a calcium channel antagonist. A fourth chemical class was consequently introduced into the rapidly expanding pharmacotherapeutic area of calcium channel antagonism.

In an attempt to categorize these differing chemical structures, Fleckenstein proposed a simple two-group system based on the specificity of the compounds (Table 1) (Fleckenstein, 1983). This classification, although it is convenient, highlights only the specificity of these compounds for the calcium channel over the sodium channel. It is apparent, however, that even the most potent compounds grouped into Class 1 differ in their profile of activity.

2.2. Physiological Effects

As has been summarized in a number of recent reviews on Ca^{2+} channel antagonists, calcium-dependent contractions in K^+-depolarized smooth muscle are more sensitive to Ca^{2+} channel antagonists than are agonist-induced contractions. A representative collection of these data are shown in Table 2; for more complete descriptions, *see* Cauvin et al. (1983). The differences in potency for agonist-induced responses probably reflect the different extent to which different tissues depend on voltage-operated channels, receptor-operated channels, and intracellular sources of calcium.

Table 1
Classification of Calcium Channel Antagonists

Class basis	Class 1	Class 2	Class 3	Reference
Specificity vs Na^+ channel	1,4-Dihydropyridines[a], 1,5-benzothiazepines[b], phenylalkylamines[c]	Diphenylalkylamines[d]	—	Fleckenstein, 1983
In vitro physiologic effects	1,4-Dihydropyridines	Phenylalkylamines, 1,5-benzothiazepines	Diphenylalkylamines, Diphenylbutyl-piperidines[e]	Spedding, 1982 Spedding and Berg, 1984
Binding	1,4-Dihydropyridines, diphenylalkylamines	Phenylalkylamines	1,5-Benzothiazepines	Glossmann et al., 1982
Binding	1,4-Dihydropyridines	Phenylalkylamines, diphenylalkylaimines, diphenylbutylpiperi-dines, 1,5-benzothia-zepines	Maitotoxin (?)	Murphy et al., 1983 Gould et al., 1983b Freedman et al., 1984b

[a] 1,4-Dihydropyridines: nitrendipine, nifedipine, nimodipine, felodipine, niludipine, ryosidine, and others.
[b] 1,5-Benzothiazepines: diltiazem.
[c] Phenylalkylamines: verapamil, gallopamil, tiapamil, and desmethoxyverapamil.
[d] Diphenylalkylamines: prenylamine, lidoflazine, cinnarizine, and flunarizine.
[e] Diphenylbutylpiperidines: pimozide, clopimozide, fluspirilene, and penfluridol.

Table 2
Effects of Calcium Channel Antagonists

Contraction				
Tissue	Drug	Stimulus	ED_{50}, *M*	Reference
Rabbit aorta	Nifedipine	Norepinephrine,	3×10^{-5}	Schumann et al., 1975
		K^+-Depolarization	1×10^{-8}	
	Gallopamil	Norepinephrine,	1×10^{-4}	Schumann et al., 1975
		K^+-Depolarization	4×10^{-8}	
	Diltiazem	Norepinephrine,	1×10^{-4}	Van Breeman et al., 1981
		K^+-Depolarization	5×10^{-7}	
Rabbit mesenteric artery	Nifedipine	Norepinephrine,	2×10^{-6}	Schumann et al., 1975
		K^+-Depolarization	1×10^{-8}	
	Gallopamil	Norepinephrine,	4×10^{-5}	Schumann et al., 1975
		K^+-Depolarization	5×10^{-8}	
	Diltiazem	Norepinephrine,	2×10^{-7}	Saida and Van Breeman, 1982
		K^+-Depolarization	5×10^{-7}	
Dog coronary artery	Verapamil	$PGF_{2\alpha}$,	$>10^{-5}$	Shimizu et al., 1980
		K^+-Depolarization	3×10^{-7}	
	Gallopamil	Norepinephrine,	1×10^{-6}	Van Breeman and Siegel, 1980
		K^+-Depolarization	1×10^{-7}	
Guinea pig ileum	Nifedipine	Acetylcholine,	5×10^{-9}	Rosenberger and Triggle, 1978
		K^+-Depolarization	3×10^{-9}	
	Gallopamil	Acetylcholine,	3×10^{-8}	Rosenberger and Triggle, 1978
		K^+-Depolarization	3×10^{-8}	

Secretion				
Rabbit ear artery	Flunarizine, diltiazem, nicardipine	Norepinephrine release	NE[a] at 10^{-6}	Rezvani et al., 1983
Guinea pig ileum	Nicardipine, gallopamil, diltiazem	Acetylcholine release	NE at 10^{-5}	Kaplita and Triggle, 1983
Rat phrenic nerve	Nitrendipine	Norepinephrine release	NE at 2×10^{-5}	Fairhurst et al., 1983
Rabbit candate nucleus	Verapamil, diltiazem	Dopamine and acetylcholine release	NE at 10^{-5}	Starke et al., 1984
PC12 Cells	Nifedipine, nitrendipine	$^{45}Ca^{2+}$ uptake	3×10^{-9} 5×10^{-9}	Toll, 1982
NG108-15	Nitrendipine, gallopamil, diltiazem	$^{45}Ca^{2+}$ uptake	7×10^{-9} 7×10^{-7} 2×10^{-6}	Freedman et al., 1984a

[a]NE: No effect.

For example, the rabbit aorta is relatively insensitive to nifedipine when induced to contract by norepinephrine, whereas the guinea pig ileum is highly and equally sensitive to nifedipine whether contractions are induced by acetylcholine or potassium depolarization (Schumann et al., 1975; Rosenberger and Triggle, 1978). The rabbit aorta is believed to contain "pure" voltage-operated channels and receptor-operated channels (Karaki and Weiss, 1984). That is, the contraction induced by norepinephrine is only caused by calcium flux through the receptor-operated channels, with no contribution from voltage-operated channels. Thus, nifedipine has only weak and nonpharmacological effects on norepinephrine-induced contractions in this tissue. By contrast, the guinea pig ileum contains a channel or channels that are responsive to both agonists and depolarizations. Both channels, if they are indeed separate, come into action with either stimulus, and thus calcium channel antagonists are equipotent against both activators. In general, verapamil and diltiazem are approximately equipotent on cardiac and smooth muscle, whereas nifedipine is much more active on smooth muscle than on cardiac muscle.

These differences in vitro are reflected in vivo and form the basis for the therapeutic indications of the various antagonists (for reviews, *see* Stone et al., 1980; Zelis and Flaim, 1982; Janis and Triggle, 1983; Schwartz and Triggle, 1984; Fleckenstein, 1977; Weiss, 1981; Scriabine et al., 1984). The dihydropyridines exert some negative inotropic effects and very little effect on arrhythmias (Dangman and Hoffman, 1980; Henry, 1980). Dihydropyridines differ from diphenylalkylamines in their selectivity for vascular smooth muscle over cardiac muscle. At concentrations that increase cardiac perfusion, nifedipine has less of a tendency to depress cardiac contractility, and thus nifedipine is more effective than verapamil as an antianginal agent (Henry, 1980). On the other hand, verapamil and gallopamil have frequency- and voltage-dependent effects on calcium fluxes (Mannhold et al., 1981; Ehara and Daufmann, 1978). They consequently have a more profound effect on atrioventricular conduction and are more effective in treating cardiac arrhythimias (Henry, 1980; Wit and Ning, 1983; Singh et al., 1980). The differences in physiological effects have led to the proposal that calcium channel antagonists fall into three classes, as shown in Table 1.

Calcium channel antagonists have little effect on stimulus-secretion coupling with a few exceptions (for an extensive review, *see* Miller and Freedman, 1984). As shown in Table 2, only heroic levels of calcium channel antagonists block responses such as $^{45}Ca^{2+}$

uptake or Ca^{2+}-dependent neurotransmitter release. One exception to this is in cultured transformed cells where $^{45}Ca^{2+}$ uptake is blocked by 1,4-dihydropyridines (Toll, 1982; Freedman et al., 1984a). Thus, there appears to be a fundamental difference between excitation–secretion coupling and excitation–contraction coupling. This is reflected in vivo because there is no therapeutic indication for calcium channel antagonists for any nervous dysfunction, nor is there any effect of 1,4-dihydropyridines on most tests of central nervous system function (Hoffmeister et al., 1982). However, as will be discussed in section 3.1.2, there are suggestions from radioligand binding studies that "novel" calcium channel antagonists may have subtle effects on the central nervous system.

Recently, electrophysiological studies have provided evidence for multiple calcium channels in both neuronal and cardiac tissue (Nowycky et al., 1985; Hess et al., 1985). In neuronal tissue, there appear to be three types of channels distinguished by their kinetic characteristics (Nowycky et al., 1985). Only one of these, the L type, is sensitive to a 1,4-dihydropyridine, Bay K 8644. Bay K 8644 stimulates calcium channels and thus is considered a calcium channel agonist (Schramm et al., 1983a,b). The T calcium channel produces a transient current at more negative membrane potentials, whereas the N channels require very large depolarizations. L and T channels are also found in cardiac cells. Since smooth and cardiac muscles are sensitive to 1,4-dihydropyridines, L channels apparently play a larger role in these tissues than in neuronal tissue. In most neuronal tissues, N or T channels may play a more prominent role. This would explain the relative insensitivity of most neuronal tissue to calcium channel blockers. Under special circumstances or in transformed cells, L channels may mediate neurotransmitter release or $^{45}Ca^{2+}$ uptake. This would account for the small component of $^{45}Ca^{2+}$ uptake that can be blocked by nitrendipine or the block of Bay K 8644-induced neurotransmitter release by nitrendipine (Turner and Goldin, 1985; Middlemiss and Spedding, 1985). A potent marine toxin, maitotoxin, apparently acts as a calcium channel activator and will release norepinephrine from nerve terminals. Only part of the effect of this compound is blocked by low levels of verapamil, and it does not appear to interact with any of the drug binding sites within the nitrendipine-sensitive calcium channel (Ohizumi et al., 1983; Miller et al., 1985; Freedman et al., 1984b). Conceivably, such a toxin may act at the N or T channels, although it also may be depolarizing the cell and thus affecting calcium channels indirectly (Kuroda et al., 1984).

3. Biochemical Characterization

3.1. ^{3}H-Dihydropyridines and Novel Antagonists

3.1.1. Characteristics of Drug Binding Sites

The high potency of 1,4-dihydropyridines coupled with their specificity makes them ideal to use as radioligands for exploring the biochemical characteristics of the voltage-operated calcium channels. Radioligand studies have allowed a description of how calcium channel antagonists of different chemical classes interact and the description of drugs whose mechanisms may be through calcium channels. This has allowed for the description of new potential sites for therapeutic intervention.

The initial reports of ^{3}H-1,4-dihydropyridine binding reported the presence of high-affinity binding sites for ^{3}H-nitrendipine to cardiac membranes and cerebral cortex membranes (Bellemann et al., 1981; Murphy and Snyder, 1982). Since then, 1,4-dihydropyridine binding sites have been reported using ^{3}H-nifedipine, ^{3}H-nimodipine, ^{3}H-nisoldipine, ^{3}H-PN 200 110, and ^{3}H-Bay K 8644 (Holck et al., 1982; Ferry et al., 1983; Janis et al., 1982, 1984a; Pan et al., 1983). In most cases, the same general results are obtained no matter which ligand is used, although differences in the number of binding sites may be seen (Ferry et al., 1983). Triggle and Janis (1984) have recently published an extensive compilation of data using radioligands. The number of tissues that contain 1,4-dihydropyridine binding sites is large, as one would expect from the ubiquity of excitation–contraction coupling. Binding sites are found in tissues such as cardiac and smooth muscle, where they form the basis for the therapeutic utility of the compounds. They are also found in tissues such as neuronal tissue and skeletal muscle, where the pharmacology of such sites is poorly defined or where no effect at all can be found (Table 3). Triggle and Janis (1984) have also compiled an extensive list of the tissues in which ^{3}H-1,4-dihydropyridine receptors are found.

The binding of ^{3}H-nitrendipine is identical in neuronal tissue, smooth muscle, or cardiac muscle, except for the number of binding sites in these tissues (Table 3; Gould et al., 1984a). A variety of 1,4-dihydropyridine derivatives show virtually identical affinities for ^{3}H-nitrendipine binding sites in these three preparations (Gould et al., 1984a; Triggle and Janis, 1984). For example, in guinea pig heart, cerebral cortex, and ileum, the order of potency is SKF-24260 > nitrendipine > PY108-068 > nisoldipine > nimodipine >

nifedipine (Gould et al., 1984a). Excellent correlations are seen when binding to neuronal tissue, smooth muscle, or heart is compared to inhibition of K^+-stimulated ileal smooth muscle contraction (Janis and Triggle, 1983; Triggle and Janis, 1984). The congruence of binding parameters suggest that the high-affinity ^{3}H-nitrendipine binding sites in heart, smooth muscle, and neuronal tissue are either identical or very similar, even though there are only limited effects of 1,4-dihydropyridines in neuronal tissues (Middlemiss and Spedding, 1985; Turner and Goldin, 1985; Freedman and Miller, 1984; Freedman et al., 1984a). By contrast, ^{3}H-nitrendipine binding sites in skeletal muscle have approximately tenfold lower affinity for 1,4-dihydropyridines and altered affinity for nondihydropyridine drugs (Gould et al., 1984a; Fairhurst et al., 1983). Thus, skeletal muscle binding sites cannot be considered pharmacologically the same as sites in brain, heart, or smooth muscle.

The binding sites are located on plasma membranes. After separating neuronal membranes by differential centrifugation, ^{3}H-nitrendipine binding sites distribute with β-adrenoceptors measured with ^{3}H-dihydroalprenolol. This suggests they are associated with synaptosomal membranes. They are also localized to areas of synaptic connection as shown by in vitro autoradiography of brain slices incubated with ^{3}H-nitrendipine and may be associated with dendrites (Gould et al., 1985, 1982a; Cortes et al., 1983; Quirion, 1983). ^{3}H-Nitrendipine binding sites also copurify with enzymes that are considered markers for sarcolemma from smooth muscle and cardiac muscle (Bolger et al., 1983; Triggle et al., 1982; Grover et al., 1984; DePover et al., 1982; Sarmiento et al., 1983). Binding with whole cells substantiates this conclusion (Marsh et al., 1983; DePover et al., 1983; Green et al., 1985). Thus binding analyses are in agreement with the physiological data that voltage-operated calcium channels are found predominantly in the plasma membrane of cells.

^{3}H-Nitrendipine binding is regulated by multivalent cations (Gould et al., 1982b; Bolger et al., 1983; Ehlert et al., 1982; DePover et al., 1982; Marangos et al., 1982). If membranes are treated with EDTA and/or EGTA to deplete endogenous divalent metal ions, binding is reduced by 90% (Gould et al., 1982b; Bolger et al., 1983). Addition of divalent or trivalent cations, but not monovalent cations, can restore binding. The effects are totally caused by a diminished number of binding sites with little or no effect on the affinity (Gould et al., 1982b; Bolger et al., 1983). Complete restoration of binding sites occurs with those divalent cations having an

Table 3
Characteristics of Calcium Channel Antagonist Binding Sites

Tissue	K_d, nM	B_{max}, fmol/mg protein	References[a]
	^{3}H-Nitrendipine		
Cardiac muscle			
Rat ventricle	0.11–0.33	80–400	1-5
Guinea pig ventricle	0.14–0.16	200–300	6,7
Dog ventricle	0.11–0.30	190–1500	8–11
Dog atria	0.14	170	8
Vascular smooth muscle			
Rat mesenteric artery	0.10	18	12
Dog mesenteric artery	0.25; 1.5	25, 62	12
Pig coronary artery	1.6	35	9
Rabbit aorta	1	54	13
Dog aorta	0.3; 4.4	20, 178	12
Bovine aorta	2.1	40–60	14
Other smooth muscle			
Rat ileum	0.26	25	1
Guinea pig ileum	0.16–0.17	94–1130	7,15
Rat myometrium	0.14	720	16
Rabbit myometrium	0.70	116	17
Rat fundus	0.13	430	16
Nervous tissue			
Rat cerebral cortex	0.11–1.0	102–103	1,4
Guinea pig cerebral cortex	0.20	166	7
Rat forebrain	0.16	92	1,4
Rat cerebral cortex	0.11	13[b]	7
Hippocampus	0.13	14[b]	3
Olfactory bulb	0.11	12[b]	18
Striatum	0.13	11[b]	18
Thalamus/hypothalmus	0.12	6[b]	18
Cerebellum	0.09	4[b]	18
Midbrain	NM[c]	<1[b]	18
Brainstem	NM[c]	<1[b]	18
Pheochromocytoma (PC12) cells	1.1	27.5	19
Skeletal muscle			
Guinea pig leg	2.28–3.6	1112–7000	7,20
Frog hindlimb	0.5	20,000	21
Rabbit hindlimb	1.5–2.5	800–50,000	21,22

Table 3 *(Continued)*
Characteristics of Calcium Channel Antagonist Binding Sites

Tissue	K_d, nM	B_{max}, fmol/mg protein	References[a]
	^{3}H-Verapamil		
Cardiac muscle			
Frog heart	4.25	50	23
Porcine left ventricle	55	880	24
Nervous tissue			
Rat cerebral cortex	94	435[b]	25
Skeletal muscle			
Rat skeletal muscle	38	550[b]	25
Rabbit T-tubule	27	5000	26
	^{3}H-Desmethoxyverapamil		
Nervous tissue			
Rat cerebral cortex	1.2; 64	3900; 20,000	27
	^{3}H-Diltiazem		
Cardiac muscle			
Porcine left ventricle	70(25°C)	300–400	28
Nervous tissue			
Rat cerebral cortex	50(30°C)	1150	29
Skeletal muscle			
Guinea pig hind limb	37(30°C)	2900	30

[a]References: (1) Ehlert et al., 1982 (2) Janis et al., 1982 (3) Marangos et al., 1982 (4) Murphy and Snyder, 1982 (5) Pan et al., 1983 (6) Bellemann et al., 1981 (7) Gould et al., 1984a (8) Sarmiento et al., 1982 (9) DePover et al., 1982 (10) Sarmiento et al., 1983 (11) Williams and Jones, 1983 (12) Triggle et al., 1982 (13) Bristow et al., 1982 (14) Williams and Tremble, 1982 (15) Bolger et al., 1983 (16) Grover et al., 1984 (17) Miller and Moore, 1983 (18) Gould et al., 1982b (19) Toll, 1982 (20) Ferry et al., 1983 (21) Fossett et al., 1983 (22) Fairhurst et al., 1983 (23) Hulthen et al., 1982 (24) Garcia et al., 1984 (25) Reynolds et al., 1983 (26) Galizzi et al., 1984a,b (27) Reynolds et al., 1985 (28) Kaczorowski et al., 1985 (29) Schoemaker and Langer, 1985 (30) Glossmann et al., 1983.

[b]pmol/g wet wet tissue.

[c]NM, Not measurable.

ionic radius, and thus a relative charge density, similar to Ca^{2+} (Gould et al., 1982b). There is an inverse linear relationship between charge density of both tri- and divalent cations and percent of maximal ^{3}H-nitrendipine binding sites (Triggle and Janis, 1984). Thus, within the site of drug action of the 1,4-dihydropyridines, there is a coordination site for Ca^{2+} or a closely related cation, such as Sr^{2+}.

As mentioned previously, the Ca^{2+} channel antagonists are a chemically diverse group of compounds (Fig. 1). There appear to be at least two drug binding sites within the calcium channel complex (Table 1; Ehlert et al., 1982; Ferry and Glossman, 1982; Bolger et al., 1983; Murphy et al., 1983). It is clear that the two (or three) sites can interact in complex fashions. Before discussing these interactions in detail, it is useful to separate the calcium channel antagonists into two groups. The first group includes all of the 1,4-dihydropyridines, both calcium agonists and antagonists. These compounds are strictly competitive inhibitors of ^{3}H-nitrendipine binding. The second group includes all the calcium channel antagonists of the other chemical classes: the phenylalkylamines, the diphenylalkylamines, and the 1,5-benzothiazepines. This group modulates the affinity of the ^{3}H-nitrendipine site allosterically via a separate binding site. The compounds in the second group fall along a spectrum of efficacy, where the efficacy is greater than zero if the affinity of ^{3}H-nitrendipine binding is lowered in the presence of the compound and is less than zero if the affinity of the ^{3}H-nitrendipine site is raised. A greater positive efficacy would lead to larger decreases in the affinity of the ^{3}H-nitrendipine site. A greater negative efficacy would lead to larger increases in the affinity of the ^{3}H-nitrendipine site. An efficacy of zero would have no effect on the affinity of the site and would thus have no effect on the measurable binding. If, however, a compound with positive or negative efficacy were present with a compound with zero efficacy, the effects of the positive or negative efficacious compound would be blocked by the zero-effect compound. Likewise, a compound with positive efficacy would reverse the allosteric modulation of a compound with negative efficacy.

Verapamil and gallopamil show stereoselective partial inhibition of ^{3}H-1,4-dihydropyridine binding, with the (−)-isomer being more potent than the (+)-isomer, consistent with studies on contractile tissues (Glossman and Ferry, 1983; Ferry and Glossman, 1982; Janis et al., 1984b; Bayer et al., 1975). It should be pointed out that binding sites in skeletal muscle have a reverse specificity [(+) more potent than (−)], illustrating another difference between skeletal muscle binding sites and sites on smooth muscle, cardiac muscle, or neuronal tissue (Triggle and Janis, 1984; Galizzi et al.,

1984a,b). Scatchard analysis of ^{3}H-nitrendipine binding to membranes in the presence of various concentrations of verapamil indicates that verapamil reduces the K_d of nitrendipine for its site, up to a maximum of 2.5-fold (Gould et al., 1984a; Ehlert et al., 1982). Moreover, maximal displacement by verapamil of ^{3}H-nitrendipine differs depending on the ^{3}H-nitrendipine concentration. These two data are best explained by a negative heterotropic interaction of verapamil and nitrendipine binding sites. Verapamil has a positive efficacy. Gallopamil, under identical conditions, causes less of a maximal effect than verapamil, but also decreases the affinity. It therefore would have a lower positive efficacy than verapamil. Tiapamil, another analog of verapamil, causes greater apparent displacement than verapamil, and thus has a greater positive efficacy (Eigenmann et al., 1981; Murphy et al., 1983). All of these compounds accelerate the dissociation rate of ^{3}H-nitrendipine, consistent with their noncompetitive effects on equilibrium binding (Murphy et al., 1983; Ehlert et al., 1982). Thus, the phenylalkylamines have a different site of action than the 1,4-dihydropyridines, and all appear to have positive efficacy. They are thus considered to be negative heterotropic allosteric regulators.

The diphenylalkylamines may similarly be classified as negative heterotropic allosteric regulators (Murphy et al., 1984; Gould et al., 1983a; Janis et al., 1984b). This is based on the three observations discussed above for the phenyalkylamines. First, incomplete displacement is seen by flunarizine and lidoflazine (Murphy et al., 1983, 1984; Janis et al., 1984b). Second, the maximal displacement varies according to the ^{3}H-nitrendipine concentration (Murphy et al., 1984). Third, the dissociation rate of ^{3}H-nitrendipine is accelerated in the presence of the diphenylalkylamines, but not in the presence of a 1,4-dihydropyridine (Murphy et al., 1983, 1984; Janis et al., 1984b).

Do these two chemical classes, both negative heterotropic regulators, act at the same site or at different sites? This issue may be resolved by hypothesizing that they do, in fact, act at the same site. If this were the case, then a compound with a lower positive efficacy should be able to reverse the effects of a compound with a higher positive efficacy. Second, the ^{3}H-nitrendipine displacement curve of the compound with a more positive efficacy should be shifted to the right by the compound with a less positive efficacy. Gallopamil has a low positive efficacy and may test this hypothesis. As predicted, gallopamil reverses the inhibition of ^{3}H-nitrendipine binding produced by both penylalkylamines (tiapamil) and diphenylalkylamines (prenylamine, flunarizine, and lidoflazine) (Murphy et al., 1983). Verapamil has a similar effect (Janis et al., 1984b). In addition, the concentration–response curve for the inhibition of ^{3}H-nitrendipine binding is right-shifted in the presence of gallopamil (Murphy et al., 1983). Thus, the diphenylalkylamines and phenylalkylamines appear to act at the same site (Table 1).

Recently, this conclusion has been challenged from functional studies with depolarized taenia (Spedding and Berg, 1984). These studies are more indirect than analyses of binding interactions, and are complicated because they were done in the presence of a calcium channel agonist. It is not clear how calcium channel agonists would affect compounds with different efficacies at this site. This conclusion is also different from a classification of calcium channel antagonist sites, also drawn from binding analyses (Table 1; Glossmann et al., 1982). These authors concluded that diphenylalkylamines should be classified with 1,4-dihydropyridines, because at lower tracer concentrations the ^{3}H-dihydropyridine is completely displaced. These basic data are consistent with a single site of action.

Diltiazem, a 1,5-benzothiazepine, has curiously been shown to have no effect, an inhibitory effect, or a stimulatory effect on ^{3}H-1,4-dihydropyridine binding (Bolger et al., 1983; DePover et al., 1982; Murphy and Snyder, 1982; Ferry et al., 1983; Yamamura et al., 1982; Schoemaker et al., 1983; Ferry and Glossman, 1982; Murphy et al., 1983; Janis et al., 1984b). However, it is generally agreed that the stimulatory effect is temperature-sensitive and that inhibition occurs only at high concentrations. The stimulation of binding shows the same pharmacology as effects on the voltage-dependent calcium channel, i.e., *d-cis*-diltiazem > 1-*cis*-diltiazem > desmethyldiltiazem (DePover et al., 1982; Glossman and Ferry, 1983). This increased binding results from a decrease in the dissociation rate and a consequent increase in the affinity of the receptor for ^{3}H-1,4-dihydropyridines. Thus, diltiazem is a positive heterotropic allosteric effector of ^{3}H-1,4-dihydropyridine binding sites. It also has negative efficacy at higher doses because it causes a decreased affinity of the binding sites for 1,4-dihydropyridines.

This positive effect has been interpreted as diltiazem acting at a different site than 1,4-dihydropyridines, phenylalkylamines, or diphenylalkylamines (Table 1; Glossman et al., 1982). Diltiazem, however, can fully reverse the effects of gallopamil, prenylamine, or tiapamil (Murphy et al., 1983, 1984; Janis et al., 1984b). This reversal is also pharmacologically relevant because *trans*-diltiazem, which has very low activity against calcium channels, cannot elicit it, and desmethyldiltiazem is less potent than *cis*-diltiazem. Finally, dose-response curves for inhibition of ^{3}H-nitrendipine binding by prenylamine and tiapamil demonstrate a large shift to the right in the presence of diltiazem (Murphy et al., 1984). By the criteria established earlier, then, diltiazem acts at the same site as the phenylalkylamines and diphenylalkylamines.

3.1.2. "Novel" Calcium Channel Antagonists

The ability of *cis*-diltiazem to reverse the inhibition of ^{3}H-1,4-dihydropyridine binding suggested a sensitive assay for compounds that act at this nondihydropyridine binding site. That is, if a compound could reverse inhibition induced by tiapamil, or if diltiazem could reverse a compound's inhibition, it may be acting at the nondihydropyridine site in the calcium channel (Murphy et al., 1983). This assay therefore has greater sensitivity and specificity than casual stimulation or inhibition of binding. A number of compounds, including dimethindene, bepridil, thioridazine, biperiden, chlorpheniramine, and mesoridazine are shown by this assay to act at drug binding sites within the calcium channel. In addition, dimethindene and thioridazine are both calcium channel antagonists in potassium-depolarized smooth muscle contractions (Murphy et al., 1983; Gould et al., 1984b). This effect is specific for compounds having structural similarities to known calcium channel antagonists because promazine, chlorimipramine, nortriptyline, fluphenazine, clozapine, pyrilamine, histamine, atropine, and a variety of other compounds are ineffective.

Thioridazine is particularly interesting since it is an example of a compound that has no efficacy of its own at the nondihydropyridine binding site. That is, it has no effect under the conditions of our assay on ^{3}H-nitrendipine binding. It can clearly reverse the inhibitory effects of tiapamil (Gould et al., 1984b). Recently, it has been shown that, under certain conditions, thioridazine can stimulate ^{3}H-nitrendipine binding and thus has a greater negative efficacy than we had anticipated. This makes it more like diltiazem than was originally thought (Thorgeirsson and Rudolph, 1984). Since the potency of thioridazine at this calcium channel site (1–2 μM) is in the same range as serum concentrations (2–3 μM), it is postulated that calcium channel antagonism could account for some of the side effects seen with thioridazine (Gould et al., 1984b). Thioridazine is also unique among the neuroleptics in that it induces apomorphine hyposensitivity following chronic administration (Goetz et al., 1984). Other neuroleptics, haloperidol for instance, induce hypersensitivity to apomorphine (Rupniak et al., 1983). Thus, one may speculate that calcium channel antagonists may show behavioral effects under certain experimental conditions.

While exploring the structure–activity relationship of the nondihydropyridine drug binding site on the calcium channel, a group of compounds with high positive efficacies and high potencies was found (Gould et al., 1983a,b; Quirion et al., 1985). The effects of these compounds, diphenylbutylpiperidines, are reversed by gallopamil, which has a lower positive efficacy. In addition, two of the compounds, pimozide and fluspirilene, are calcium channel antagonists in potassium-depolarized smooth muscle (Spedding, 1982; Gould et al., 1983b). These compounds are also D_2-dopamine receptor antagonists, but are atypical in that they block prolactin release at concentrations slightly higher than their potency at D_2 receptors (Denef et al., 1979; MacLeod and Lamberts, 1978). These effects have been postulated to be mediated by calmodulin antagonism. This seems unlikely since diphenylbutylpiperidines are 2–3 orders of magnitude more potent against ^{3}H-nitrendipine binding than against calmodulin-dependent phosphodiesterase activation (Spedding, 1982; Spedding and Berg, 1984; Gould et al., 1983a). Pimozide, penfluridol, and fluspirilene are almost as potent at inhibiting ^{3}H-nitrendipine binding as they are at dopamine receptors (Gould et al., 1983b). Thus, at clinical doses, nondihydropyridine drug binding sites within the calcium channel may also be occupied. The diphenylbutylpiperidines are unique in the symptoms they relieve. Classic neuroleptics like phenothiazines and butyrophenones are much less effective at the negative symptoms of schizophrenia, flattened affect, and social withdrawal than are the diphenylbutylpiperidines (Crow, 1980; Haas and Beckman, 1982; Lapierre, 1978; Frangos et al., 1978). These drugs may then be unique among antischizophrenic drugs in having actions on neuronal calcium channels.

3.2. ^{3}H-Phenylalkylamines

The ^{3}H-verapamil binding site may be studied directly using ^{3}H-verapamil, ^{3}H-gallopamil, or ^{3}H-desmethoxy-verapamil (Hulthen et al., 1982; Reynolds et al., 1983; Garcia et al., 1984; Gallizzi et al., 1984a,b; Reynolds et al., 1985). Binding sites are found in cerebral cortex, skeletal muscle, and heart (Table 3). As expected from the ^{3}H-1,4-dihydropyridine binding studies, the ^{3}H-phenylalkylamine binding is regulated allosterically by 1,4-dihydropyridines

(Gallizi et al., 1984a,b; Garcia et al., 1984; Reynolds et al., 1985). The effects of diltiazem vary with the tissue. In skeletal muscle and cerebral cortex, ^{3}H-nitrendipine sites are in a 1:1 ratio with ^{3}H-verapamil sites and diltiazem; gallopamil and bepridil appear competitive (Gallizi et al., 1984a,b). This is consistent with a single site of action for phenylalkylamines and 1,5-benzothiazepines. In crude cardiac membranes, ^{3}H-verapamil sites and ^{3}H-nitrendipine sites are in a 4:1 ratio. If purified sarcolemmal vesicles are prepared, the sites are in a 1:1 ratio. Diltiazem gives complete displacement in purified membranes, but not in crude membranes, whereas nitrendipine only partially displaces in both membrane preparations. A membrane fraction with ^{3}H-verapamil sites that do not respond to either diltiazem or 1,4-dihydropyridines can be prepared, but these sites have not been shown to be relevant to calcium channel activity (Garcia et al., 1984). In cardiac membranes, ^{3}H-verapamil sites are heterogeneous, and one population of sites has the characteristics of the non-dihydropyridine site as defined with ^{3}H-nitrendipine (Murphy et al., 1983). To date, new putative blockers of calcium channels have not been tested in these assays.

3.3. ^{3}H-Diltiazem

^{3}H-Diltiazem binding sites have been detected in skeletal muscle, heart, and cerebral cortex (Table 3; Glossman et al., 1983; Kaczorowski et al., 1985; Schoemaker and Langer, 1985). As one could expect from the studies with ^{3}H-1,4-dihydropyridines, binding is temperature-dependent. In skeletal muscle, the B_{max} increases with decreasing temperature with no change in the K_d (Glossmann et al., 1983). In cerebral cortex, the K_d increased with decreasing temperature (Schoemaker and Langer, 1985). Binding also shows allosteric regulation by 1,4-dihydropyridines, but phenylalkylamines and diphenylalkylamines appear competitive (Schoemaker and Langer, 1985; Garcia, et al., 1985; Glossmann et al., 1983). Putative new blockers of calcium channels have not been explored in these studies.

4. Summary

The history of the development of calcium channel antagonists highlights a spectrum of drug discovery techniques. The original observations by Fleckenstein were based on traditional in vitro physiological measurements of smooth and cardiac muscle contractility, coupled with in vivo extrapolations. Compounds were discovered simply by looking for vasorelaxant activity. Fleckenstein's contribution was his postulation of calcium channel antagonism as a "pharmacodynamic principle" and demonstration that nifedipine and prenylamine worked by this principle. Then, by using these and related compounds, the binding sites for these drugs were characterized. It was postulated that compounds having similar structures to known calcium channel antagonists might also be antagonists. This was confirmed by the diphenylbutylpiperidines. These compounds may then exert unique therapeutic effects through these sites in tissues previously thought to be insensitive to calcium channel antagonists.

What does the future hold for drugs working at calcium channels? Several possibilities present themselves.

1. Potent calcium channel agonists like Bay K 8644 have obvious implications for positive inotropic effects and increased smooth muscle contractility. Agonists at the phenylalkylamine/benzothiazepine site have not yet been described. Since antagonists show different therapeutic utility, agonists may similarly show differential effects on cardiac muscle, vascular smooth muscle, or pacemaker activity.
2. Affective disorders may be responsive to new classes of calcium channel agonists/antagonists like diphenylbutylpiperdines.
3. Bay K 8644 has profound behavioral effects unlike the 1,4-dihydropyridines that are antagonists (Bolger et al., 1985; Hoffmeister et al., 1982). Effects were consistent with both CNS stimulation (increased sensitivity to auditory stimuli, convulsions) and depression (ataxia, ptosis). Thus disorders of the CNS may be responsive to agonists or antagonists at calcium channels.
4. Calcium channels in brain or heart that are not blocked by available calcium channel antagonists may be assayed for by K^+-depolarization and subsequent uptake of $^{45}Ca^{2+}$ or neurotransmitter release in synaptosomes. These channels are at least pharmacologically distinct, and their presence has been verified electrophysiologically. They may represent a new generation of targets for calcium channel effectors.

5. If maitotoxin does, in fact, have a direct effect on calcium channels, it represents a site of action distant from 1,4-dihydropyridines, phenylalkylamines, 1,5-benzothiazepines, or diphenylalkylamines.

A variety of molecular targets thus exist within voltage-operated calcium channel. Exploration using drug binding sites should increase the armamentarium of compounds effective on this channel.

References

Allen, G. S., Ahn, H. S., Preziosi, T. J., Battye, R., Boone, S. C., Chou, S. N., Kelly, D. L., Weir, B. K., Crabbe, R. A., Lavik, P. J., Rosenbloom, S. B., Dorsey, F. C., Ingram, C. R., Mellits, D. E., Bertsch, L. A., Boisvert, D. P. J., Hundley, M. B., Johnson, R. K., Strom, J. A., and Transou, C. R. (1983) Cerebral arterial spasm. A controlled trial of nimodipine in patients with subarachnoid hemorrhage. *N. Eng. J. Med.* **308**, 619–624.

Bartoletti, M. and Labo, G. (1981) Clinical and manometric effects of nifedipine in patients with esophageal achalasia. *Gastroenterology* **83**, 963–969.

Bayer, R. Kaufmann, R., and Mannhold, R. (1975) Inotropic and electrophysiological actions of verapamil and D600 in mammalian myocardium. *Naunyn Schmiedebergs Arch. Pharmacol.* **290**, 69–80.

Bellemann, P., Ferry, D., Lubbecke, F., and Glossmann, H. (1981) [^{3}H]-Nitrendipine, a potent calcium antagonist, binds with high affinity to cardiac membranes. *Arzneimittelforsch.* **31**, 2064–2067.

Berridge, M. J. and Irvine, R. F. (1984) Inositol trisphosphate, a novel second messenger in cellular signal transduction. *Nature* **312**, 315–321.

Bolger, G. T., Gengo, P., Klockowski, R., Luchowski, E., Siegl, H., Janis, R. A., Triggle, A. M., and Triggle, D. J. (1983) Characterization of binding of the Ca^{2+}-channel antagonist, [^{3}H]-nitrendipine, to guinea pig ileal smooth muscle. *J. Pharmacol. Exp. Ther.* **225**, 291–310.

Bolger, G. T., Weissman, B. A., and Skolnick, P. (1985) The behavioral effects of the calcium agonist BAY K 8644 in the mouse: Antagonism by the calcium antagonist nifedipine. *Naunyn Schmiedebergs Arch. Pharmacol.* **328**, 373–377.

Bolton, T. B. (1979) Mechanism of action of transmitters and other substances on smooth muscle. *Physiol. Rev.* **59**, 606–718.

Bristow, M. R., McAuley, B., Ginsburg, R., Minobe, W. A., and Baisch, M. (1982) Does [^{3}H]nitrendipine bind to sarcolemmal slow channels? *Circulation* **66 (II)**, 95.

Burgess, G. M., Godfrey, P. P., McKinney, J. S., Berridge, M. J., Irvine, R. F., and Putney, Jr., J. W. (1984) The second messenger linking receptor activation to internal Ca release in liver. *Nature* (Lond.) **309**, 63–66.

Cauvin, C., Loutzinhiser, R., and Van Breemen, C. (1983) Mechanisms of calcium antagonist-induced vasodilation. *Ann. Rev. Pharmacol. Toxicol.* **23**, 373–396.

Cerrina, J., Denjean, A., Alexandre, G., Lockhard, A., and Duroux, P. (1981) Inhibition of exercise-induced asthma by a calcium antagonist nifedipine. *Am. Rev. Respir. Dis.* **123**, 156–160.

Cortes, R., Supavilai, P., Karobath, M., and Palacios, J. M. (1983) The effects of lesions in the rat hippocampus suggest the association of calcium channel blocker binding sites with specific neuronal population. *Neurosci. Lett.* **42**, 249–254.

Crow, T. J. (1980) Molecular biology of schizophrenia: More than one disease process? *Br. Med. J.* **280**, 66–67.

Dangman, K. H. and Hoffman, B. F. (1980) Effects of nifedipine on electrical activity of cardiac cells. *Am. J. Cardiol.* **46**, 1059–1067.

Denef, D., Van Neuten, J. M., Leysen, J. E., and Janssen, P. A. J. (1979) Evidence that pimozide is not a partial agonist of dopamine receptors. *Life Sci.* **25**, 217–225.

DePover, A., Matlib, M. A., Lee, S. W., Dupe, G. P., Grupp, I. L., Grupp, G., and Schwartz, A. (1982) Specific binding of [^{3}H]-nitrendipine to membranes from coronary arteries and heart in relation to pharmacological effects. Paradoxical stimulation by diltiazem. *Biochem. Biophys. Res. Commun.* **108**, 112–117.

DePover, A., Lee, S. W., Matlib, M. A., Whitmer, K., Davis, B. A., Powell, T., and Schwartz, A. (1983) [^{3}H]-Nimodipine specific binding to cardiac myocytes and subcellular fractions. *Biochem. Biophys. Res. Commun.* **113**, 185–191.

Ehara, T. and Daufmann, R. (1978) The voltage- and time-dependent effects of (−)-verapamil on the slow inward current in isolated cat ventricular myocardium. *J. Pharmacol. Exp. Ther.* **207**, 49–55.

Ehlert, F. J., Roeske, W. R., Itoga, E., and Yamamura, H. I. (1982) The binding of [^{3}H]-nitrendipine to receptors for calcium channel antagonists in the heart, cerebral cortex, and ileum of rats. *Life Sci.* **30**, 2191–2202.

Eigenmann, R., Blaber, L., Nakamura, K., Thorens, S., and Haeusler, G. (1981) Tiapamil, a new calcium antagonist. 1. Demonstration of calcium antagonistic activity and related studies. *Arzneimittelforsch.* **31**, 1393–1401.

Emanuel, M. B. (1979) Specific calcium antagonists in the treatment of peripheral vascular disease. *Angiology* **30**, 454–469.

Fairhurst, A. S., Thayer, S. A., Colker, J. E., and Beatty, D. A. (1983) A calcium antagonist drug binding site in skeletal muscle sarcoplasmic reticulum: Evidence for a calcium channel. *Life Sci.* **32**, 1331–1339.

Ferry, D. R. and Glossmann, H. (1982) Evidence for multiple receptor sites within the putative calcium channel. *Naunyn Schmiedebergs Arch. Pharmacol.* **321**, 80–83.

Ferry, D. R., Goll, A., and Glossmann, H. (1983) Differential labelling of putative skeletal muscle calcium channels by [^{3}H]nifedipine, [^{3}H]nimodipine and [^{3}H]PN 200 110. *Naunyn Schmiedebergs Arch. Pharmacol.* **323**, 276–277.

Fleckenstein, A. (1971) Specific Inhibitors and Promoters of Calcium Action in the Excitation-Contraction Coupling of Heart Muscle and Their Role in the Prevention or Production of Myocardial Lesions, in *Calcium and the Heart* (Harris, P. and Opie, L., eds.) Academic, New York.

Fleckenstein, A. (1977) Specific pharmacology of calcium in myocardium, cardiac pacemakers, and vascular smooth muscle. *Ann. Rev. Pharmacol. Toxicol.* **17**, 149–166.

Fleckenstein, A. (1981) Fundamental Actions of Calcium Antagonists on Myocardial and Cardiac Pacemaker Cell Membranes, in *New Perspectives on Calcium Antagonists* (Weiss, G. B., ed.) American Physiological Society, Bethesda, Maryland.

Fleckenstein, A. (1983) *Calcium Antagonism in Heart and Smooth Muscle,* Wiley-Interscience, New York.

Fleckenstein, A. (1984) Calcium Antagonism: History and Prospects for a Multifaceted Pharmacodynamic Principle, in *Calcium Antagonists and Cardiovascular Disease* (Opie, L. H., ed.) Raven, New York.

Fleckenstein, A. and Grun, G. (1969) Reversible blockade of excitation–contraction coupling in rat's uterine smooth muscle by means of organic calcium antagonists (Iproveratril, D600, prenylamine). *Pflugers Arch. Physiol.* **307**, R26.

Fleckenstein, A., Tritthart, H., Fleckenstein, B., Herbst, A., and Grun, G. (1969) A new group of competitive Ca antagonists (Iproveratril, D600, prenylamine) with highly potent inhibitory effects on excitation–contraction coupling in mammalian myocardium. *Pflugers Arch.* **391**, R12.

Forman, A., Andersson, K.-E., and Ulmsten, U. (1981) Inhibition of myometrial activity by calcium antagonists. *Seminars in Perinatology* **5**, 288–294.

Fossett, M., Jaimovich, E., Delpont, E., and Lazdunski, M. (1983) [^{3}H]Nitrendipine receptors in skeletal muscle. Properties and preferential localization in transverse tubules. *J. Biol. Chem.* **258**, 6086–6092.

Fozzard, H. A., Chapman, R. A., and Friedlander, I. R. (1985) Measurement of intracellular calcium ion activity with neutral exchanger ion sensitive microelectrodes. *Cell Calcium* **6**, 57–68.

Frangos, H., Zissis, N. P., Leontopoulos, I., Diamantas, N., Tsitouridis, S., Gavril, I., and Tsolis, K. (1978) Double-blind therapeutic evaluation of fluspirilene compared with fluphenazine decanoate in chronic schizophrenics. *Acta Psychiatr. Scand.* **57**, 436–446.

Freedman, S. B. and Miller, R. J. (1984) Calcium channel activation: A different type of drug action. *Proc. Natl. Acad. Sci. USA* **81**, 5580–5583.

Freedman, S. B., Dawson, G., Villereal, M. L., and Miller, R. J. (1984a) Identification and characterization of voltage sensitive calcium channels in neuronal clonal cell lines. *J. Neurosci.* **4**, 1453–1467.

Freedman, S. B., Miller, R. J., Miller, D. M., and Tindall, D. R. (1984b) Interactions of maitotoxin with voltage-sensitive calcium channels in cultured neuronal cells. *Proc. Natl. Acad. Sci. USA* **81**, 4582–4585.

Galizzi, J. P., Fosset, M., and Lazdunski, M. (1984a) [^{3}H]Verapamil binding sites in skeletal muscle transverse tubule membranes. *Biochem. Biophys. Res. Commun.* **118**, 239–245.

Galizzi, J. P., Fosset, M., and Lazdunski, M. (1984b) Properties of receptors for the Ca^{2+}-channel blocker verapamil in transverse-tubule membranes of skeletal muscle. *Eur. J. Biochem.* **144**, 211–215.

Garcia, M. L., Trumble, M. J., Reuben, J. P., and Kaczorowski, G. J. (1984) Characterization of verapamil binding sites in cardiac membrane vesicles. *J. Biol. Chem.* **259**, 15013–15016.

Garcia, M. L., King, V. F., and Kaczorowski, G. (1985) Interaction of diltiazem binding sites with dihydropyridine and verapamil receptors in cardiac sarcolemmal membrane vesicles. *Fed. Proc.* **44**, 715.

Gelmers, H. J. (1983) Nimodipine, a new calcium antagonist, in the prophylactic treatment of migraine. *Headache* **23**, 106–109.

Glossmann, H. and Ferry, D. R. (1983) Molecular Approach to the Calcium Channel, in *New Calcium Antagonists* (Fleckenstein, A., Hashimoto, K., Herrmann, M., Schwartz, A., and Seipel, L., eds.) Gustav Fischer Verlag, New York.

Glossmann, H., Ferry, D. R., Lubbecke, F., Mewes, R., and Hofmann, F. (1982) Calcium channels: Direct identification with radioligand binding studies. *Trends Pharmacol. Sci.* **3**, 431–437.

Glossmann, H., Linn, T., Rombusch, M., and Ferry, D. R. (1983) Temperature-dependent regulation of *d-cis*-[^{3}H]diltiazem binding to Ca^{2+} channels by 1,4-dihydropyridine channel agonists and antagonists. *FEBS Lett.* **160**, 226–232.

Goetz, C. G., Carvey, P. M., Tanner, C. M., and Klawans, H. L. (1984) Neuroleptic-induced dopamine hyposensitivity. *Life Sci.* **34**, 1475–1479.

Gould, R. J., Murphy, K. M. M., and Snyder, S. H. (1982a) Autoradiographic visualization of [^{3}H]nitrendipine binding sites in rat brain: Localization to synaptic zones. *Eur. J. Pharmacol.* **81**, 517–519.

Gould, R. J., Murphy, K. M. M., and Snyder, S. H. (1982b) [^{3}H]Nitrendipine-labelled calcium channels discriminate inorganic calcium agonists and antagonists. *Proc. Natl. Acad. Sci. USA* **79**, 3656–3660.

Gould, R. J., Murphy, K. M. M., and Snyder, S. H. (1983a) Studies on voltage-operated calcium channels using radioligands. *Cold Spring Harbor Symp. Quant. Biol.* **48**, 355–362.

Gould, R. J., Murphy, K. M. M., Reynold, I. J., and Snyder, S. H. (1983b) Antischizophrenic drugs of the diphenylbutylpiperidine type act as calcium channel antagonists. *Proc. Natl. Acad. Sci. USA* **80**, 5122–5125.

Gould, R. J., Murphy, K. M. M., and Snyder, S. H. (1984a) Tissue heterogeneity of calcium channel antagonist binding sites labelled by [^{3}H]nitrendipine. *Mol. Pharmacol.* **25**, 235–241.

Gould, R. J., Murphy, K. M. M., Reynolds, I. J., and Snyder, S. H. (1984b) Calcium channel blockade: Possible explanation for thioridazine's peripheral side effects. *Am. J. Psych.* **141**, 352–357.

Gould, R. J., Murphy, K. M. M., and Snyder, S. H. (1985) Autoradiographic localization of calcium channel antagonist receptors in rat brain with [^{3}H]nitrendipine. *Brain Res.* **330**, 217–223.

Green, F. J., Farmer, B. B., Wiseman, G. L., Jose, M. J. L., and Watanabe, A. M. (1985) Effect of membrane depolarization on binding of ^{3}H-nitrendipine to rat myocytes. *Circ. Res.* **56**, 576–585.

Grover, A. K., Kwan, C. Y., Luchowski, E., Daniel, E. E., and Triggle, D. J. (1984) Subcellular distribution of [^{3}H]nitrendipine binding in smooth muscle. *J. Biol. Chem.* **259**, 2223–2226.

Haas, S. and Beckmann, H. (1982) Pimozide versus haloperidol in acute schizophrenia. A double blind controlled study. *Pharmacopsychiatria* **15**, 70–74.

Haas, H. and Hartfelder, G. (1962) α-Isopropyl-α-(*N*-methyl-homoveratryl)-γ-aminopropyl-3,4-dimethoxy-phenylacetonitril, eine Substanz mit Coronargefasserweiternden Eigenschaften. *Arzneimittelforsch.* **12**, 549–558.

Henry, P. D. (1980) Comparative pharmacology of calcium antagonists: Nifedipine, verapamil and diltiazem. *Am. J. Cardiol.* **46**, 1047–1058.

Hess, P., Lansman, J. B., and Tsien, R. W. (1985) A novel type of cardiac calcium channel in ventricular cells. *Nature* (Lond.) **316**, 443–446.

Hoffmeister, F., Benz, U., Heise, H., Krause, P., and Neuser, V. (1982) Behavioral effects of nimodipine in animals. *Arzneimitelforsch.* **32**, 347–360.

Holck, M., Thorens, S., and Haeusler, G. (1982) Characterization of [^{3}H]nifedipine binding sites in rabbit myocardium. *Eur. J. Pharmacol.* **85**, 305–315.

Hulthen, U. L., Landmann, R., Burgisser, E., and Buhler, F. R. (1982) High-affinity binding sites for [^{3}H]verapamil in cardiac membranes. *J. Cardiovasc. Pharmacol.* **4**, S291–S93.

Janis, R. A. and Triggle, D. J. (1983) New developments in Ca^{2+} channel antagonists. *J. Med. Chem.* **26**, 775–785.

Janis, R. A., Maurer, S. C., Sarmiento, J. G., Bolger, G. T., and Triggle, D. J. (1982) Binding of [^{3}H]nimodipine to cardiac and smooth muscle membranes. *Eur. J. Pharmacol.* **82**, 191–194.

Janis, R. A., Rampe, D., Sarmiento, J. G., and Triggle, D. J. (1984a) Specific binding of a calcium channel activator, [^{3}H]Bay K 8644, to membranes from cardiac muscle and brain. *Biochem. Biophys. Res. Comm.* **121**, 317–323.

Janis, R. A., Sarmiento, J. G., Maurer, S. C., Bolger, G. T., and Triggle, D. J. (1984b) Characteristics of the binding of [^{3}H]nitrendipine to rabbit ventricular membranes: Modification by other Ca^{2+} channel-antagonists and the Ca^{2+} channel antagonist Bay K 8644. *J. Pharmacol. Exp. Ther.* **221**, 8–15.

Kaczorowski, G., King, V. F., and Garcia, M. L. (1985) Characterization of diltiazem binding sites in cardiac sarcolemmal membrane vesicles. *Fed. Proc.* **44**, 169–192. 715.

Kambara, H., Fujimoto, K., Wakabayashi, A., and Kawai, C. (1981) Primary pulmonary hypertension: Beneficial therapy with diltiazem. *Am. Heart J.* **101**, 230–231.

Kaplita, P. V. and Triggle, D. J. (1983) Actions of Ca^{2+} antagonists on the guinea-pig ileal myenteric plexus preparation. *Biochem. Pharmacol.* **32**, 65–68.

Karaki, H. and Weiss, G. B. (1984) Calcium channels in smooth muscle. *Gastroenterology* **87**, 960–970.

Kass, R. S. and Tsien, R. W. (1975) Multiple effects of calcium antagonists on plateau currents in cardiac Purkinje fibers. *J. Gen. Physiol.* **66**, 169–192.

Kohlhardt, M., Bauer, B., Krause, H., and Fleckenstein, A. (1972) Differentiation of the transmembrane Na and Ca channels in mammalian cardiac fibers by the use of specific inhibitors. *Pflugers Arch.* **335**, 309–322.

Kuroda, Y., Yoshii, M., Tsunoo, A., Yasumoto, T., and Narahashi, T. (1984) Recording of calcium current in gigaohm sealed neuroblastoma cells and the mechanism of action of maitotoxin. *Neurochem. Res.* **9**, 1172.

Landmark, K., Sire, S., Thaulow, E., Amlie, J. P., and Nitter-Hauge, S. (1982) Hemodynamic effects of nifedipine and propranolol in patients with hypertropic obstructive cardiomyopathy. *Br. Heart J.* **48**, 19–26.

Lapierre, Y. D. (1978) A controlled study of penfluridol in the treatment of chronic schizophrenia. *Am. J. Psychiatry* **135**, 956–959.

Lee, K. S. and Tsien, R. W. (1983) Mechanism of calcium channel blockade by verapamil, D600, diltiazem and nitrendipine in single dialyzed heart cells. *Nature* (Lond.) **302**, 790–794.

Lindner, E. (1960) Phenyl-propyl-diphenyl-propyl-amin, eine neue Substanz mit coronargefasserweiternder Wirkung. *Arzneimittelforsch.* **10**, 569–573.

MacLeod, R. M. and Lamberts, S. W. J. (1978) The biphasic regulation of prolactin secretion by dopamine agonists-antagonists. *Endocrinology* **103**, 200–203.

Mannhold, R., Zierden, P., Bayer, R., Rodenkirchen, R., and Steiner, R. (1981) The influence of aromatic substitution on the negative inotropic action of verapamil in the isolated cat papillary muscle. *Arzneimittelforsch.* **31**, 773–780.

Marangos, P. J., Patel, J., Miller, C., and Martino, A. M. (1982) Specific calcium antagonist binding in brain. *Life Sci.* **31**, 1575–1585.

Marsh, J. D., Loh, E., LaChance, D., Barry, W. H., and Smith, T. W. (1983) Relationship of binding of a calcium channel blocker to inhibition of contraction in intact cultured embryonic chick ventricular cells. *Circ. Res.* **53**, 539–543.

Metzger, H., Stern, H. O., Pfitzer, G., and Ruegg, J. C. (1982) Calcium antagonists affect calmodulin-dependent contractility of a skinned smooth muscle. *Arzneimittelforsch.* **32**, 1425–1427.

Middlemiss, D. N. and Spedding, M. (1985) A functional correlate for the dihydropyridine binding site in rat brain. *Nature* (Lond.) **314**, 94–96.

Miller, R. J. and Freedman, S. B. (1984) Are dihydropyridine binding sites voltage sensitive calcium channels? *Life Sci.* **34**, 1205–1221.

Miller, W. C. and Moore, J. B. (1983) [^{3}H]Nitrendipine binding in uterine smooth muscle. *Pharmacologist* **25**, 520.

Miller, D. M., Davin, W. T., and Tindall, D. R. (1985) Effects of maitotoxin on guinea pig ileum contraction partially ameliorated by calcium channel antagonists. *Toxicon* **23**, 34.

Murphy, K. M. M. and Snyder, S. H. (1982) Calcium antagonist receptor binding sites labeled with [^{3}H]nitrendipine. *Eur. J. Pharmacol.* **77**, 201–202.

Murphy, K. M. M., Gould, R. J., Largent, B. L., and Snyder, S. H. (1983) A unitary mechanism of calcium antagonist drug action. *Proc. Natl. Acad. Sci. USA* **80**, 860–864.

Murphy, K. M. M., Gould, R. J., and Snyder, S. H. (1984) Regulation of [^{3}H]Nitrendipine Binding: A Single Allosteric Site for Verapamil, Diltiazem and Prenylamine, in *Nitrendipine* (Scriabine, A., Vanov,

S., and Deck, K., eds.) Urban and Schwarzenberg, Baltimore, Maryland.

Nagao, T., Sato, M., Iwasawa, Y., Takada, T., Ishida, R., Nakajima, H., and Kiyomoto, A. (1972) Studies on a new 1,5-benzothiazepine derivative (CRD-401). III. Effects of optical isomers of CRD-401 on smooth muscle and other pharmacological properties. *Jap. J. Pharmacol.* **22**, 467–478.

Nagao, T., Ikeo, T., and Sato, M. (1977) Influence of calcium ions on responses to diltiazem in coronary arteries. *Jap. J. Pharmacol.* **27**, 330–332.

Nakajima, H., Hoshiyama, M., Yamashita, K., and Kiyomoto, A. (1975) Effect of diltiazem on electrical and mechanical activity of isolated cardiac ventricular muscle of guinea pig. *Jap. J. Pharmacol.* **25**, 383–392.

Narducci, F., Bassotti, G., Gaburri, M., Farroni, F., and Morelli, A. (1985) Nifedipine reduces the colonic motor response to eating in patients with the irritable colon syndrome. *Am. J. Gastroenterol.* **80**, 317–319.

Nowycky, M. C., Fox, A. P., and Tsien, R. W. (1985) Three types of neuronal calcium channel with different calcium agonist sensitivity. *Nature* (Lond.) **316**, 440–443.

Ohizumi, Y., Kajiwara, A., and Yasumoto, T. (1983) Excitatory effect of the most potent marine toxin, maitotoxin, on the guinea pig vas deferens. *J. Pharmacol. Exp. Ther.* **227**, 199–204.

Pan, M., Janis, R. A., and Triggle, D. J. (1983) Comparison of the equilibrium and kinetic binding characteristics of tritiated Ca^{2+} channel inhibitors, nisoldipine, nitrendipine, and nifedipine. *Pharmacologist* **25**, 523.

Quirion, R. (1983) Autoradiographic localization of a calcium channel antagonist, [^{3}H]nitrendipine, binding site in rat brain. *Neurosci. Lett.* **36**, 267–271.

Quirion, R., Lafaille, F., and Nair, N. P. V. (1985) Comparative potencies of calcium channel antagonists and antischizophrenic drugs on central and peripheral calcium channel binding sites. *J. Pharm. Pharmacol.* **37**, 437–440.

Reuter, H. (1967) The dependence of the slow inward current on external calcium concentration in Purkinje fibers. *J. Physiol.* (Lond.) **192**, 479–492.

Reuter, H. (1968) Slow inactivation of currents in cardiac Purkinje fibers. *J. Physiol.* (Lond.) **197**, 233–253.

Reynolds, I. J., Gould, R. J., and Snyder, S. H. (1983) [^{3}H]Verapamil binding sites in brain and skeletal muscle: Regulation by calcium. *Eur. J. Pharmacol.* **95**, 319–321.

Reynolds, I. J., Snowman, A. M., and Snyder, S. H. (1985) [^{3}H]Methoxyverapamil ([^{3}H]D-600) and [^{3}H]desmethoxyverapamil ([^{3}H]D-888) label multiple receptors in brain and heart. *Soc. Neurosci. Abstr.* **11**, 516.

Rezvani, A., Huidobro-Toro, J. P., and Way, E. L. (1983) Effect of 4-aminopyridine and verapamil on the inhibitory action of normorphine on the guinea-pig ileum. *Eur. J. Pharmacol.* **86**, 111–115.

Richter, J. E., Spurling, T. J., Cordova, C. M., and Castell, D. O. (1984) Effects of oral calcium blocker, diltiazem, on esophageal contraction: Studies in volunteers and patients with "nutcracker esophagus." *Dig. Dis. Sci.* **29**, 649–656.

Ringer, S. (1883) A further contribution regarding the influence of the different constituents of the blood on the contraction of the heart. *J. Physiol.* (Lond.) **4**, 29–42.

Rodeheffer, R. J., Rommer, J. A., Wigley, F., and Smith, G. R. (1983) Controlled double-blind trial of nifedipine in the treatment of Raynaud's phenomenon. *N. Eng. J. Med.* **308**, 880–883.

Rosenberger, L. B. and Triggle, D. J. (1978) Calcium, Calcium Translocation and Specific Calcium Antagonists, in *Calcium and Drug Action* (Weiss, G. B., ed.) Plenum, New York.

Rupniak, N. M. J., Jenner, P., and Marsden, C. D. (1983) The effect of chronic neuroleptic administration on cerebral dopamine receptor function. *Life Sci.* **32**, 2289–2311.

Saida, K. and Van Breemen, C. (1982) Inhibiting effect of diltiazem on intracellular Ca^{2+} release in vascular smooth muscle. *Blood Vessels* **20**, 105–108.

Saikawa, T., Nagamoto, Y., and Atrita, M. (1977) Electrophysiologic effects of diltiazem, a new slow channel inhibitor, on canine cardiac fibers. *Jap. Heart J.* **18**, 235–245.

Sandahl, B., Ulmsten, U., and Anderssen, K. E. (1979) Trial of the calcium antagonist nifedipine in the treatment of primary dysmenorrhoea. *Arch. Gynecol.* **227**, 147–151.

Sarmiento, J. G., Janis, R. A., Colvin, R. A., Maurer, S. C., and Triggle, D. J. (1982) Comparison of calcium channels in canine arterial and ventricular myocardium using [^{3}H]nitrendipine binding. *Fed. Proc.* **41**, 1707A.

Sarmiento, J. G., Janis, R. A., Colvin, R. A., Triggle, D. J., and Katz, A. M. (1983) Binding of the calcium channel blocker, nitrendipine to its receptor in purified sarcolemma from canine cardiac ventricle. *J. Mol. Cell. Cardiol.* **15**, 135–137.

Sato, M., Nagao, T., Yamaguchi, I.,, Nakajima, H., and Kiyomoto, A. (1971) Pharmacological studies on a new 1,5-benzothiazepine derivative (CRD-401). I. Cardiovascular actions. *Arzneimittelforsch.* **21**, 1338–1343.

Schoemaker, H. and Langer, S. Z. (1985) [^{3}H]Diltiazem binding to calcium channel antagonists recognition sites in rat cerebral cortex. *Eur. J. Pharmacol.* **111**, 273–277.

Schoemaker, H., Itoga, E., Boles, R. G., Roeske, W. R., Ehlert, F. W., Kito, S., and Yamamura, H. I. (1983) Temperature Dependence and Kinetics of [^{3}H]Nitrendipine Binding in the Rat Brain, in *Nitrendipine* (Scriabine, A., Vanov, S., and Deck, K., eds.) Urban and Schwarzenberg, Baltimore, Maryland.

Schramm, M., Thomas, G., Towart, R., and Frankowiak, G. (1983a) Activation of calcium channels by novel 1,4-dihydropyridines: A new mechanism for positive inotropics or smooth muscle stimulants. *Arzneimittelforsch.* **33**, 1268–1272.

Schramm, M., Thomas, G., Towart, R., and Frankowiak, G. (1983b) Novel dihydropyridines with positive inotropic action through activation of Ca^{2+} channels. *Nature* **303**, 535–537.

Schumann, H. J., Gorlitz, B. D., and Wagner, J. (1975) Influence of papaverine, D600 and nifedipine on the effects of noradrenaline and calcium on the isolated aorta and mesenteric artery of the rabbit. *Naunyn Schmiedebergs Arch. Pharmacol.* **289**, 409–418.

Schwartz, A. and Triggle, D. J. (1984) Cellular action of calcium channel blocking drugs. *Ann. Rev. Med.* **35**, 325–339.

Scriabine, A., Vanov, S., and Deck, K., eds. (1984) *Nitrendipine.* Urban and Schwarzenberg, Baltimore, Maryland.

Schmizu, K., Ohta, T., and Noda, N. (1980) Evidence for greater susceptibility of isolated dog cerebral arteries to Ca antagonists than peripherial arteries. *Stroke* **11**, 261–266.

Singh, B. N., Collett, J. T., and Chew, C. Y. (1980) New perspectives in the pharmacologic therapy of cardiac arrhythmias. *Prog. Cardiovasc. Dis.* **22**, 243–301.

Singh, B. N., Nademanee, K., and Baky, S. H. (1983) Calcium antagonists. Clinical use in the treatment of arrhythmias. *Drugs* **25**, 125–133.

Spedding, M. (1982) Assessment of "Ca^{2+}-antagonist" effects of drugs in K^+-depolarized smooth muscle. Differentiation of antagonist subgroups. *Naunyn Schmiedebergs Arch. Pharmacol.* **318**, 234–240.

Spedding, M. and Berg, C. (1984) Interactions between a "calcium channel agonist," Bay K 8644, and calcium antagonists differentiate calcium antagonist subgroups in K^+-depolarized smooth muscle. *Naunyn Schmiedebergs Arch. Pharmacol.* **328**, 69–75.

Spivack, C., Ocken, S., and Frishman, W. H. (1983) Calcium antagonists. Clinical use in the treatment of systemic hypertension. *Drugs* **25**, 154–177.

Starke, K., Spath, L., and Wichmann, T. (1984) Effects of verapamil, diltiazem and ryosidine on the release of dopamine and acetylcholine in rabbit caudate nucelus slices. *Naunyn Schmiedebergs Arch. Pharmacol.* **325**, 124–130.

Stone, P. H., Antman, E. M., Muller, J. E., and Braunwald, E. (1980) Calcium channel blocking agents in the treatment of cardiovascular disorders. II. Hemodynamic effects and clinical applications. *Ann. Intern. Med.* **93**, 886–904.

Theroux, P., Taeymans, Y., and Waters, D. D. (1983) Calcium antagonists. Clinical use in the treatment of angina. *Drugs* **25**, 178–195.

Thorgeirsson, G. and Rudolph, S. A. (1984) Diltiazem-like effect of thioridazine on the dihydropyridine binding sites of the calcium channel of rat myocardial membranes. *Biochem. Biophys. Res. Commun.* **121**, 657–663.

Toll, L. (1982) Calcium antagonists. High-affinity binding and inhibition of calcium transport in a clonal cell line. *J. Biol. Chem.* **257**, 13189–13192.

Triggle, D. J. and Janis, R. A. (1984) Calcium channel antagonists: New perspectives from the radioligand binding assay. *Mod. Meth. Pharmacol.* **2**, 1–28.

Triggle, C. R., Agrawal, D. K., Bolger, G. T., Daniel, E. E., Kwan, C. Y., Luchowski, E. M., and Triggle, D. J. (1982) Calcium channel antagonist binding to isolated vascular smooth muscle membranes. *Can. J. Physiol. Pharmacol.* **60**, 1738–1741.

Tsien, R. Y., Pozzan, T., and Rink, T. J. (1984) Measuring and manipulating cytosolic Ca^{2+} with trapped indicators. *Trends Biochem. Sci.* **9**, 263–266.

Turner, T. J. and Goldin, S. M. (1985) Calcium channels in rat brain synaptosomes: Identification and pharmacological characterization. High affinity blockade by organic Ca^{2+} channel blockers. *J. Neurosci.* **5**, 841–849.

Ulmsten, U., Anderssen, K. E., and Wingerup, L. (1980) Treatment of premature labor with the calcium antagonist nifedipine. *Arch. Gynecol.* **229**, 1–5.

Van Breeman, C. and Siegel, B. (1980) The mechanism of α-adrenergic activation of the dog coronary artery. *Circ. Res.* **46**, 426–429.

Van Breemen, C., Hwang, O., and Meisheri, K. D. (1981) The mechanism of inhibitory action of diltiazem on vascular smooth muscle contractility. *J. Pharmacol. Exp. Ther.* **218**, 459–463.

Vitek, M. and Trautwein, W. (1971) Slow inward current and action potential in cardiac Purkinje fibers. The effect of Mn ions. *Pfleugers Arch.* **323**, 204–218.

Weiss, G. B., ed. (1981) *New Perspectives on Calcium Antagonists*, American Physiological Society, Bethesda, Maryland.

Williams, L. T. and Jones, L. R. (1983) Specific binding of the calcium antagonist, [^{3}H]nitrendipine, to subcellular fractions isolated from canine myocardium—evidence for high affinity binding to ryanodine-sensitive sarcoplasmic reticulum. *J. Biol. Chem.* **258**, 5344–5347.

Williams, L. T. and Tremble, P. (1982) Binding of a calcium antagonist, [^{3}H]nitrendipine, to high affinity sites in bovine aortic smooth muscle and canine cardiac membranes. *J. Clin. Invest.* **70**, 209–212.

Wit, A. L. and Ning, W. (1983) Effects of the Slow Channel Blockers Nifedipine and Verapamil on the Electrical Activity of the Sinoatrial and Atrioventricular Nodes, in *Calcium Antagonists: The State of The Art and Role in Cardiovascular Disease*. (Hoffman, B. F., ed.) College of Physicians of Philadelphia, Philadelphia, Pennsylvania.

Yamamoto, H. and Van Breemen, C. (1985) Inositol 1,4,5-trisphosphate releases calcium from skinned, cultured smooth muscle cells. *Biochem. Biophys. Res. Commun.* **130**, 270–274.

Yamamura, H. I., Schoemaker, H., Boles, R. G., and Roeske, W. R. (1982) Diltiazem enhancement of [^{3}H]nitrendipine binding to calcium channel associated drug receptor sites in rat brain synaptosomes. *Biochem. Biophys. Res. Commun.* **108**, 640–646.

Zelis, R. and Flaim, S. F. (1982) Calcium blocking drugs for angina pectoris. *Ann. Rev. Med.* **33**, 465–478.

Index